**Enhanced Radio Access Technologies for Next Generation Mobile Communication**

# Enhanced Radio Access Technologies for Next Generation Mobile Communication

*Edited by*

Yongwan Park
*Yeung Nam University, Korea*

*and*

Fumiyuki Adachi
*Tohoku University, Japan*

A C.I.P. Catalogue record for this book is available from the Library of Congress.

ISBN 978-1-4020-5531-7 (HB)
ISBN 978-1-4020-5532-4 (e-book)

---

Published by Springer,
P.O. Box 17, 3300 AA Dordrecht, The Netherlands.

*www.springer.com*

*Printed on acid-free paper*

# CONTENTS

CHAPTER 1

# OVERVIEW OF MOBILE COMMUNICATION

YONGWAN PARK[1] AND FUMIYUKI ADACHI[2]

[1] *Department of Information and Communication Engineering, Yeungnam University, 214-1 Dae-dong, Gyeongsan-si, Gyeongsanbuk-do, Korea*

[2] *Graduate School of Engineering, Tohoku University, 6-6-05 Aza-Aoba, Aramaki, Aoba-ku, Sendai 980-8579, Japan*

**Abstract:** Following chapter introduces the mobile communication, gives a short history of wireless communication evolution, and highlights some application scenarios predestined for the use of mobile devices. Cellular and wireless based systems related to different generations of mobile communication, including GSM, IS-95, PHS, AMPS, D-AMPS, cdma2000 and WCDMA are also described by this Chapter. Much attention in this chapter is given to express the wireless based networks, such as Wi-Fi and WiBro/WiMax, and wireless broadcasting systems, including DMB, DVB-H, and ISDB-T. We conclude the chapter with the future vision of mobile communication evolution

**Keywords:** mobile communication; wireless communication; first generation (1G); second generation (2G); thirdgeneration (3G); IMT-2000; UMTS; WCDMA; cdma2000; TDSCDMA; IEEE 802.11; WiFi-; IEEE 802.15; Bluetooth; UWB; WiBro; WiMax; wireless broadcasting; DMB; DVB-H; ISDB-T; OFDMA; MC DS-CDMA

## 1. INTRODUCTION TO MOBILE COMMUNICATION SYSTEM

To this day, there have been three different generations of mobile communication networks. First-generation of (1G) wireless telephone technologies are the analog cell phone standards that were introduced in the 80s and continued until being replaced by **2G** digital cell phones in 1990s. Example of such standards are **NMT** (Nordic Mobile Telephone), used in Nordic countries, **NTT** system in Japan, and the **AMPS** (Advanced Mobile Phone System) operated in the United States. The second-generation **(2G)** technology is based on digital cellular technology. Examples of the 2G are the Global System for Mobile Communications **(GSM),** Personal Digital Cellular (**PDC**), and North American version of CDMA standard (**IS-95**). The third generation **(3G)** started in October 2001 when **WCDMA** network was launched in Japan. The services associated with 3G provide the ability to transfer both voice data (a telephone call) and non-voice data (such as downloading information, exchanging email, and instant messaging).



*Y. Park and F. Adachi (eds.), Enhanced Radio Access Technologies for Next Generation Mobile Communication*, 1–37.

| Technology | 1G | 2G | 2.5G | 3G | 3.5G | 4G |
|---|---|---|---|---|---|---|
| Standard | AMPS, TACS, NMT, ETC. | TDMA, CDMA, GSM, PDC | GPRS, EDGE, 1×RTT | WCDMA, CDMA2000 | HSDPA WiBro(Mobile WiMax) | Single standard |
| Implementation | 1984 | 1991 | 1999 | 2002 | 2006 | 2010 |
| Data Rate | 1.9Kbps | 14.4Kbps | 384Kbps | 2Mbps | 10 ~ 50Mbps | 100Mbps ~ 1Gbps |
| Multiplexing | FDMA | TDMA, CDMA | TDMA, CDMA | CDMA | CDMA, OFDMA | CDMA, OFDMA, ? |
| Service | Analog voice, synchronous data to 9.5 Kbps | Digital voice, Short messages | Higher capacity, packetized data | Higher capacity, broadband data up to 2Mbps | Portable Internet, High speed Wireless Internet, multimedia | Higher capacity, completely IP oriented, multimedia, data up to 1Gbps |

*Figure 1.* Mobile communication generations

**Figure 1** illustrates a brief overview on each generation. More detail information about mobile communication evolution steps is given in section 2.

The advances in cellular systems, wireless LANs, wireless MANs, personal area networks (PANs), and sensor networks are bound to play a significant role in the people communication manner in the future. It is expected that in the following years most of the access part of the Internet will be wireless. Increasing capacity and data rate of mobile communication systems enable to develop extended applications and services. **Figure 2** demonstrates some application environments and modern services focusing on South Korea and Japan's markets and technology trends. The current and awaited mobile services in these countries can be viewed as follows:

**E-mail:** This is a killer application regardless of the mobile network generation. The e-mail applications both send a message to other mobile phone or to anyone who has an Internet e-mail address. Mobile terminals also can receive e-mail. Low cost and fully compatibility with normal Internet e-mail makes this service popular among the mobile Internet users.

**Web Browsing:** Although mobile browsing is not popular everywhere today, it is very likely that within next ten years from now, mobile phone users will connect to Internet and use a mobile browser as an everyday tool. But this requires that the mobile browsing user experience improves: connection speed, number of services, and usability must increase, while cost per byte must decrease. While 2G networks allow predominantly text-based HTML browsing, 2.5G and 3G mobile terminals with TFT displays with 262,144 colors enables mobile users to browse Internet contents with high quality.

Two candidates aimed to enabling the Web browsing application to be built with wireless technology. One of them is Wireless Application Protocol (WAP) which was designed to provide services equivalent to a Web browser with some mobile-specific additions, being specifically designed to address the limitations of very small portable devices. The Japanese *i-mode* system is the other major competing wireless data protocol. WAP was hyped at the time of its introduction, leading users to expect WAP to have the performance of the Web. In terms of speed, ease of use, appearance, and interoperability, the reality fell far short of expectations. This led

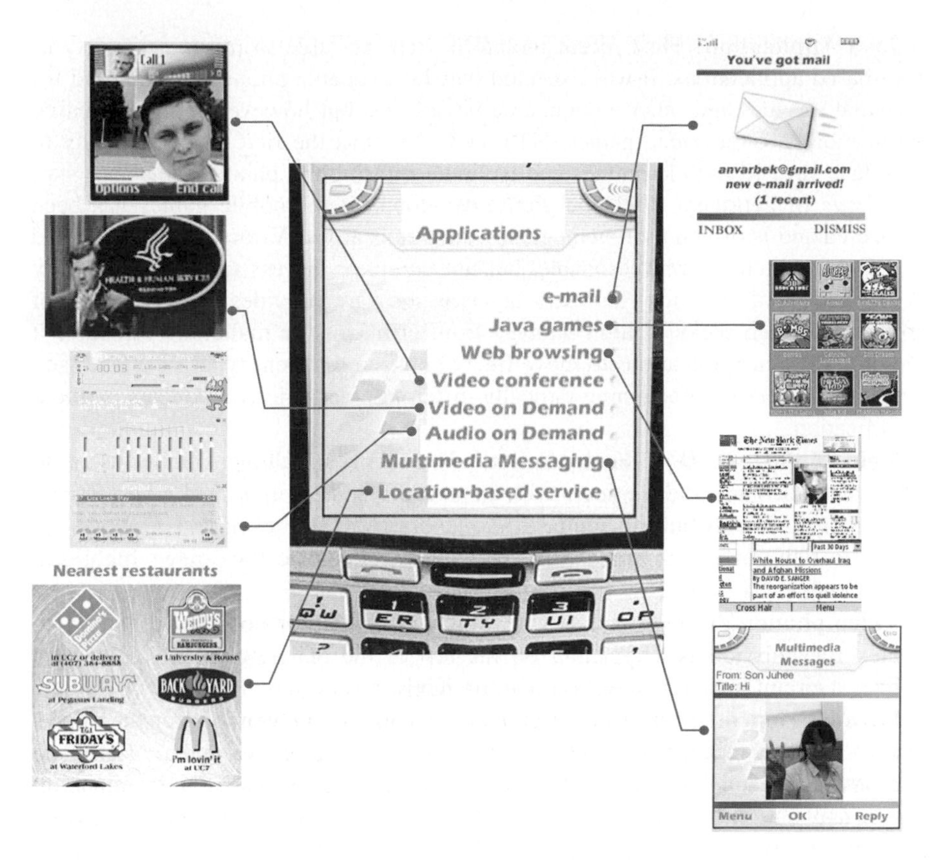


*Figure 2.* Mobile applications

to the wide usage of sardonic phrases such as "Worthless Application Protocol", "Wait And Pay", and so on. While WAP did not succeed, *i-mode* soon became a tremendous success. *i-mode* phones have a special *i-mode* button for the user to access the start menu. There are numerous official sites – and even more unofficial ones – that can be made available by anyone, using HTML and with access to a standard Web server. As of June 2005, *i-mode* has 45 million customers in Japan and over 5 million in the rest of the world.

**Multimedia Messaging Service (MMS)** is a technology for transmitting not only text messages, but also various kinds of multimedia content (e.g. images, audio, and/or video clips) over wireless telecommunications networks. MMS-enabled mobile phones enable mobile users to compose and send messages with one or more multimedia parts. Mobile phones with built-in or attached cameras, or with built-in MP3 players are very likely to also have an MMS messaging client – a software program that interacts with the mobile subscriber to compose, address, send, receive, and view MMS messages.

**Java Application:** Most recent mobile devices are able to run wide variety of Java-based applications. It was expected that Java capable phones will be used for financial services and other e-commerce businesses, but however main Java-based applications are the video games. NTT DoCoMo was the first carrier globally to introduce Java to mobile phones and for games on mobile phones. Since the start of *i-mode* in February 1999, the global development of mobile games has been pioneered and is driven by *i-mode* games. Java runs atop a Virtual Machine (called the KVM) which allows reasonable, but not complete, access to the functionality of the underlying phone. This extra layer of software provides a solid barrier of protection which seeks to limit damage from erroneous or malicious software. It also allows Java applications to move freely between different types of phone (and other mobile device) containing radically different electronic components, without modification.

**Videoclip/Music Download:** Current 3G networks allow mobile users to download video and audio content with enhanced speeds of up to 384 Kbps. Recent mobile devices with built in multimedia players and high resolution displays can access to rich content of video clips, movie trailers, music files, news highlights and so on.

**Video phone:** Visual phone service which is capable of both audio and video duplex transmission is a typically on the top of the 3G networks. This service utilizes a circuit switch connection with 64 Kbps.

**Location-dependent services:** Contemporary mobile networks offer the opportunity to employ recently developed position-determining devices and to offer many new and interesting location-dependent services. In many cases it is important for an application to know something about the location or the user might need location information for further activities. In 2001 Japanese company NTT DoCoMo launched the first location-dependent Web browsing service. The service delivers mobile users a broad range of location-specific Web content. The location estimation accuracy depends on cell size and the associated base station. The mobile user can gain access to cell-range information such as restaurants, hotels, shopping centers, and download relevant maps. On April 2004 Korean SK Telecom also launched the commercial Location-Based Service, called "Becktermap". Unlike existing Location-Based Services that show the location by downloading a complete map like a photo, the Becktermap service directly draws a map with a specific location on the cellular screen. It does this by downloading its configuration information from base stations or a Global Positioning System. This service includes weather conditions at the location, discount information at department stores, nearby restaurant information, and the changing location information of the pedestrian.

Future location based service systems will use both GPS and network information, and will support the interoperability between outdoor (GPS, Cellular, etc.) and indoor (based on WLAN, UWB, etc.) localization and tracking systems.

**Figure 3** shows the mobile communication services and applications evolution towards 3G. Today's mobile users already comprise some, but future users will comprise many mobile communication systems and mobility aware applications.

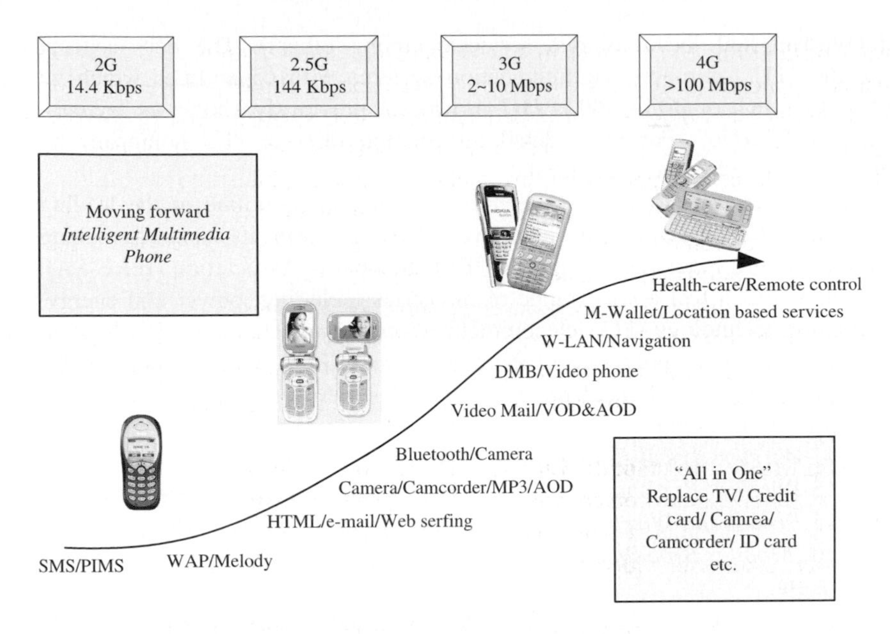


*Figure 3.* Mobile applications and services evolution

Music, news, road conditions, weather and financial reports, business information, infotainment and others are received via digital audio broadcasting (DAB) with 1.5 Mbps. DMB (Digital Multimedia Broadcasting) allows to transmit data, radio and TV to mobile devices. For personal communication a UMTS phone might be available offering voice and data connectivity with 384 Kbps. Satellite communications can be used for remote areas, while the current position of mobile user is determined using GPS.

In the next generation the cell phone will be an important mobile platform for daily life tools. The machine-to-machine services such as remote control of vendor machines, home-security, commuter pass, delivery tracking, and telemetry are now becoming commercially available, and it is reasonable to expect that this application area will grow into a significant component of next generation services.

The major standardization bodies that play an important role in defining the specifications for the mobile technology are:

- ITU (*International Telecommunication Union*): International organization within the United Nations, where governments and the private sector coordinate global telecom networks and services. One of the sectors of ITU, *ITU-T* produces the quality standards covering all the fields of telecommunications. More than 1500 specialists from telecommunication organizations and administrations around the world participate in the work of the Radiocommunication Sector of ITU (*namely ITU-R*). ITU's *IMT-2000* (*International Mobile Telecommunications-2000*) global standard for 3G wireless communications has opened the way to enabling innovative applications and services (e.g. multimedia entertainment, infotainment and location-based services, among others). The new concept from the ITU for mobile communication systems with capabilities which go further than that of IMT-2000 is *IMT-Advanced*, previously known as "*systems beyond IMT-2000*". For more detail information refer to ITU homepage by http://www.itu.int/home/index.html.

- IEEE (*Institute of Electrical and Electronics Engineers*) is one of the leading standards-making organizations in the world. IEEE performs its standards making and maintaining functions through the IEEE Standards Association (IEEE-SA). IEEE standards affect a wide range of industries including: power and energy, information technology (IT), telecommunications, nanotechnology, information assurance, and many more. One of the more notable IEEE standards is the *IEEE 802 LAN/MAN* group of standards which includes the:
  - 802.3 *Ethernet standard,*
  - 802.11 *Wireless Local Area Networks (Wi-Fi),*
  - 802.15 *Wireless Personal Area Networks (Bluetooth, ZigBee, Wireless USB),*
  - 802.16 *Broadband Wireless Access (WiMax, Mobile WiMax/WiBro),*
  - 802.20 *Mobile Broadband Wireless Access (suspended until 1 October 2006),* etc.

For more information about *IEEE* and *802 LAN/MAN* group refer to http://www.ieee.org/ and http://www.ieee802.org/ Web pages, respectively.

- ETSI (*European Telecommunication Standard Institute*) is an independent, non-profit, standardization organization of the telecommunications industry (equipment makers and network operators) in Europe, with worldwide projection. ETSI has been successful in standardizing the GSM cell phone system and the TETRA professional mobile radio system. Owing to the technical and commercial success of the GSM, this body plays an important role in the development of 3G mobile systems. See http://www.etsi.org/ for detailed information.
- *ARIB* (The Association of Radio Industries and Businesses) was chartered by the Minister of Posts and Telecommunications of Japan as a public service corporation on May 15, 1995. Established in response to several trends such as the growing internationalization of telecommunications, the convergence of telecommunications and broadcasting, and the need for promotion of radio-related industries, this body is playing an important role in the 3G development. ARIB Web page located at http://www.arib.or.jp/english/.
- TTA (*Telecommunications Technology Association*) is a Korean IT standards organization that develops new standards and provides one-stop services for the establishment of IT standards as well as testing and certification for IT products. One of the successful standards approved by TTA is the TTA PG302, the standard for 2.3 GHz Portable Internet (*WiBro*). For further information see TTA organization Web site at http://tta.or.kr/English/.
- 3GPP (*Third Generation Partnership Project*) was created to maintain overall control of the specification design and process for 3G networks. The scope of 3GPP is to make a globally applicable 3G mobile phone system specification within the scope of the ITU's IMT-2000 project. The 3GPP is an

international collaboration of a number of telecommunications standards bodies to standardize UMTS (*Universal Mobile Telecommunications System*). The original scope of 3GPP was to produce globally applicable Technical Specifications and Technical Reports for a 3rd Generation Mobile System based on evolved GSM core networks and the radio access technologies that they support. The current Organizational Partners are Japanese (*ARIB and TTC*), Chinese (*CCSA*), European (*ETSI*), American (*ATIS*) and Korean (*TTA*). 3GPP Web site located at http://www.3gpp.org/ .

- 3GPP2 is the other major 3G standardization organization, which promotes the cdma2000 system. In the world of IMT-2000, this proposal is known as IMT-MC. The major difference between 3GPP and 3GPP2 approaches into the air specification development is that 3GPP has specified a completely new air interface without any constraints from the past, whereas 3GPP2 has specified a system that is backward compatible with IS-95 systems. Official Web page of 3GPP2 organization is http://www.3gpp2.org/.

Next in this chapter we discuss the aforesaid mobile communication generations in detail and describe the services and applications suitable for mobile communication systems.

## 2. EVOLUTION OF MOBILE COMMUNICATION SYSTEMS

For a better understanding of today's wireless communication systems and developments, we will present a short history of wireless communications. The name, which is closely connected with the success of wireless communication, is that **Guglielmo Marconi**. In 1895, he gave the first demonstration of **wireless telegraphy**. Six years later in 1901 the first transatlantic transmission followed. The first radio broadcast took place in 1906 when **Reginald A. Fessenden** transmitted voice and music for Christmas. Within the next years huge work has been made, and in 1915 the first **wireless transmission** was set up between New York and San Francisco. Since, all this done using long wave transmission, sender and receiver still needed huge antennas and high transmission power (up to 200 kW).

The situation was resolutely changed with the **discovery of short waves** in 1920 by Marconi. Since then became possible to send short radio waves around the world bouncing at the ionosphere. After the Second World War governments started to invest in development of wireless communication projects. In 1958 Germany launches the first analogue wireless network named **A-Netz**, using 160 MHz carrier frequency. Connection setup was only possible from the mobile station, no handover, i.e., changing of the base station, was possible. System had coverage of 80 percent and 11,000 customers. In 1972 **B-Netz** followed in Germany, using the same 160 MHz. This network could initiate the connection setup from a station in the fixed telephone network, but, the current location of the mobile receiver had to be known. In 1979, B-Netz had 13,000 customers and needed a heavy sender and receiver, typically built into cars.

At the same time, the Northern European countries of Denmark, Finland, Norway and Sweden agreed upon the **Nordic Mobile Telephone (NMT)** system. NMT is based on analog technology (first generation or 1G) and two variants exist: **NMT-450** and **NMT-900**. The numbers indicate the frequency bands uses. NMT-900 was introduced in 1986 because it carries more channels than the previous NMT-450 network. The cell sizes in an NMT network range from 2 km to 30 km. With smaller ranges the network can serve more simultaneous callers; for example in a city the range can be kept short for better service. NMT used full duplex transmission, allowing for simultaneous receiving and transmission of voice. Car phone versions of NMT used transmission power of up to 15 watt (NMT-450) and 6 watt (NMT-900), handsets up to 1 watt. NMT had automatic switching (dialing) and handover of the call built into the standard from the beginning. Additionally, the NMT standard specified billing as well as national and international roaming.

In 1979 NTT introduced the analog mobile phone system using frequency division multiplexing (FDMA) and operating at 800MHz band. NTT system aimed to provide nationwide service by introducing the cellular architecture, location registration and handoff. In 1983 Bell Labs officially introduced the analog mobile phone system standard **Advanced Mobile Phone System (AMPS),** using FDMA and working at 850 MHz. Though analog is no longer considered advanced at all, AMPS introduced the relatively seamless cellular switching technology, that made the original mobile radiotelephone practical, and was considered quite advanced at the time. Using FDMA, each cell site would transmit on different frequencies, allowing many cell sites to be built near each other. However it had the disadvantage that each site did not have much capacity for carrying calls. It also had a poor security system which allowed people to force a phone's serial code to use for making illegal calls.

The boundary line between 1G and 2G systems is obvious: it is the analog/digital split. The 2G systems have much higher capacity than the 1G systems. One frequency channel is simultaneously divided among several users, either by code or time division. There are four main standards for 2G: *Global System for Mobile* (GSM), *Digital AMPS* (D-AMPS), code-division multiple access (CDMA, IS-95), and *Personal Digital Cellular* (PDC).

**PDC** is the Japanese 2G standard. Originally it was known as *Japanese Digital Cellular (**JDC**)*, but the name was changed to PDC to make system more attractive outside Japan. However, this renaming did not bring about the desired result, and this standard is commercially used only in Japan. PDC operates in two frequency bands, 800 MHz and 1,500 MHz. It has both analog and digital modes. PDC has been very popular system in Japan.

Another, popular Japanese 2G system is **Personal Handy-phone System (PHS)**, also marketed as the **Personal Access System (PAS)**, is a mobile network system operating in the 1880-1930 MHz frequency band. PHS is, essentially, a cordless telephone with the capability to handover from one cell to another. PHS cells are small, with transmission power a maximum of 500mW and range typically measures in tens or at most hundreds of meters, as opposed to the multi-kilometer ranges of

GSM. Originally developed by NTT Laboratory in Japan in 1989 and far simpler to implement and deploy than competing systems like PDC or GSM, the commercial services have been started by 3 PHS operators (NTT-Personal, DDI-Pocket and ASTEL) in Japan in 1995. However, the service has been pejoratively dubbed as the "poor man's cellular" due to its limited range and roaming capabilities in Japan. Recently, PHS has been reconsidered again in Japan as a cost-effective solution to providing broadband services of data rate up to 64Kbps, which is much faster than any other 2G systems. Also in other Asian countries, e.g., China, PHS has been deployed in addition to 2G cellular systems.

In accordance with the general idea of European Union, the European countries decided to develop a pan-European phone standard in 1982. The new system aimed to:

- use a new spectrum at 900 MHz;
- allow roaming throughput Europe;
- be fully digital;
- offer voice and data service.

The "**Groupe Speciale Mobile**" **(GSM)** was founded for this new development. From 1982 to 1985 discussions were held to decide between building an analog or digital system. After multiple field tests, a digital system was adopted for GSM. The next task was to decide between a narrow or broadband solution. In May 1987, the narrowband time division multiple access (TDMA) solution was chosen. In 1989, ETSI took over control and by 1990 the first GSM specification was completed, amounting to over 6,000 pages of text. Commercial operation began in 1991 with Radiolinja in Finland. GSM differs significantly from its predecessors in that both signaling and speech channels are digital, which means that it is considered a second generation (2G) mobile phone system. This first version GSM, now called global system for mobile communication, works at 900 MHz and uses 124 full-duplex channels. GSM offers full international roaming, automatic location services, authentication, encryption on the wireless link, and a relatively high audio quality. GSM is by far the most successful and widely used 2G system. Originally designed as a Pan-European standard, it was quickly adopted all over the world.

It was soon discovered that the analog AMPS in the US and digital GSM at 900 MHz in Europe are not sufficient for the high user densities in cities. These triggered off the search for more able systems. While the Europeans agreed to use the GSM in the new 1800 MHz band (**DCS 1800**), in the US, different companies developed three different new, more bandwidth-efficient technologies to operate side-by-side with AMPS in the same frequency band. This resulted in three incompatible systems, the analog narrowband AMPS (IS-88), and the two digital systems D-AMPS (IS-136) and CDMA (IS-95).

**D-AMPS** (also known as US-TDMA) is used in the Americas, Israel, and in some Asian countries. D-AMPS uses existing AMPS channels and allows for smooth transition between digital and analog systems in the same area. Capacity was increased over the preceding analog design by dividing each 30 kHz channel pair into three time slots (TDMA) and digitally compressing the voice data, yielding three times the call capacity in a single cell. A digital system also made calls more secure because analog scanners could not access digital signals.

The first **CDMA**-based digital cellular standard **IS-95** (Interim Standard 95) is pioneered by Qualcomm. The brand name for IS-95 is ***cdmaOne***. IS-95 is also known as TIA-EIA-95. CDMA or "code division multiple access" is a digital radio system that transmits streams of bits (PN Sequences). CDMA permits several users to share the same frequencies. Unlike TDMA, a competing system used in GSM, all transmitters can be active all the time, because network capacity does not directly limit the number of active users. Since larger numbers of users can be served by smaller numbers of cell-sites, CDMA-based standards have a significant economic advantage over TDMA-based standards, or the oldest cellular standards that used FDMA.

In 1993 South Korea adopts CDMA, although some experts worried Korea would lag behind with the launch of the then-untested CDMA network, while the world was commercializing the GSM standard. The decision to adopt CDMA technology turned a new page in Korea's telecommunications history. In January 1996, Korea successfully launched the world's first commercial operation of CDMA network in Seoul and its neighboring cities. Since then, CDMA has become the fastest-growing of all wireless technologies, with over 100 million subscribers worldwide. In addition to supporting more traffic, CDMA brings many other benefits to carriers and mobile users, including better voice quality, broader coverage and stronger security. IS-95 is the only CDMA standard so far to be operated commercially as a 2G system.

Note that quite often when the 2G is discussed, *digital cordless systems* are also mentioned. In 1991, ETSI adopted the standard **Digital European cordless telephone (DECT)** for digital cordless telephony. DECT works at a spectrum of 1880–1900 MHz with a range of 100–500m. 120 duplex channels can carry up to 1.2 Mbps for data transmission. Several new features, such as voice encryption and authentication, are built-in. Today, DECT has been renamed *digital enhanced cordless telecommunications*.

2.5 Generation (2.5G) is a designation that broadly includes all advanced upgrades for the 2G networks. 2.5G provides some of the benefits of 3G (e.g. it is packet-switched) and can use some of the existing 2G infrastructure in GSM and CDMA networks. **Figure 4** demonstrates the evolution of cellular based systems from 2G towards 4G. **General Packet Radio Service (GPRS)** is a 2.5G technology used by GSM operators. Some protocols, such as **EDGE (Enhanced Data Rates for Global Evolution)** for GSM and **CDMA2000 1x-RTT** for CDMA, can qualify as "3G" services (because they have a data rate of above 144 Kbps), but are considered by most to be 2.5G services because they are several times slower than "true" 3G services.

With **GPRS** technology, the data rates can be pushed up to 115 Kbps, or even higher. It provides moderate speed data transfer, by using unused TDMA channels in the GSM network. Originally there was some thought to extend GPRS to cover

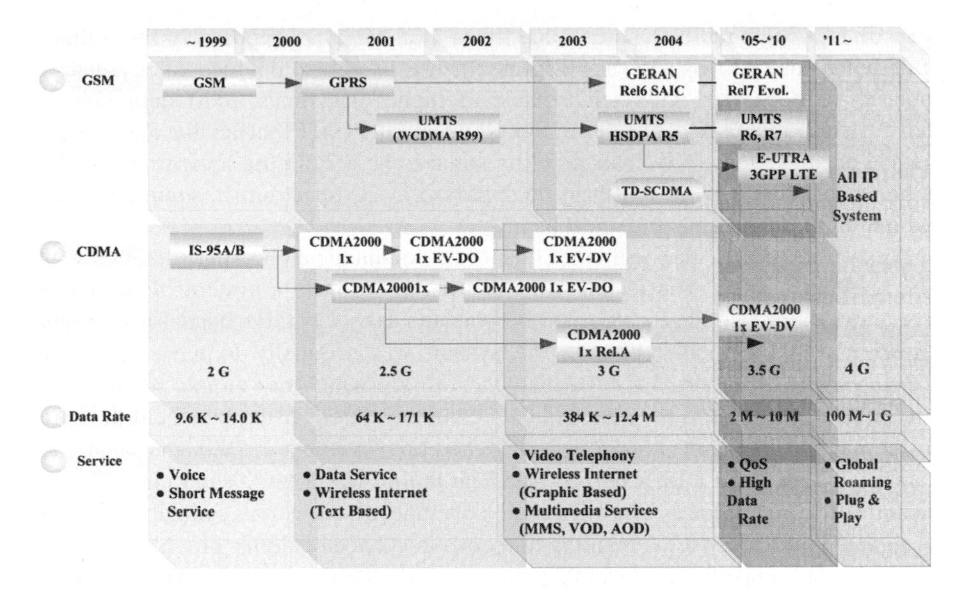


*Figure 4.* Mobile communication systems evolution towards 4G

other standards, but instead those networks are being converted to use the GSM standard, so that it is the only kind of network where GPRS is in use. First it was standardized by ETSI but now that effort has been handed onto the 3GPP. GPRS is packet switched, and thus it does not allocate the radio resources continuously but only when there is something to be sent. A consequence of this is that packet switched data has a poor bit rate in busy cells. The theoretical limit for packet switched data is approx. 160.0 Kbps (using 8 time slots). A realistic bit rate is 30–80 Kbps, because it is possible to use max 4 time slots for downlink. GPRS is especially suitable for non-real-time applications, such as e-mail and Web surfing. It is not well suited for real-time applications, as the resource allocations in GPRS is connection based and thus it cannot guarantee an absolute maximum delay.

A change to the radio part of GPRS called **EDGE** (sometimes called EGPRS or Enhanced GPRS) allows higher bit rates of between 160 and 236.8 Kbps (theoretical maximum is 473.6 Kbps for 8 timeslots). Although EDGE requires no hardware changes to be made in GSM core networks, base stations must be modified. EDGE compatible transceiver units must be installed and the base station subsystem (BSS) needs to be upgraded to support EDGE. New mobile terminal hardware and software is also required to decode/encode the new modulation and coding schemes and carry the higher user data rates to implement new services.

**CDMA2000 1xRTT**, the core CDMA2000 wireless air interface standard, is known by many terms: 1x, 1xRTT, IS-2000, CDMA2000 1X, and cdma2000 (lowercase). The designation “1xRTT” (1 times Radio Transmission Technology) is used to identify the version of CDMA2000 radio technology that operates in a pair of 1.25-MHz radio channels (one times 1.25 MHz, as opposed to three times 1.25 MHz in 3xRTT as shown in **Figure 5**). 1xRTT almost doubles voice capacity over IS-95 networks. Although capable of higher data rates, most deployments have limited the peak data rate to 144 Kbps. While 1xRTT officially qualifies as 3G technology, 1xRTT is considered by some to be a 2.5G (or sometimes 2.75G) technology. This has allowed it to be deployed in 2G spectrum in some countries which limit 3G systems to certain bands.

Year 1998 marked the beginning of mobile communication using satellites with the **Iridium** system. The Iridium satellite constellation is a system of 66 active communication satellites in low earth orbit and uses 1.6 GHz band for communication with the mobile phone. The system was originally to have 77 active satellites, and was named for the element iridium, which has atomic number 77. Iridium allows worldwide voice and data communications using handheld devices. Iridium communications service was launched on November 1, 1998 and went into bankruptcy on August 13, 1999. Its financial failure was largely due to insufficient demand for the service. The increased coverage of terrestrial cellular networks (e.g. GSM) and the rise of roaming agreements between cellular providers proved to be fierce competition. Nowadays the system is being used extensively by the U.S. Department of Defense for its communication purposes through the DoD Gateway in Hawaii. The commercial Gateway in Tempe, Arizona provides voice, data and paging services for commercial customers on a global basis. Typical customers include maritime, aviation, government, the petroleum industry, scientists, and frequent world travelers. Iridium Satellite LLC claims to have approximately 142,000 subscribers as of December 31, 2005.

In 1999 IEEE published several powerful WLAN standards. One of them is **802.11b Wi-Fi** standard offering 11 Mbps at 2.4 GHz. 802.11b products appeared on the market very quickly, since 802.11b is a direct extension of the DSSS (Direct-sequence spread spectrum) modulation technique defined in the original

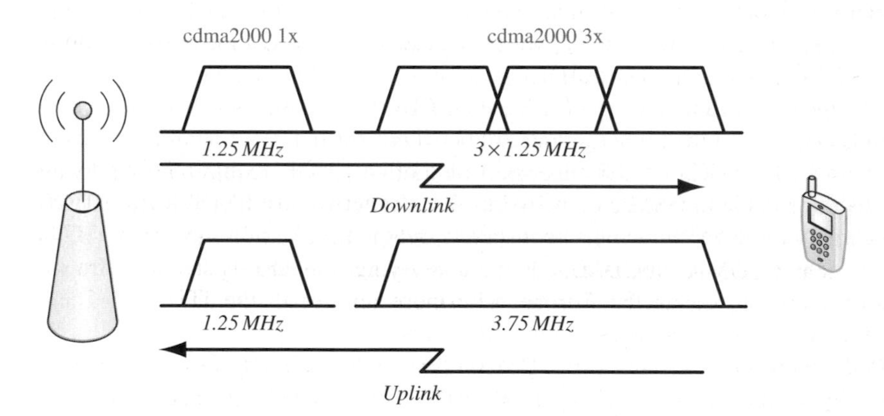


*Figure 5.* Relationship between 1x and 3x modes in spectrum usage

standard. Hence, chipsets and products were easily upgraded to support the 802.11b enhancements. The dramatic increase in throughput of 802.11b (compared to the original standard) along with substantial price reductions led to the rapid acceptance of 802.11b as the definitive wireless LAN technology. The same spectrum is used by **Bluetooth**, a short-range technology to set-up wireless personal area networks (PANs) with gross data rates less than 1 Mbps. Bluetooth is an industrial specification for PANs, also known as **IEEE 802.15.1**. Bluetooth provides a way to connect and exchange information between devices like personal digital assistants (PDAs), mobile phones, laptops, PCs, printers and digital cameras via a secure, low-cost, globally available short range radio frequency.

The rapid development of mobile communication systems was one of the most notable success stories of the 1990s. The 2G systems began their operation at the beginning of the decade, and since then they have been expanding and evolving continuously. In 2000 there were 361.7 million GSM and more than 100 million CDMA subscribers worldwide. Main disadvantage of 2G systems was that the standards for developing the networks were different for different parts of the world. Hence, it was decided to have a network that provides services independent of the technology platform and whose network design standards are same globally. Thus, **3G** was born. To understand the background to the differences between 2G and 3G systems, we need to look at the new requirements of the 3G systems which are listed below:

- Bit rates up to 2 Mbps;
- Variable bit rate to offer bandwidth on demand;
- Multiplexing of services with different quality requirements on a single connection, e.g. speech, video and packet data;
- Delay requirements from delay-sensitive real time traffic to flexible best-effort packet data;
- Quality requirements from 10% frame error rate to $10^{-6}$ bit error rate;
- Co-existence of 2G and 3G systems and inter-system handovers for coverage enhancements and load balancing;
- Support asymmetric uplink and downlink traffic, e.g. Web browsing causes more loading to downlink than to uplink;
- High spectrum efficiency;
- Co-existence of FDD and TDD modes.

ITU started the process of defining the standard for 3G systems, referred to as **IMT-2000**. In 1998 Europeans agreed on the **Universal Mobile Telecommunications System (UMTS)** as the European proposal for the 3G systems. UMTS uses Wideband-CDMA (**WCDMA**) as the underlying standard, is standardized by the 3GPP, and represents the European/Japanese answer to the ITU IMT-2000 requirements for 3G systems.

**IMT-2000** offers the capability of providing value-added services and applications on the basis of a single standard. The system envisages a platform for distributing converged fixed, mobile, voice, data, Internet, and multimedia services. One of its key visions is to provide seamless global roaming, enabling users to move across borders while using the same number and handset. IMT-2000 also aims to provide seamless delivery of services, over a number of media (satellite, fixed, etc...). It is expected that IMT-2000 will provide higher transmission rates: a minimum speed of 2Mbps for stationary or walking users, and 348 Kbps in a moving vehicle.

## 3. MODERN CELLULAR COMMUNICATION SYSTEMS

In 2001 the 3G systems started with the **FOMA** service in Japan, with several field trials in Europe and with **cdma2000** in South Korea. The first country which introduced 3G on a large commercial scale was Japan. In 2005 about 40% of subscribers use 3G networks only, and 2G is on the way out in Japan. It is expected that during 2006 the transition from 2G to 3G will be largely completed in Japan, and upgrades to the next 3.5G stage with maximum around 14 Mbps data rate is underway.

3G technologies are an answer to the ITU's IMT-2000 specification. Originally, 3G was supposed to be a single, unified, worldwide standard, but in practice, there are two main competing technologies, WCDMA and cdma2000. Also there is another 3G standard called TD-SCDMA, developing by the Chinese Academy of Telecommunications Technology (CATT).

This section describes the stated above 3G systems, lists their main parameters and gives information about their evolutions, like HSDPA/HSUPA for WCDMA and 1x EV-DO/EV-DV for cdma2000 systems.

### 3.1 WCDMA/HSDPA/HSUPA

WCDMA (Wideband Code Division Multiple Access) is a type of 3G cellular network. WCDMA is the technology behind the 3G UMTS standard and is allied with the 2G GSM standard. WCDMA was developed by NTT DoCoMo as the air interface for their 3G network. Later NTT DoCoMo submitted the specification to the ITU as a candidate for the international 3G standard known as IMT-2000. The ITU eventually accepted WCDMA as part of the IMT-2000 family of 3G standards. Later WCDMA was selected as the air interface for UMTS, the 3G successor to GSM.

WCDMA is a wideband Direct-sequence Code Division Multi Access (DS-CDMA) system. Compared to the first DS-CDMA based standard, IS-95, WCDMA uses a three times larger bandwidth equal to 5 MHz, as a result using 3.84 Mcps chip rate. Higher chip rate of 3.84 Mcps enables higher bit rate and provides more multipath diversity than the chip rate of 1.2288 Mcps (IS-95), especially in urban cells. In order to support high bit rates up to 2 Mbps, WCDMA supports the use of variable spreading factor and multicode connections.

WCDMA supports highly variable user data rates and the Bandwidth on Demand (BoD) is well supported. Although, the user data rate is constant during each 10 ms frame, the data capacity among the users can change from frame to frame. WCDMA utilizes fast closed loop power control in both uplink and downlink. Fast power control in the downlink improves link performance and enhances downlink capacity. However, this requires new functionalities in the mobile, such as SIR (signal-to-interference ratio) estimation and outer-loop power control. Also, WCDMA supports both Frequency Division Duplex (FDD) and Time Division Duplex (TDD) operation modes. In the FDD mode, separate 5 MHz carrier frequencies are used for uplink and downlink respectively, whereas in TDD only 5 MHz is time-shared between the uplink and downlink. WCDMA supports the operation of asynchronous base stations, so that there is no need for a global time reference such as GPS. Deployment of indoor and micro base stations is easier when no GPS signal needs to be received.

In standardization forums, WCDMA technology has emerged as the most widely adopted 3G air interface. Its specification has been created in 3GPP. Within 3GPP, WCDMA is called Universal Terrestrial Radio Access (UTRA) FDD and TDD, the name WCDMA being used to cover both FDD and TDD operation. Further, experience from 2G systems like GSM and cdmaOne has enabled improvements to be incorporated in WCDMA. Focus has also been put on ensuring that as much as possible of WCDMA operators' investments in GSM equipment can be reused. Examples are the re-use and evolution of the core network, the focus on co-siting and the support of GSM handover. In order to use GSM handover the subscribers need dual mode handsets.

Inter-frequency handovers are considered important in WCDMA, to maximize the use of several carriers per base station. In cdmaOne inter-frequency measurements are not specified, making inter-frequency handovers more difficult. Also, WCDMA includes transmit diversity mechanism to improve the downlink capacity to support asymmetric capacity requirements between downlink and uplink.

WCDMA supports up to 1920 Kbps data transfer rates (and not 2 Mbps as previously expected), although at the moment users in the real networks can expect performance up to 384 Kbps – in Japan, its evolved version High Speed Down Link Packet Access (HSDPA) will be deployed in 2006 to provide mobile users with higher rate packet services than WCDMA. HSDPA and High Speed Up Link Packet Access (HSUPA) will enable high-speed wireless connectivity comparable to wired broadband. HSDPA/HSUPA enables individuals to send and receive email with large file attachments, play real-time interactive games, receive and send high-resolution pictures and video, download video and music content or stay wirelessly connected to their office PCs – all from the same mobile device.

HSDPA refers to the speed at which individuals can receive large data files, the "downlink." In this respect it extends WCDMA in the same way that EV-DO extends CDMA2000. HSDPA provides a smooth evolutionary path for UMTS networks allowing for higher data capacity (up to 14.4 Mbps in the downlink). It is an evolution of the WCDMA standard, designed to increase the available data rate by a factor of 5 or more. HSDPA defines a new WCDMA channel, the high-speed downlink shared channel (HS-DSCH) that operates in a different way from existing WCDMA channels, but is only used for downlink communication to the mobile.

HSUPA (high-speed uplink packet access) refers to the speed at which individuals can send large data files, the "uplink." HSUPA extremely increases upload speeds up to 5.76 Mbps. HSUPA is expected to use an uplink enhanced dedicated channel (E-DCH) on which it will employ link adaptation methods similar to those employed by HSDPA. Similarly to HSDPA there will be a packet scheduler, but it will operate on a request-grant principle where the MSs request a permission to send data and the scheduler decides when and how many MSs will be allowed to do so. In HSUPA, unlike in HSDPA, soft and softer handovers will be allowed for packet transmissions. Similar to HSDPA, HSUPA is considered **3.75G.**

HSDPA considerably improves the 3G end-user data experience by enhancing downlink performance. HSDPA significantly reduces the time it takes a mobile user to retrieve broadband content from the network. A reduced delay is important for many applications such as interactive games. In general, HSDPA allows a more efficient implementation of "interactive" and "background" Quality of Service (QoS) classes as standardized by 3GPP. HSDPA high data rates also improve the use of streaming applications, while lower roundtrip delays will benefit Web browsing applications. In addition, HSDPA's improved capacity opens the door for new and data-intensive applications that cannot be fully supported with Release 99 because of bandwidth limitations.

### 3.2 cdma2000/1xEV-DO/1xEV-DV

The other significant 3G standard is **cdma2000**, which is an outgrowth of the earlier 2G CDMA standard IS-95. cdma2000's primary proponents are outside the GSM zone in the Americas, Japan and Korea. cdma2000 is managed by 3GPP2, which is separate and independent from UMTS's 3GPP. The various types of transmission technology used in cdma2000 include 1xRTT, cdma2000-1xEV-DO and 1xEV-DV. cdma2000 offers data rates of 144 Kbps to over 3 Mbps. It has been adopted by the International Telecommunication Union - ITU. Arguably the most successful introduction of cdma2000 3G systems is South Korean SK Telecom, which has more than 20 million 3G subscribers. In October 2000, they debuted the world's first commercial CDMA 1x service; and in February 2002, they released the first commercial CDMA 1xEV-DO service, which achieves data rates up to 2.4 Mbps.

Same as IS-95 cdma2000 1x uses one times the chip rate of 1.2288 Mcps. However, in addition, the cdma2000 also supports Spreading Rate 3 (or 3x), which is used when higher data rate transmissions are required. Spreading Rate 3 has two implementation options: DSSS (Direct-sequence spread spectrum) or MCSS (multicarrier spread-spectrum).

On the downlink of the MC system three narrowband 1x carriers, each with 1.25 MHz, are bundled to form a multicarrier transmission with approximately 3.75 MHz (3x) bandwidth. On the uplink, cdma2000 3x system uses the DSSS option, which allows the mobile to directly spread its data over a wider bandwidth using a chip rate of 3.6864 Mcps. To harmonize with other 3G systems such as UMTS WCDMA, a Spreading Rate 3 signal can have 625 kHz of guard band on each side resulting in a total 5 MHz RF bandwidth. Although currently, there do not seem to be commercial commitments for actual adopting the MC mode, but instead the focus has been more on the further development of narrowband operation, wider bandwidth options such as 6x, 9x, and 12x are under consideration for even higher data rate applications.

Launched in South Korea in **2002, cdma2000 1xEV-DO** (1x Evolution-Data Optimized, originally 1x Evolution-Data Only), is an evolution of cdma2000 1x with High Data Rate (HDR) capability added and where the forward link is time-division multiplexed. This 3G air interface standard is denoted as IS-856. 1xEV-DO is capable of delivering data ar speeds comparable to wireline broadband. By dividing radio spectrum into separate voice and data vhannels, cdma2000 1xEV-DO, which uses a 1.25 MHz data channel, improves network efficiency and eliminates the chance that an increase in voice traffic would cause data speeds to drop.

cdma2000 1xEV-DO in its latest revision, Rev. A, supports downlink data rates up to 3.1 Mbps and uplink data rates up to 1.8 Mbps in a radio channel dedicated to carrying high-speed packet data. 1xEV-DO Rev. A was first deployed in Japan and will be deployed in North America in 2006. The Rev. 0 that is currently deployed in North America has a peak downlink data rate of 2.5 Mbps and a peak uplink data rate of 154 Kbps.

**cdma2000 1xEV-DV** (1x Evolution-Data/Voice), is another piece of the 3G CDMA roadmap. Promising efficient, high speed packet data capabilities added to cdma2000 1x circuit-switched voice capability, cdma2000 1xEV-DV supports downlink (forward link) data rates up to 3.1 Mbps and uplink (reverse link) data rates of up to 1.8 Mbps. 1xEV-DV can also support concurrent operation of legacy 1x voice users, 1x data users, and high speed 1xEV-DV data users within the same radio channel.

In 2005, Qualcomm put the development of EV-DV on an indefinite halt, due to lack of carrier interest, mostly because both Verizon Wireless and Sprint are using EV-DO.

## 3.3 TD-SCDMA

**TD-SCDMA** (Time Division-Synchronous Code Division Multiple Access) is a 3G mobile telecommunications standard, being pursued in the People's Republic of China by the Chinese Academy of Telecommunications Technology (CATT).

TD-SCDMA uses TDD, in contrast to the FDD scheme used by WCDMA. By dynamically adjusting the number of timeslots used for downlink and uplink, the system can more easily accommodate asymmetric traffic with different data rate requirements on downlink and uplink than FDD schemes. Since it does not require paired spectrum for downlink and uplink, spectrum allocation flexibility is also increased. Also, using the same carrier frequency for uplink and downlink means that the channel condition is the same on both directions, and the base station can

*Table 1.* Modern cellular systems main parameter

| Parameter | WCDMA | cdma2000 | TD-SCDMA |
|---|---|---|---|
| Multiple access | DS-CDMA | 1x: DS-CDMA; **3x:** MC-CDMA | TDMA, CDMA, FDMA |
| Carrier spacing | 5 MHz | 1x: 1.25 MHz; 3x: 3.75 MHz | 1.6 MHz |
| Chip rate | 3.84 Mcps | 1x: 1.2288 Mcps; 3x: 3.6864 Mcps | 1.28 Mcps |
| Data rate | up to 1920 Kbps (up to 10 Mbps using HSDPA) | 153.6 Kbps, up to 2.4 Mbps with EV-DO and 5.2 Mbps with EV-DV | up to 2 Mbps |
| Duplexing method | FDD/TDD | FDD | TDD |
| Power control frequency | 1500 MHz | 800 Hz in uplink/downlink | Downlink and uplink |
| BS synchronization | Asynchronous | Synchronous | Synchronous |
| Frame length | 10 ms | 5 ms, 10 ms, 20 ms | 10 ms |
| Spreading factors | Variable SF from 4 to 512 | 4~256 UL | 1, 2, 4, 8 and 16 |
| Data modulation | QPSK/ dual-channel QPSK | BPSK/QPSK | QPSK or 8PSK |
| Antenna processing | DL transmit diversity (Space-Time Coding) | DL transmit diversity (Space-Time Spreading) | Smart antenna with beamforming |

deduce the downlink channel information from uplink channel estimates, which is helpful to the application of beamforming techniques.

TD-SCDMA also uses TDMA in addition to the CDMA used in WCDMA. This reduces the number of users in each timeslot, which reduces the implementation complexity of multiuser detection and beamforming schemes, but the non-continuous transmission also reduces coverage (because of the higher peak power needed), mobility (because of lower power control frequency) and complicates radio resource management algorithms.

The "S" in TD-SCDMA stands for "synchronous", which means that uplink signals are synchronized at the base station receiver, achieved by continuous timing adjustments. This reduces the interference between users of the same timeslot using different codes by improving the orthogonality between the codes, therefore increasing system capacity, at the cost of some hardware complexity in achieving uplink synchronization. The standard has been adopted by 3GPP since Rel-4, known as "UTRA TDD 1.28Mcps Option".

We conclude this section by listing the main parameters of modern cellular based networks in **Table 1**.

## 4. WIRELESS DATA SERVICES

While many of the classical mobile phone systems converged to IMT-2000 systems (with cdma2000 and WCDMA/UMTS), the Wireless Local Area Networks (WLAN) area developed more or less independently.

**WLAN** is expected to continue to be an important form of connection in many business areas. The market is expected to grow as the benefits of WLAN are recognized. It is estimated that the WLAN market will have been 0.3 billion US dollars in 1998 and 1.6 billion dollars in 2005. So far WLANs have been installed in universities, airports, and other major public places. Decreasing costs of WLAN equipment has also brought it to many homes.

Early development of WLANs included industry-specific solutions and proprietary protocols, but at the end of the 1990s these were replaced by standards, primarily the various versions of IEEE 802.11 (Wi-Fi). An alternative ATM-like 5 GHz standardized technology, HIPERLAN, has not succeeded in the market, and with the release of the faster 54 Mbps 802.11a (5 GHz) and 802.11g (2.4 GHz) standards, almost certainly never will.

In this section we discuss the most succeed wireless network standards such as IEEE 802.11a/b/g Wi-Fi and IEEE 802.15 family standards including Bluetooth, ZigBee and Wireless USB. Much attention in this chapter is paid to high speed wireless Internet services such as WiBro and WiMax.

### 4.1 IEEE 802.11 Family Standards

IEEE 802.11, the Wi-Fi standard, denotes a set of WLAN standards developed by working group 11 of the IEEE LAN/MAN Standards Committee (IEEE 802). The

802.11 family currently includes six over-the-air modulation techniques that all use the same protocol. The most popular (and prolific) techniques are those defined by the *b*, *a*, and *g* amendments to the original standard. 802.11b and 802.11g standards use the 2.4 GHz band. Because of this choice of frequency band, 802.11b and 802.11g equipment can incur interference from microwave ovens, cordless telephones, Bluetooth devices, and other appliances using this same band. The 802.11a standard uses the 5 GHz band, and is therefore not affected by products operating on the 2.4 GHz band.

The **802.11a** amendment to the original standard was ratified in 1999. The 802.11a standard uses the same core protocol as the original standard, operates in 5 GHz band, and uses a 52-subcarrier orthogonal frequency-division multiplexing (OFDM) with a maximum raw data rate of 54 Mbps, which yields realistic net achievable throughput in the mid-20 Mbps. The data rate is reduced to 48, 36, 24, 18, 12, 9 then 6 Mbps if required.

Since the 2.4 GHz band is heavily exploited, using the 5 GHz band gives 802.11a the advantage of less interference. However, this high carrier frequency also brings disadvantages. It restricts the use of 802.11a to almost line of sight, necessitating the use of more access points; it also means that 802.11a cannot penetrate as far as 802.11b since it is absorbed more readily, other things (such as power) being equal.

802.11a products started shipping in 2001, lagging 802.11b products due to the slow availability of the 5 GHz components needed to implement products. 802.11a was not widely adopted overall because 802.11b was already widely adopted, because of 802.11a's disadvantages, because of poor initial product implementations, making its range even shorter, and because of regulations.

**802.11b** products appeared on the market very quickly, since 802.11b is a direct extension of the DSSS modulation technique defined in the original standard. Hence, chipsets and products were easily upgraded to support the 802.11b enhancements. The dramatic increase in throughput of 802.11b (compared to the original standard) along with substantial price reductions led to the rapid acceptance of 802.11b as the definitive wireless LAN technology.

802.11b is usually used in a point-to-multipoint configuration, wherein an access point communicates via an omni-directional antenna with one or more clients that are located in a coverage area around the access point. Typical indoor range is 30 m at 11 Mbps and 90 m at 1 Mbps. Extensions have been made to the 802.11b protocol (e.g., channel bonding and burst transmission techniques) in order to increase speed to 22, 33, and 44 Mbps, but the extensions are proprietary and have not been endorsed by the IEEE. Many companies call enhanced versions "802.11b+". These extensions have been largely obviated by the development of 802.11g, which has data rates up to 54 Mbps and is backwards-compatible with 802.11b.

In June 2003, a third modulation standard was ratified: **802.11g**. This flavor works in the 2.4 GHz band (like 802.11b) but operates at a maximum raw data rate of 54 Mbps, or about 24.7 Mbps net throughput like 802.11a. 802.11g hardware will work with 802.11b hardware. Details of making *b* and *g* work well together occupied much of the lingering technical process. In older networks, however, the presence of an 802.11b participant significantly reduces the speed of an 802.11g network. The modulation scheme used in 802.11g is OFDM for the data rates of 6, 9, 12, 18, 24, 36, 48, and 54 Mbps. The maximum range of 802.11g devices is slightly greater then that of 802.11b devices, but the range in which a client can achieve full (54 Mbps) data rate speed is much shorter than that of 802.11b. The 802.11g standard swept the consumer world of early adopters starting in January 2003, well before ratification. The corporate users held back and Cisco and other big equipment makers waited until ratification. By summer 2003, announcements were flourishing. Most of the dual-band 802.11a/b products became dual-band/tri-mode, supporting a, b, and g in a single mobile adaptor card or access point. Despite its major acceptance, 802.11g suffers from the same interference as 802.11b in the already crowded 2.4 GHz range. Devices operating in this range include microwave ovens, Bluetooth devices, and cordless telephones.

In January 2004 IEEE announced that it had formed a new **802.11n** Task Group to develop a new amendment to the 802.11 standard for wireless local-area networks. The real data throughput is estimated to reach a theoretical 540 Mbit/s (which may require an even higher raw data rate at the physical layer), and should be up to 100 times faster than 802.11b, and well over 10 times faster than 802.11a or 802.11g. It is projected that 802.11n will also offer a better operating distance than current networks.

802.11n builds upon previous 802.11 standards by adding MIMO (multiple-input multiple-output). MIMO uses multiple transmitter and receiver antennas to allow for increased data throughput through spatial multiplexing and increased range by exploiting the spatial diversity, perhaps through coding schemes like Alamouti coding.

The Enhanced Wireless Consortium (EWC) was formed to help accelerate the IEEE 802.11n development process and promote a technology specification for interoperability of next-generation WLAN products.

According to the IEEE 802.11 Working Group Project Timelines, the 802.11n standard is not due for final approval until July 2007.

**Table 2** summarizes this section and list main parameters of 802.11 standards family.

*Table 2.* IEEE 802.11 family standards major parameters

| | 802.11 | 802.11a | 802.11b | 802.11g | 802.11n |
|---|---|---|---|---|---|
| Release Date | 1997 | 1999 | 1999 | 2003 | Expected mid.2007 |
| Frequency band | 2.4 GHz | 5 GHz | 2.4 GHz | 2.4 GHz | 2.4 GHz |
| Data Rate (typical) | 1 Mbps | 25 Mbps | 6.5 Mbps | 25 Mbps | 200 Mbps |
| Data rate (max) | 2 Mbps | 54 Mbps | 11 Mbps | 54 Mbps | 540 Mbps |
| Range (indoor) | 30 meters | 10 meters | 30 meters | 30 meters | 50 meters |
| Transmission | FHSS, DSSS, IR | DSSS with CCK | OFDM | OFDM, CCK DSSS+ DB(Q)PSK | MIMO-OFDM |

## 4.2 IEEE 802.15 Family Standards

15th working group of the IEEE 802 specializes in Wireless PAN (Personal Area Network) standards. It includes four task groups, numbered from 1 to 4.

Task group 1 or **802.15.1** derived a WPAN standard based on **Bluetooth** specification, which is the simple choice for convenient, wire-free, short-range communication between devices. It is a globally available standard that wirelessly connects mobile phones, portable computers, cars, stereo headsets, MP3 players, and more (**Figure 6**). Bluetooth was designed to fill a range of use cases or applications. To improve interoperability, Bluetooth Profiles were written to make sure that the application level works the same way across different manufacturers' products.

Bluetooth radios operate in the unlicensed ISM band at 2.4 GHz using 79 channels between 2.402 GHz to 2.480 GHz (23 channels in some countries). The range for Bluetooth communication is 10 meters with a power consumption of 0dBm (1mW).

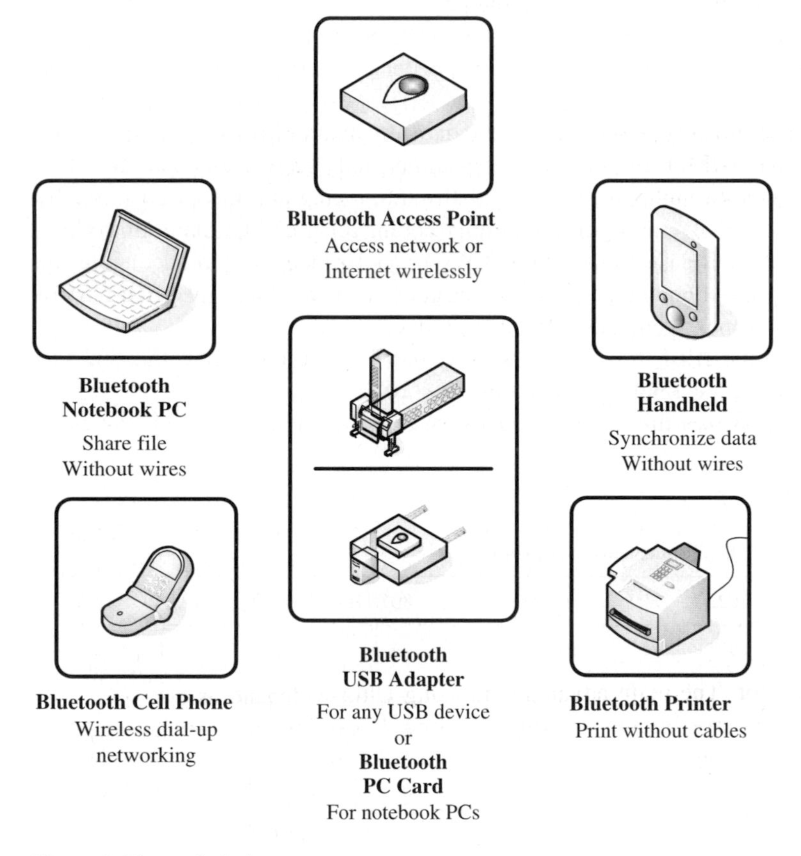


*Figure 6.* Bluetooth devices

This distance can be increased to 100 meters by amplifying the power to 20dBm. The Bluetooth radio system is optimized for mobility.

The name Bluetooth was born from the 10th century king of Denmark, King Harold Blaatand (whose surname is sometimes written as Bluetooh), who engaged in diplomacy which led warring parties to negotiate with each other. The inventors of the Bluetooth technology thought this a fitting name for their technology which allowed different devices to talk to each other.

The Bluetooth specification was first developed by Ericsson (now Sony Ericsson), and was later formalized by the Bluetooth Special Interest Group (**SIG**). The SIG was formally announced on May 20, 1999. It was established by Sony Ericsson, IBM, Intel, Toshiba and Nokia, and later joined by many other companies as Associate or Adopter members.

Bluetooth technology already plays a part in the rising Voice over IP (VOIP) scene, with Bluetooth headsets being used as wireless extensions to the PC audio system. As VOIP becomes more popular, and more suitable for general home or office users than wired phone lines, Bluetooth may be used in Cordless handsets, with a base station connected to the Internet link.

In March 2006, the Bluetooth Special Interest Group (SIG) announced its intent to work with UWB (ultra-wideband) manufacturers to develop a next-generation Bluetooth technology using UWB technology and delivering UWB speeds. This will enable Bluetooth technology to be used to deliver high speed network data exchange rates required for wireless VOIP, music and video applications.

The IEEE **802.15.3** High Rate Task Group (TG3) for WPANs is chartered to draft and publish a new standard for high-rate (20Mbit/s or greater) WPANs. Besides a high data rate, the new standard will provide for low power, low cost solutions addressing the needs of portable consumer digital imaging and multimedia applications. Another member of 802.15 family is the IEEE 802.15 High Rate Alternative PHY Task Group (TG3a) or **802.15.3a** is working to define a project to provide a higher speed **Ultra-wideband (UWB)** PHY enhancement amendment to 802.15.3 for applications which involve imaging and multimedia.

**Ultra-Wideband (UWB)** is a recently allocated unlicensed spectrum (3.1–10.6 GHz) that provides an efficient use of scarce radio bandwidth while enabling both high data rate personal-area network wireless connectivity as well as long-range, low data rate applications. UWB was previously defined as an impulse radio, but the industry now views it as an available bandwidth set with an emissions limit that enables coexistence without harmful interference.

Due to its extremely short range, UWB is limited to the same sort of devices that Bluetooth is used for. The main advantage to using Ultra-wideband as opposed to Bluetooth is, as the name implies, bandwidth speed. Excepting any interference, a UWB device could theoretically achieve transfer speeds of up to 1 Gbps (today's Bluetooth devices have a theoretical limit of 3Mbps). The ranges of applications for these kinds of speeds are staggering even given the range limitations of UWB.

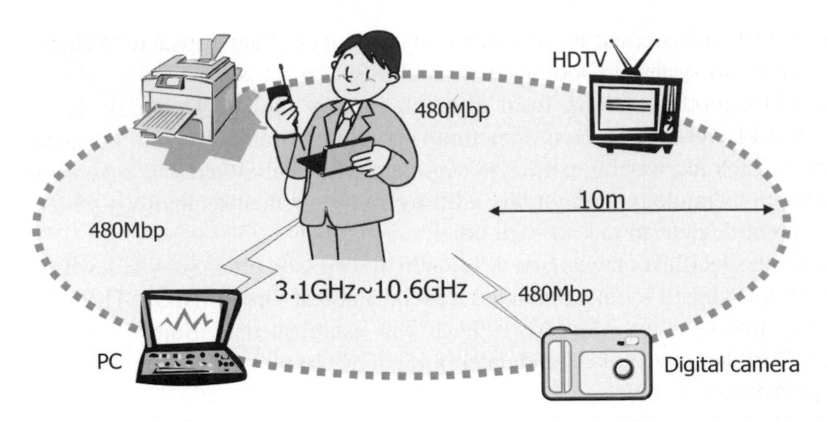


*Figure 7.* UWB applications example

As **Figure 7** indicates, UWB is a potential market includes a broad spectrum of products and applications. One typical scenario is promising wireless data connectivity between a host and associated peripherals such as keyboards, mouse, printer, scanner, and so on. A UWB link functions as a 'cable replacement' with transfer data rate requirements that range from 1000 Kbps for wireless mouse to 100 Mbps for rapid file sharing or download of images/graphic files. Additional driver applications relate to streaming of digital media content between consumer electronics appliances, such as digital TVs, VCRs, CD/DVD players, MP3 players and so on. In summary UWB is seen as having potential for applications that to date have not been fulfilled by other wireless short-range technologies currently available, such as, 802.11 LANs or Bluetooth PANs.

One of the technologies fully utilizing the advantages of UWB is the **Wireless USB (WUSB)**. WUSB is a new wireless extension to USB intended to combine the speed and security of wired technology with the ease-of-use of wireless technology. WUSB is based on ultra wideband wireless technology defined by WiMedia (***IEEE 802.15.3a***), which operates in the range of 3.1–10.6 GHz.

Wireless USB supports the 480 Mbps data rate over a distance of two meters. If the speed is lowered to 110 Mbps, UWB will go a longer distance (up to 10 meters). WUSB supports so-called dual-role devices, which in addition to being a WUSB client device, can function as a host with limited capabilities. For example, a digital camera could act as a client when connected to a computer, and as a host when transferring pictures directly to a printer.

WUSB will be used in devices that are now connected via regular USB cables, such as game controllers, printers, scanners, digital cameras, MP3 players, hard disks and flash drives, but it is also suitable for transferring parallel video streams.

4th and last member of IEEE 802.15 family is the **IEEE 802.15.4** was chartered to investigate a low data rate solution with multi-month to multi-year battery life and very low complexity. This standard specifies operation in the unlicensed 2.4 GHz, 915 MHz and 868 MHz ISM bands. The raw, over-the-air data rate is

250 Kbps per channel in the 2.4 GHz band, 40 Kbps per channel in the 915 MHz band, and 20 Kbps in the 868 MHz band. Transmission range is between 10 and 75 meters.

ZigBee is the most succeed technology based on 802.15.4 standard. ZigBee's current focus is to define a general-purpose, inexpensive, self-organizing, mesh network that can be used for industrial control, embedded sensing, medical data collection, smoke and intruder warning, building automation, interactive toys, smart badges, remote controls, and home automation, etc. (**Figure 8**). The resulting network will use very small amounts of power so individual devices might run for a year or two using the originally installed battery.

We summarize the IEEE 802.15 based standards major parameters in **Table 3**.

### 4.3 WiBro/Mobile WiMax (IEEE 802.16e)

In February 2002, Korean government allocated 100 MHz bandwidth of 2.3GHz spectrum band for **WiBro** (Wireless Broadband) system. WiBro allows subscribers to use high-speed Internet more cheaply and more widely, even when moving at speeds of about 60 km (37 miles) per hour. WiBro base stations will offer an aggregate data throughput of 30 to 50 Mbps and cover a radius of 1–5 km allowing for the use of portable Internet usage within the range of a base station. From testing during the APEC Summit in Pusan in late 2005, the actual range and bandwidth were quite a bit lower than these numbers. The technology will also offer Quality of Service. The inclusion of QoS allows for WiBro to stream video content and other loss-sensitive data in a reliable manner. The WiBro system was developed as a regional and potentially international alternative to 3.5G systems, which delivers

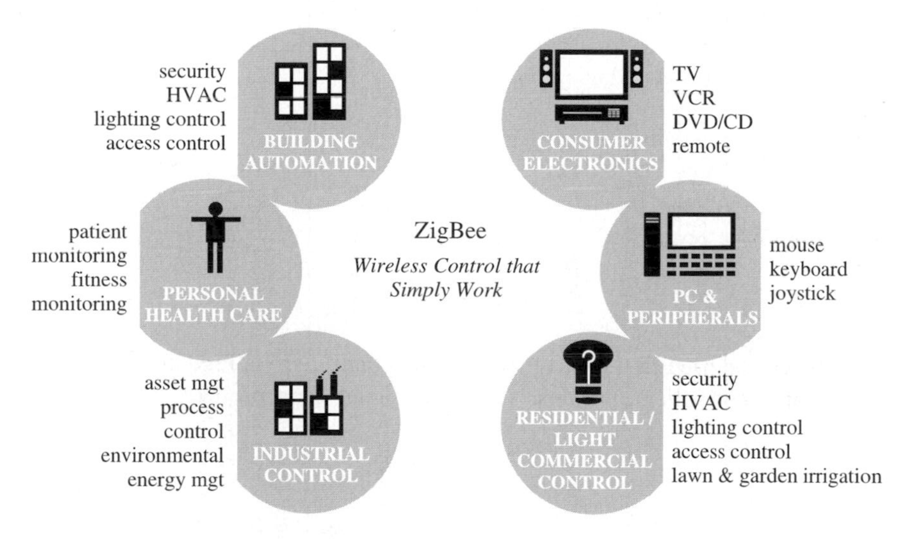


*Figure 8.* ZigBee applications example

*Table 3.* 802.15 based standards

| Key features | Bluetooth 802.15.1[a] | 802.15.3[b] | 802.15.3a UWB/HR[c] | 802.15.4 ZigBee[d] |
|---|---|---|---|---|
| Status of standard | Approved | Approved | Under discussion | Approved |
| Operating Frequency | 2.4-2.4835 GHz ISM band | 2.4-2.4835 GHz ISM band | 3.1-10.6 GHz | 868-868.6 MHz<br>2.4-2.4835 GHz ISM band |
| Max. data rate | 1 Mbps | QPSK: 11Mbps<br>64QAM:55mbps | 110Mbps (<10m)<br>200Mbps (4m)<br>480Mbps (2m) | 250Kbps<br>40Kbps<br>20Kbps |
| Max. range | 10m (opt. 100m) | 10 m | 10 m | 30 m |
| Modulation | GFSK | D-QPSK,<br>16-,32-,64QAM | BPSK, QPSK | BPSK, QPSK |
| Spreading | DS-FH | N/A | Multiband OFDM or DS-SS | DS-SS |
| Max. transmit power | 0 dBm<br>20 dBm for 100m | 100 mW | -41.3 dBm/MHz<br>0.562 mW | 20 mW |
| Cost | $5 | Unknown | $20~ | $2.50 |

[a] http://www.bluetooth.com
[b] http://www.ieee802.org/15/pub/TG3.html
[c] http://www.ieee802.org/15/pub/TG3a.html
[d] http://www.zigbee.com

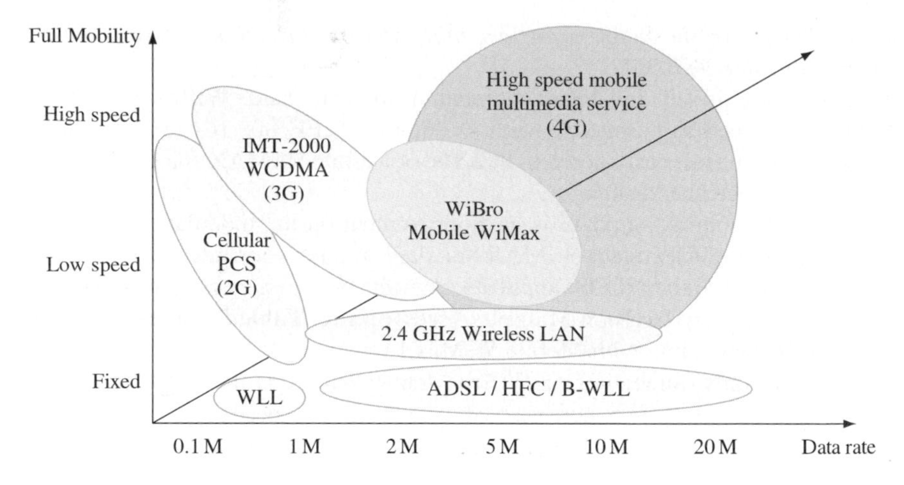


*Figure 9.* WiBro service location

superior spectral efficiency and end-user throughput than today's 3G networks, and acts as a transition to 4G (**Figure 9**).

**Table 4** lists the major parameters and radio access requirements for WiBro system. WiBro adopts OFDMA/TDD for multiple-access and duplex schemes, and aims to provide a high data rate wireless Internet access with PSS (Personal Subscriber Stations) under the stationary or mobile environment, regardless of the place and time. WiBro supports the various types of wireless multimedia applications and various types of multimedia-enabled terminals such as handsets, notebook, PDA or smart phone. Depending on urban environment WiBro service supports different cell types with different cell radius. Pico- and micro-cell radiuses equal to 100 m and 400 m, respectively, whereas macro-cell service coverage is up to 1 km. System supports mobile users at a velocity of up to 60 km/h, although last trials showed that this parameter can be upgraded up to 100 km/h.

These all appear to be the stronger advantages over another wireless broadband access standard called **WiMax (Worldwide Interoperability for Microwave Access)**. Formed in April 2001 to promote conformance and interoperability of the standard IEEE 802.16, the WiMax Forum describes WiMAX as "*a standards-based*

*Table 4.* WiBro system major parameters and radio access requirements

| Major system parameters | | Radio access requirements | |
|---|---|---|---|
| Duplexing | TDD | Frequency reuse factor | 1 |
| Multiple Access | OFDMA | Mobility | <100 Km/h |
| Frequency band | 2.3 GHz | Service coverage | <1 Km |
| System Bandwidth | 10 MHz | Throughput | Max. DL/UL=3/1 Mbps<br>Min. DL/UL=512/128 Kbps |
| Sampling frequency | 10 MHz | Handoff | <150 ms |

*technology enabling the delivery of last mile wireless broadband access as an alternative to cable and DSL."*

Given the lack of self developed momentum as a standard, WiBro has joined WiMAX and agreed to harmonize with the similar IEEE 802.16-2005 standard, formerly named but still best known as **802.16e** or **Mobile WiMAX** standard, which is approved in December, 2005.

The WiMAX mobility standard is an improvement on the modulation schemes stipulated in the original (fixed) WiMAX standard. It allows for fixed wireless and mobile Non Line of Sight (NLOS) applications primarily by enhancing the OFDMA (Orthogonal Frequency Division Multiple Access). From **Table 5** you can see that WiBro is fully compatible with Mobile WiMax.

Currently the only competitor of WiBro/Mobile Wimax is the HSDPA, which enables WCDMA operators to increase peak data download speeds fivefold. **Table 6** compares the main parameters of both, HSDPA and Mobile WiMax systems. It is

*Table 5.* Comparison between WiMax and WiBro systems

| Items | WiMax (fixed) 802.16d | 802.16e | |
|---|---|---|---|
| | | Mobile WiMax | WiBro |
| Frequency (GHz) | 3.5, 5.8 | 2.3, 2.5, 3.5, etc. | 2.3 |
| Bandwidth (MHz) | 3.5, 7, 10, 14 | 3.5, 7, 8.75, 10, 14 | 10 |
| Duplex method | TDD/FDD | TDD | TDD |
| Multiple access | TDMA | OFDMA | OFDMA |

*Table 6.* HSPDA vs. Mobile WiMAX

| Attribute | HSDPA | Mobile WiMAX |
|---|---|---|
| Base standard | WCDMA | IEEE 802.16e |
| Duplex method | FDD | TDD (FDD opt.) |
| DL multiple access | CDM-TDM | OFDM |
| UL multiple access | CDMA | OFDMA |
| Channel BW | 5 MHz | Scalable: 4.375, 5, 8.75, 10 MHz |
| BS-to-BS distance | 2.8 km | 2.8 km |
| Modulation DL | QPSK/16QAM | QPSK/16QAM/64QAM |
| Modulation UL | BPSK/QPSK | QPSK/16QAM |
| Coding | CC, Turbo | CC, Turbo |
| Peak data rate | DL:14 Mbps UL: 2 Mbps | 46(1:1)~54(3:1) Mbps (DL/UL combined (32,14), (46,8)) |
| H-ARQ | Fast 6-channel Asycnhronous CC | Multi-channel Asynchronous CC |
| Scheduling | Fast in DL | Fast in UL and DL |
| Handoff | Network initiated hard | Network optimized hard |
| Tx diversity and MIMO | Simple open & closed loop | STBC, SM |

expected that in areas where HSDPA becomes widely available, like Western Europe, and where well-suited spectrum for 802.16e is rare, the window of opportunity for mobile WiMax will be quite limited. However the scalable architecture, high data throughput and low cost deployment make WiBro/Mobile WiMAX a leading solution for wireless broadband services. The high data throughput enables efficient data multiplexing and low data latency. Attributes essential to enable broadband data services including data, streaming video and VoIP with high quality of service (QoS).

In June 2006 SK Telecom and KT launched the world first commercial WiBro service in Korea. Market research results shows that by the end of year 2006 it is expected about 610 thousand subscribers join the WiBro network, and within the next few years number of WiBro users increases rapidly. The scalable architecture, high data throughput and low cost deployment make WiBro/Mobile WiMAX a leading solution for wireless broadband services. The high data throughput enables efficient data multiplexing and low data latency. Attributes essential to enable broadband data services including data, streaming video and VoIP with high quality of service (QoS).

## 5. WIRELESS BROADCASTING SERVICES

The rushing trend of digital revolution has resulted in personalized and mobile communication service and created a new stream of personalized and mobile TV broadcasting service. Mobile digital broadcast TV combines the two best-selling consumer products in history—TVs and mobile phones. Mobile digital TV (DTV) is already becoming a delivery mechanism for TV broadcasts, bringing new interactivse content to a new generation of customers that wants both communications and entertainment – all in one place.

Like all new technologies, there are several different standards for mobile DTV around the world. These include three primary open standards developed by industry associations with contributions from multiple players in the mobile DTV marketplace:

- DMB (digital media broadcast) has deployed today in Korea with several handsets already in-market to support the standard and is expanding to Europe and other parts of Asia.
- DVB-H (digital video broadcast-handheld) is quickly gaining ground with trials in Europe, the U.S. and parts of Asia.
- ISDB-T (integrated services digital broadcast-terrestrial) is the standard in Japan.

**Digital Multimedia Broadcasting** – a digital transmission system for sending data, radio and TV to mobile devices such as mobile phones. It can operate via satellite (**S-DMB**) or terrestrial (**T-DMB**) transmission (**Figure 10**). DMB is based on the Eureka 147 DAB standard.

In May 2005, SK Telecom (South Korea) launched a satellite DMB (S-DMB) service that delivers high-quality video broadcasts to a mobile phone or car-based

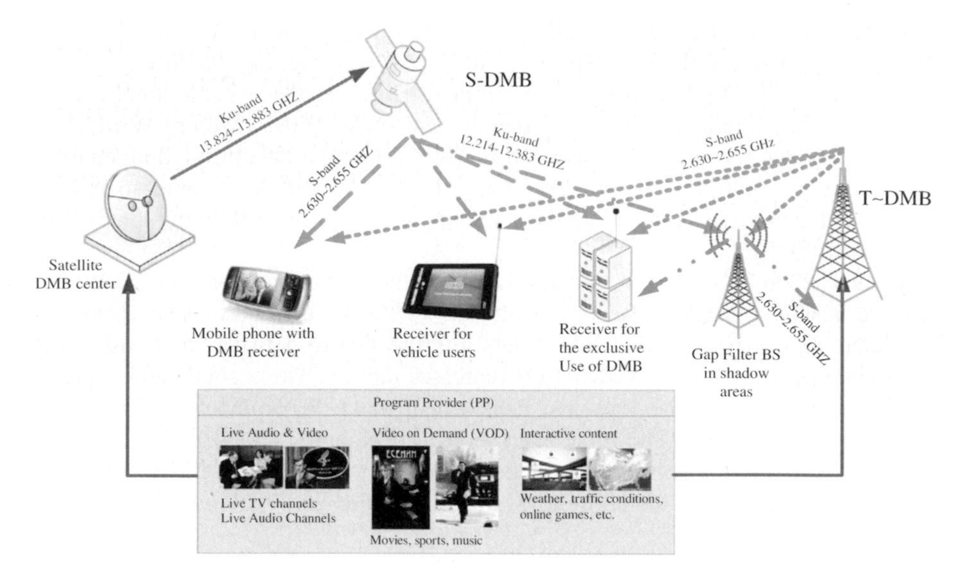


*Figure 10.* S-DMB and T-DMB services working principle

video entertainment system. SK Telecom is initially delivering 11 video channels, 25 audio channels, and 3 data channels. In S-DMB service customers can receive the signal transmitted by the satellite directly from most areas on the ground. However, there are some shadow areas such as subways, tunnels, inside buildings, etc. However, the signal receiving areas can be extended by installing Gap Fillers (Base Stations) in those shadow areas.

S-DMB utilizes Ku-Band (13.824~13.883GHz) between the Signal Transmission Center and satellite (the 144th degree of east longitude), and S-Band (2.630~2.655 GHz, 25MHz) is utilized between satellite (Or Gap Filler) and the terminals. Further, Ku-Band (12.214~12.239GHz) is used between satellite and Gap Filler (Base Station). The S-DMB service adopts the same Code Division Multiplexing (CDM) technology as the mobile phone service. Thus it is the most appropriate for signal reception in a mobile environment. This can also guard against multiple channel interferences that cause reductions in signal receiving quality within the mobile environment.

T-DMB is an ETSI standard (TS 102 427 and TS 102 428). Currently, DMB is in use in a number of countries. South Korea, in particular, started T-DMB service in December 1, 2005. Some T-DMB trials are currently planned around Europe:

- Germany will launch T-DMB for the world cup 2006
- France currently makes a trial in Paris
- Switzerland and Italy prepare a trial for 2006
- UK launch a trial for 2006

Main competitor of DMB system in mobile TV field is the **DVB-H** system. DVB-H is a technical specification for bringing broadcast services to handheld receivers and was formally adopted as ETSI standard EN 302 304 in November 2004 and has some similarities with the competing mobile TV standard DMB. DVB-H is the latest development within the set of DVB transmission standards. DVB-H technology adapts the successful DVB-T system for digital terrestrial television to the specific requirements of handheld, battery-powered receivers. DVB-H can offer a downstream channel at high data rates which can be used standalone or as an enhancement of mobile telecoms networks which many typical handheld terminals are able to access anyway.

DVB-H is designed to work in the following bands:

- VHF-III (174–230 MHz, or a portion of it)
- UHF-IV/V (470–830 MHz, or a portion of it)
- L –band (1.452–1.492 GHz)

DVB-H can coexists with DVB-T in the same multiplex. DVB-H trials are now underway in Helsinki, Berlin, Oxford, Pittsburgh, Paris, Madrid, Sydney, South Africa, The Hague, Delhi, Bern and Erlangen. Commercial launches of DVB-H services are expected in 2006 in Finland, Italy, Albania and in the USA. In Germany, DVB-H will be launched nationwide in 2007.

Another system that provides a digital TV to mobile users is Japanese **ISDB-T**. ISDB-T was adopted in commercial transmissions in Japan in December 2003. It comprises a market of about 100 million television sets. ISDB-T had 10 million subscribers by the end of April 2005. ISDB-T was pointed out as the most flexible of all for better answering the necessities of mobility and portability. It is most efficient for mobile and portable reception.

ISDB-T is characterized by the following features:

- ISDB-T can transmit a HDTV (High Definition Television) channel and a mobile phone channel within the 6 MHz bandwidth usually reserved for TV transmissions.
- ISDB-T allows switching to two or three SDTV (Standard Definition Television) channels instead of one HDTV channel (multiplexing SDTV channels).
- ISDB-T provides interactive services with data broadcasting.
- ISDB-T supports Internet access as a return channel that works to support the data broadcasting. Internet access is also provided on mobile phones.
- ISDB-T allows HDTV to be received on moving vehicles at over 100 km/h.
- 1seg is a mobile terrestrial digital audio/video broadcasting service in Japan. The 1seg can be received on mobile phones moving at a speed over 400 km/h.

Mobile DTV is coming to a phone. It's true that the technology will likely first take off in urban centers with heavy commuters and with teenagers and the younger population. But merging a mobile phone with a TV is something that everyone can understand. And with our universal hunger for information and connectivity, mobile DTV presents the perfect opportunity for users to stay informed and up to date on what is happening in the news, with their favorite sports team and even their favorite reality TV show or soap opera.

We finalize this section by comparing the main parameters of major digital broadcasting services in **Table 7**.

*Table 7*. Comparison of beyond 3G and 4G systems

| Key features | WiBro | 3G LTE | IEEE 802.20 | 4G |
|---|---|---|---|---|
| Spectrum | 2.3 GHz | 2.5~2.6 GHz | | 3~5 GHz |
| Bandwidth | 10 MHz (20 MHz) | 5MHz, 10MHz, 15MHz, 20MHz | 5MHz, 10MHz, 15MHz, 20MHz | 5~100 MHz |
| Multiple Access | OFDMA/TDD | OFDMA/FDD, TDD;SC-FDMA, | OFDMA/FDD, TDD | OFDMA, MC-CDMA, ??? |
| Service | Portable Internet/ High-speed Wireless Internet | High-speed mobile service | High-speed mobile service | Ubiquitous Broadband Convergence |
| Peak Data Rate | 30/50 Mbps@ 3 km/h | 100 Mbps@ 3 km/h | 100 Mbps@ 3 km/h | 120 Mbps@ 100 km/h<br>1/3 Gbps@ 3km/h |
| Mobility | ~120 km/h | ~350 km/h | ~350 km/h | ~350 km/h |
| Remarks | WiBro-II: better performance compared to 3G LTE and 802.20 | Similar to 802.20 | Supported by QUALCOMM | High coverage, QoS |
| Service starting time | WiBro: 09.2005<br>WiBro Evol.: 12. 2007. | July 2007 | Suspected until Oct. 2006 | 2010~2015 |

| Parameter | DMB | MediaFLO | DVB-H | ISDB-T |
|---|---|---|---|---|
| Deploying countries | South Korea, Europe | USA | USA, Europe | Japan |
| Video/Audio codec | H.264/MPEG4 | H.264/MPEG2 | H.264/MPEG2 | H.264/MPEG2 |
| Channel BW/Rate | 6MHz/9.2Mbps | 5~8MHz/ 11.2Mbps | 8MHz/ 15Mbps | 6MHz/ 23Mbps |
| Modulation | OFDM | OFDM | OFDM | OFDM |
| Available channels | 11 TV and 25 radio | 19 TV and 10 radio | 8 TV and 12 radio | -/- |

## 6. FUTURE MOBILE COMMUNICATION SYSTEMS

While the current third generation systems still heavily rely on classical telephone technology in the network infrastructure, future systems will offer users the choice of many different networks based on the Internet. No one knows exactly when and how this common platform will be available. Companies have to make their money with 3G systems first. However, already Japanese company NTT DoCoMo is talking about the introduction of **4G**. 4G technology stands to be the future standard of wireless devices. The NTT DoCoMo is testing 4G communications under brand name super 3G (or 3.9G) at 100 Mbps while moving, and 1 Gbps while stationary. NTT DoCoMo plans on releasing the first commercial network in 2010. **Figure 11** shows a mobile communication roadmap. Despite the fact that current wireless devices seldom utilize full 3G capabilities, there is a basic attitude that if the pipeline is provided then services for it will follow. In general, a generation is defined by the result of technology changes over a 10–15 year time frame. Thus,

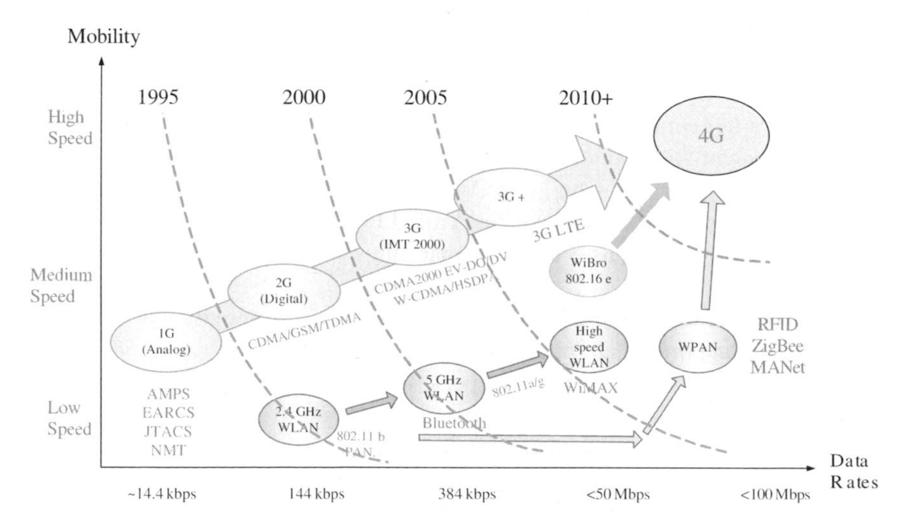


*Figure 11.* Mobile communication roadmap

4G would refer to whatever is deployed in the 2010–2015 period, assuming 3G deployment spans the 2000–2005 period. Typically, this means a new air-interface with higher data rates in the least, and some see change in the way data transport is handled end-to-end.

Currently, different international organizations such as, ITU-R, ITU-T, 3GPP, 3GPP2 are discussing visions of systems beyond 3G. The major drivers for these systems are the increasing demand for personal mobile communications with respect to performance, applications, traffic, and easy to use. Users are expecting mobile IP applications comparable to the home, a variety of services and applications with a wide spread of usage patterns, data rates and traffic volume. Despite there are different organizations working on 4G development, however all of them define the key issues of 4G systems as accessing information anywhere, anytime, with a seamless connection to a wide range of information, data, pictures, audio, video, and so on.

The term 4G is used broadly to include several types of broadband wireless access communication systems, not only cellular systems. The 4G systems not only will support the next generation of mobile services, but also will support the fixed wireless networks. 4G systems will have broader bandwidth, higher data rate, and smoother and quicker handoff and will focus on seamless service across a multitude of wireless systems and networks.

The future infrastructures of 4G will consist of a set of various networks using IP as a common protocol so that users will be able to choose every application and environment. Currently All-IP network has been tipped as the most probable technology to be synonymous with 4G systems.

The first motivation for an All-IP network is that currently it is the best architectural choice to enable a wide variety of innovative and commercially lucrative services. A second reason is that the relative technical simplicity of the basic IPs, as well as the ability of IP itself to act as a unifying abstraction or waist that hides and divides the complexity of the protocol stack, also helps to make application development for IP networks easier. This has resulted in the rapid development and deployment of a large number of Internet applications, such as, e-mail, messaging, gaming and content distribution. Simplified All-IP network example shown in **Figure 12**.

ITU is currently proposed a new concept called IMT-Advanced, which is responsible for the overall system aspects with capabilities which go further than that of IMT-2000.

The new capabilities of these IMT-Advanced systems are envisaged to handle a wide range of supported data rates according to economic and service demands in multi-user environments with target peak data rates of up to approximately 100 Mbps for high mobility such as mobile access and up to approximately 1 Gbps for low mobility such as nomadic/local wireless access, which enables the favorable conditions to develop the multimedia based services. For example with 1 Gbps data rate mobile subscribers can download the 100 mp3 music files in 2.4 seconds, one 800 Mb movie file in 5.6 seconds, and 20Mbps rate HDTV signal in 12.5 seconds.

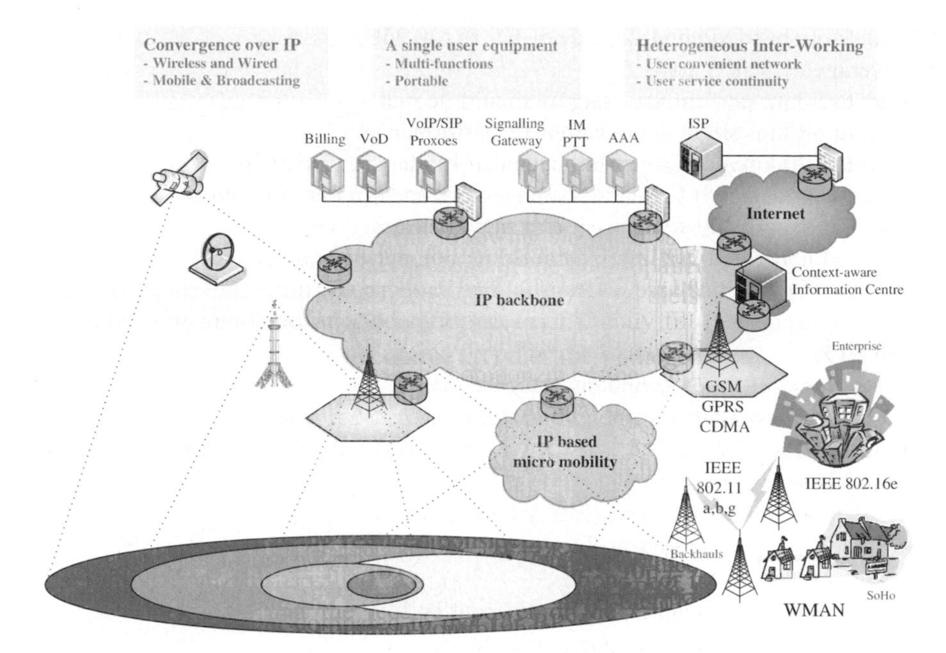


*Figure 12.* All-IP network example

To support this wide variety of services, it may be necessary for IMT-Advanced to have different radio interfaces and frequency bands for mobile access for highly mobile users and for new nomadic/local area wireless access.

One of the strong migration paths toward IMT-Advanced is 3G LTE (Long Term Evolution) concept proposed by 3GPP. LTE aims to ensure the continued competitiveness of the 3GPP technologies for the future, and focuses on enhancement of UTRA and optimization of the UTRAN architecture. Followings are the LTE requirements for 4G systems:

- Radio Access Network Latency: Significantly reduced latency
- Mobility:
  - Optimized for low mobile speed, higher mobile speed supported with high performance (up to 350 km/h)
  - Voice and other real-time services via Packet Service Domain with at least same quality as supported by 3GPP R6 over the whole of the speed range
- Co-existence and interworking with 3GPP Radio Access Technology
- Mobility between 3GPP and non-3GPP systems
- Peak data rate:
  - max 100 Mbps@20 MHz (DL) and 50 Mbps@20 MHz (UL)
  - DL: 2 receiver antennas with up to 16-QAM, UL: 1 transmitting antenna with 16-QAM
  - Spectrum efficiency: 3 to 4 times 3GPP r6 HSDPA (5bps/Hz), 2 to 3 times R6 enhanced UL (2.5 bps/Hz)

  - Scalable bandwidth: 1.25, 2.5, 5, 10, 15, 20 MHz
- Coverage
  - up to 5 km: performance targets should be met
  - up to 30 km: slight degradations of performance
  - up to 100 km: should not be precluded by the specifications
- Efficient (enhanced) E-UTRAN architecture and smooth migration
- Efficient support for transmission of higher layers
- Reduced complexity: minimization of the number of options.

Although, currently there are several solutions proposed by different vendors to meet the 4G requirements, only few of them can be defined as predominant candidates (**Figure 13**).

Below some of these technologies are described:

- *Multiple access schemes*: OFDMA, MC-CDMA, SC-FDMA
  - Strong desire of OFDMA for DL and SC-FDMA for UL by most companies
- *Modulation and coding*
  - Adaptive Modulation and Coding (AMC).
  - Hybrid Automatic Repeat Request (H-ARQ)
- *Duplex technique*
  - Hybrid Division Duplexing (considering both, TDD and FDD schemes)
- *Inter-cell interference avoiding* at cell edge
  - Interference randomization proposed by RITT, CATT, and others.
  - Interference cancellation method proposed by Panasonic and ETRI
  - Interference coordination/avoidance methods by Siemens, Alcatel and ETRI.
- *MIMO* (Multi Antenna Techniques)
  - 31 proposed MIMO technologies by different vendors (Samsung, Qualcomm, Motorola, NTT DoCoMo, etc.)

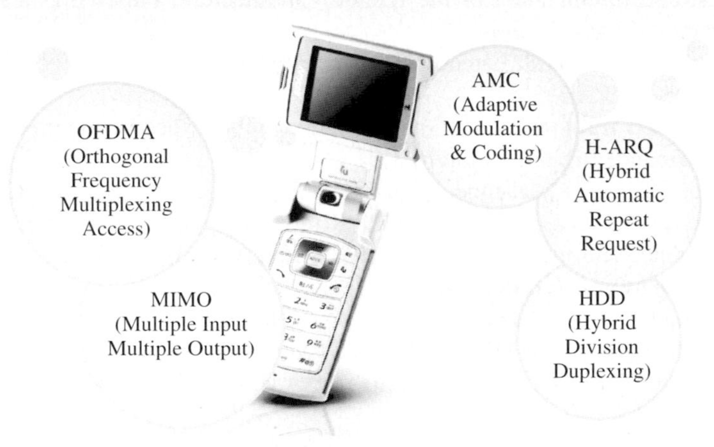


*Figure 13.* Solutions for next generation mobile communications

- *Cell search schemes*
  - Hot topic and intensive discussion currently
- *Bandwidth scalability*: SCH/BCH position at 20 MHz bandwidth
- *Random access* schemes: synchronized RACH, non-synchronized RACH, TFSTD (Time-Frequency Switching Transmit Diversity) for RACH.
  - Hot topic and intensive discussion currently.

We finalize this chapter by giving the main parameters of beyond 3G and 4G systems in **Table 7**

## REFERENCES


1. Agrawal D.P., Zeng Q., 2003, *Introduction to Wireless and Mobile Systems,* Brook/Cole, Pacific Grove.
2. Alamouti S.M., 2006, *WhyMAX: Technology Overview and Comparative Performance of Mobile WiMAX,* Wireless Broadband World Forum, Seoul.
3. Etoh M., 2005, *Next Generation Mobile Systems 3G and Beyond,* Wiley, Wiltshire.
4. Fazel K., Kaiser S., 2003, *Multi-Carrier and Spread Spectrum Systems,* Wiley, Wiltshire.
5. Glisik S.G., 2006, *Advanced Wireless Networks. 4G Technologies,* Wiley, Wiltshire.
6. Harada H., Prasad R., 2002, *Simulation and Software Radio for Mobile Communications,* Artech House, London.
7. Holma H., Toskala A., 2004, *WCDMA for UMTS,* 3rd ed., Wiley, Cornwall.
8. Hong D., 2004, *2.3 GHz Portable Internet (WiBro) for Wireless Broadband Access,* ITU Telecom Asia 2004 Forum, Seoul.
9. Korhonen J., 2003, *Introduction to 3G Mobile Communications,* 2nd ed., Artech House, Norwood.
10. Lee J.S., Miller L.R., 1998, *CDMA Systems Engineering Handbook,* Artech House, London.
11. Marks R., 2006, *IEEE 802.16 WirelessMAN Standard for Broadband Wireless Metropolitan Area Networks: Evolution to 802.16e and Beyond,* Wireless Broadband World Forum, Seoul.
12. Mishra A.R., 2004, *Fundamentals of Cellular Network Planning and Optimisation. 2G/2.5G/3G... Evolution to 4G,* Wiley, Wiltshire.
13. Nee R., Prasad R., 2000, *OFDM Wireless Multimedia Communications,* Artech House, London.
14. Rappaport T.S., 2002, *Wireless Communications. Principles and Practice,* 2nd ed., Prentice Hall, Upper Saddle River.
15. Resnick R, 2006, *Mobile WiMAX "The Global Platform for Wireless Broadband Services",* Wireless Broadband World Forum, Seoul.
16. Schiller J., 2003, *Mobile Communications,* 2nd ed., Addison-Wesley, Kent.
17. Schulze N, Luders C., 2005, *Theory and Applications of OFDM and CDMA Wideband Wireless Communications,* Wiley, Wiltshire.
18. Tachikawa K., 2002, *W-CDMA Mobile Communication System,* Wiley, Comwall.
19. Webb W., 1998, *Understanding Cellular Radio,* Artech House, London.
20. Wee K.J, 2006, *Principles for the Standardization and Harmonization of IMT-Advanced,* Wireless Broadband World Forum, Seoul.
21. Wikipedia The Free Encyclopedia, http://www.wikipedia.org
22. Wilkinson N., 2002, *Next Generation Network Services,* Wiley, Guildford.
23. Yang S.C., 2004, *3G CDMA2000 Wireless System Engineering,* Artech House, Norwood.
24. Yang S.C., 1998, *CDMA RF System Engineering,* Artech House, London.
25. Zigangirov K.Sh., 2004, *Theory of Code Division Multiple Access Communication,* IEEE press and Wiley-Interscience, Piscataway.

CHAPTER 2

# RADIO ACCESS TECHNIQUES

YONGWAN PARK[1] AND JEONGHEE CHOI[2]

[1] *Department of Information and Communication Engineering, Yeungnam University, 214-1 Dae-dong, Gyeongsan-si, Gyeongsanbuk-do, Korea, 712-749*

[2] *School of Computer and Communications Engineering, Daegu Univeristy, Jinryang, Gyeongsan-si, Gyeongsanbuk-do, Korea, 712-714*

**Abstract:** This chapter describes the key features of radio access techniques. Introducing the main multiple access methods such as FDMA, TDMA, however much attention in this chapter is paid to CDMA. Especially Section 2 gives the detailed information on CDMA concept, spreading process and spreading codes used by different CDMA standards. Also key factors of CDMA based systems such as handover processing, power control, frequency reuse factor and system capacity are also given by this section. At the end of the chapter we discuss the basics of multi-carrier transmission, introducing such multi-carrier modulation schemes as OFDM, OFDMA, MC-CDMA, MC DS-CDMA and MC CS-CDMA

**Keywords:** TDMA, FDMA, CDMA, PN sequence, Walsh code, spreading, despreading, frequency reuse factor, sectorization gain, power control, handover, MRC, EGC, Rake receiver, multi-carrier, OFDM, OFDMA, MC-CDMA, MC DS-CDMA, MC CS-CDMA

## 1. INTRODUCTION TO MULTIPLE ACCESS TECHNIQUES

In communications, a channel access method is used to share a communications channel or physical communications medium between multiple users. The goal in the design of cellular systems is to be able to handle as many calls as possible in a given bandwidth with some reliability. In a mobile cellular system, each mobile station (MS) can distinguish a signal from, the serving base station (BS) and differentiate the signals from neighboring BSs. To accommodate a number of users, many traffic channels need to be made available. In principle there are three different ways to allow access to the channel: frequency, time, and code division multiplexing and is addressed by three multiple access techniques. Next we introduce these techniques – frequency division multiple access (FDMA), time division multiple access (TDMA), and code division multiple access (CDMA).



*Y. Park and F. Adachi (eds.), Enhanced Radio Access Technologies for Next Generation Mobile Communication*, 39–79.

## 1.1 Frequency Division Multiple Access

If a system employs different carrier frequencies to transmit the signal for each user, it is called FDMA system. FDMA subdivides the total bandwidth into $N_c$ narrowband sub-channels which are available during the whole transmission time. This requires bandpass filters with sufficient stop band attenuation. The concept of FDMA is shown in

**Figure 1**, three different users are assigned different frequency bands, transmitting own data at the same time duration. In order to cope with frequency deviations of local oscillators and to minimize interference from adjacent channels, a sufficient guard band is left between two adjacent spectra.

**Figure 2** shows the simple example of how FDMA works. Couples of users are talking each other at the same time, but separated in frequency domain. While message "OK" is transmitted by the Frequency Channel 1, message "ME" is transmitted by Frequency Channel 2.

The main advantages of FDMA are in its low required transmit power, robustness against multipath fading and easy frequency planning. However, its drawbacks in cellular system might be the implementation of $N_c$ modulators and demodulators at the BS, sensitivity to narrow band interference and bandwidth leakage due to guard bands.

FDMA has been widely adopted in analog systems for portable and automobile telephones. FDMA technique also successfully exploited in GSM cellular system. 124 channels, in GSM 900 and 374 channels in GSM 1800, each 200 kHz wide, are used for FDMA.

## 1.2 Time Division Multiple Access

TDMA is a popular multiple access technique, which is used in several international standards. TDMA splits a single carrier wave into time slots and distributes the slots among multiple users. As shown in

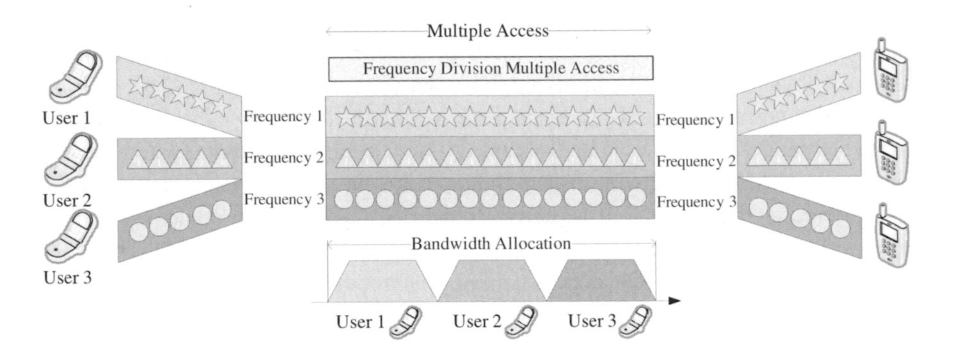


*Figure 1.* Concept of FDMA

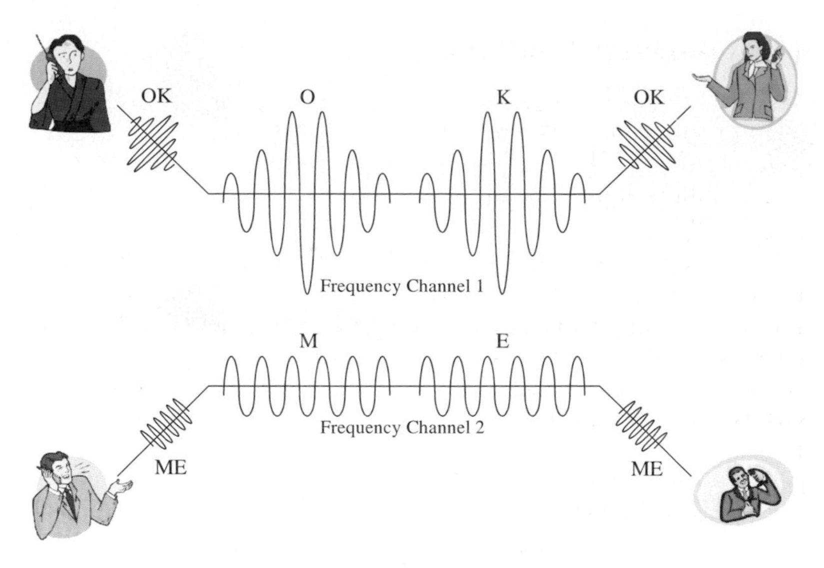


*Figure 2.* Example of FDMA working principle

**Figure 3**, three different users sharing the same frequency spectrum, but separated in time domain. The users transmit in rapid succession, one after the other, each using their own timeslot. This allows multiple users to share the same transmission medium (e.g. radio frequency) whilst using only the part of its bandwidth they require.

**Figure 4** demonstrates TDMA working principle, two couple of users talking via using of same frequency channel, but separated by different time slots. While

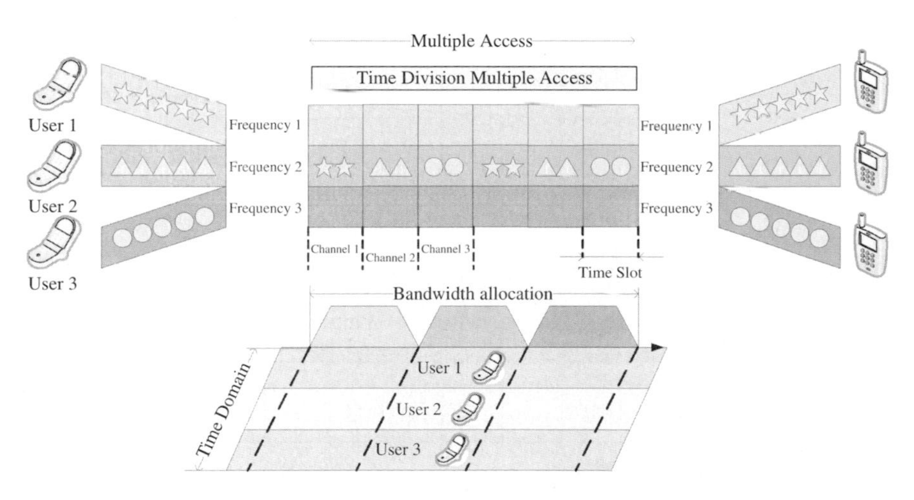


*Figure 3.* Concept of TDMA

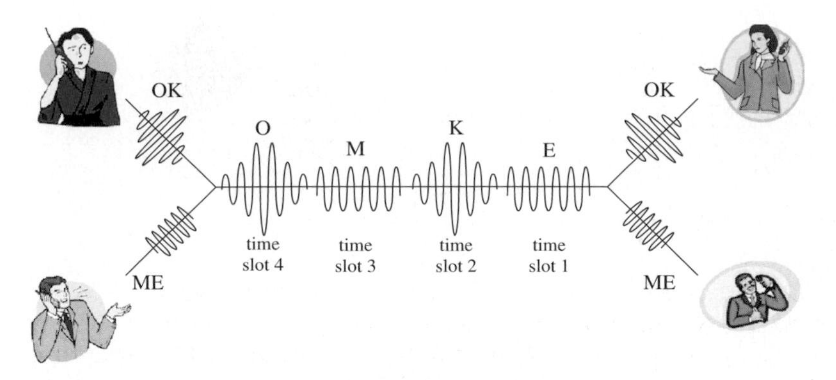


*Figure 4.* Example of TDMA working principle

message "OK" occupies 2nd and 4th time slots, message "ME" is transmitted during the 1st and 3rd time slots.

TDMA system may be in two modes: FDD (Frequency division duplex) and TDD (Time division duplex). FDD and TDD are the techniques used to separate uplink and downlink channels signals. Channel structures of TDMA/FDD and TDMA/TDD systems are illustrated in **Figures 5** and **6**, respectively.

Time division duplex has a strong advantage in the case where the asymmetry of the uplink and downlink data speed is variable. As the amount of uplink data increases, more bandwidth can be allocated to that and as it shrinks it can be taken away. Another advantage is that the uplink and downlink radio paths are likely to be very similar in the case of a slow moving system. This means that techniques such as beamforming work well with TDD systems.

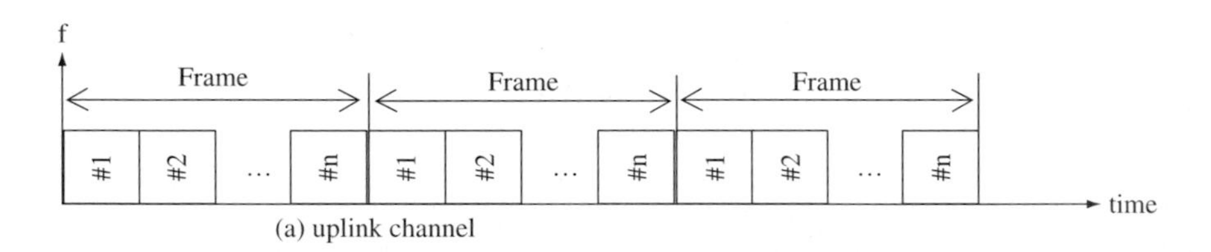


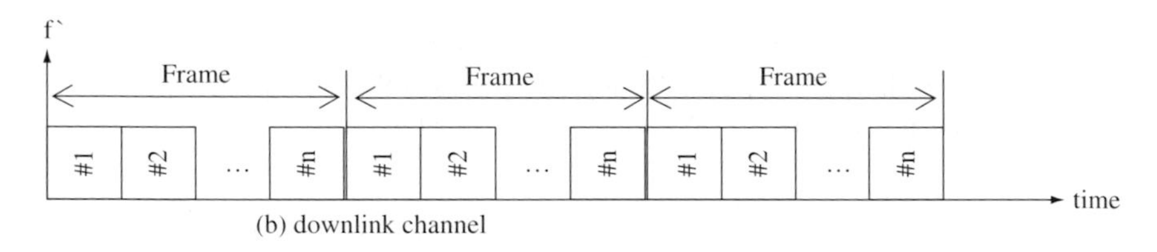


*Figure 5.* Structure of uplink and downlink channels in a TDMA TDD system

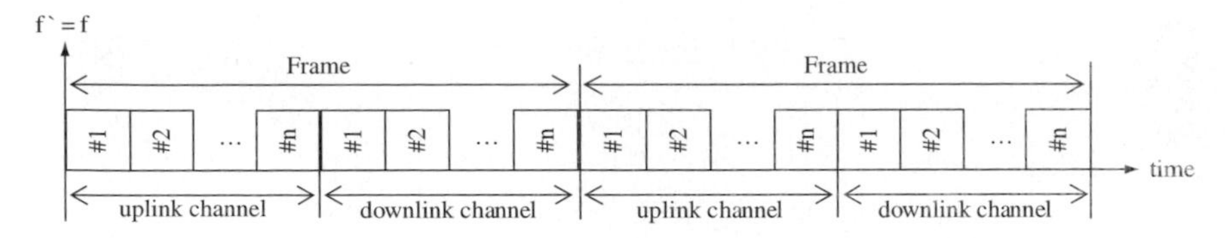


*Figure 6.* Structure of uplink and downlink channels in a TDMA TDD system

In FDD the uplink and downlink sub-bands are said to be separated by the "frequency offset". Frequency division duplex is much more efficient in the case of symmetric traffic. In this case TDD tends to waste bandwidth during switchover from transmit to receive, has greater inherent latency, and may require more complex and more power-hungry circuitry.

Another advantage of FDD is that it makes radio planning easier and more efficient since base stations do not "hear" each other (as they transmit and receive in different sub-bands) and therefore will normally not interfere each other. With TDD systems, care must be taken to keep guard bands between neighboring base stations (which decreases spectral efficiency) or to synchronize base stations so they will transmit and receive at the same time (which increases network complexity and therefore cost, and reduces bandwidth allocation flexibility as all base stations and sectors will be forced to use the same uplink/downlink ratio).

A major advantage of TDMA is that the radio part of the mobile only needs to listen and broadcast for its own timeslot. For the rest of the time, the mobile can carry out measurements on the network, detecting surrounding transmitters on different frequencies. Disadvantages of TDMA system are high transmit power, sensitivity to multipath and difficulty in frequency planning.

TDMA technique mostly used in digital cellular standards, such as, GSM, PDC, D-AMPS, and others. TDMA is also used extensively in satellite systems, local area networks and physical security systems. While early efforts to incorporate TDMA into UMTS failed, leaving UMTS as a purely CDMA technology, China still continue pursuing TD-SCDMA (time division synchronous CDMA) for the next generation mobile system.

## 1.3 Code Division Multiple Access

CDMA is a form of multiplexing and a method of multiple access that does not divide up the channel by time (as in TDMA), or frequency (as in FDMA), but instead encodes data with a certain code associated with a channel and uses the constructive interference properties of the signal medium to perform the multiplexing.

As shown in **Figure 7** in CDMA system, different spread-spectrum codes are assigned to each user, and multiple users share the same frequency, at the same time, but are separated from each other via use of a set of orthogonal codes, thus, codes 1010 and 1100 are assigned to User 1 and User 2, respectively, whereas User 3 exploits code 1001 for spreading. In the receiver side original data can be

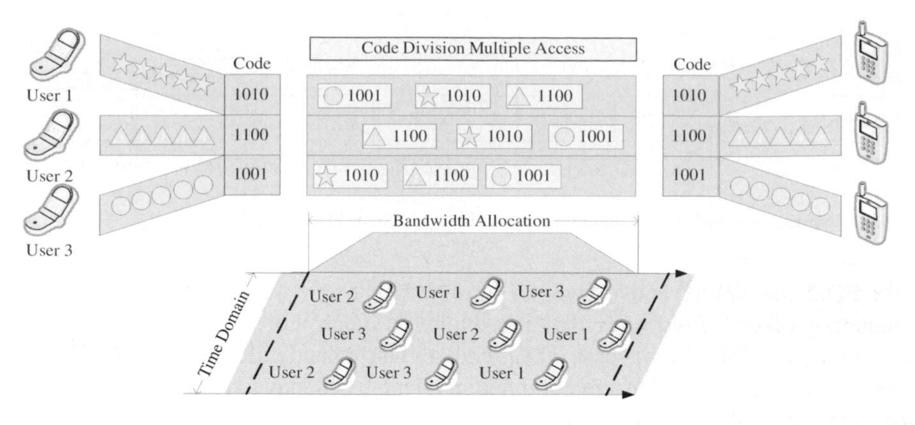


*Figure 7.* Concept of CDMA

recovered only if the receiver knows the code that used to spread received data. Otherwise, received signals will be identified as interference or noise from other users.

**Figure 8** shows the example of CDMA working principle. Both of the messages, "OK" and "ME" transmitted simultaneously in the same frequency channel, but separated by different codes.

CDMA system is based on spread spectrum technology, which provides some advantages which are primarily of interest in secure communication systems, e.g., low probability of intercept or robustness to jamming. In a spread spectrum communication system users employ signals which occupy a significantly larger bandwidth than the symbol rate. Bandwidth spreading is accomplished before the transmission through the use of a code that independent of the transmitted data. The same code is used to demodulate the data at the receiving end.

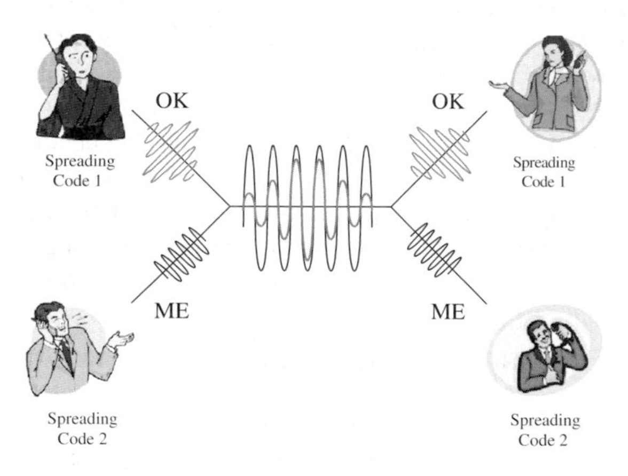


*Figure 8.* Example of CDMA working principle

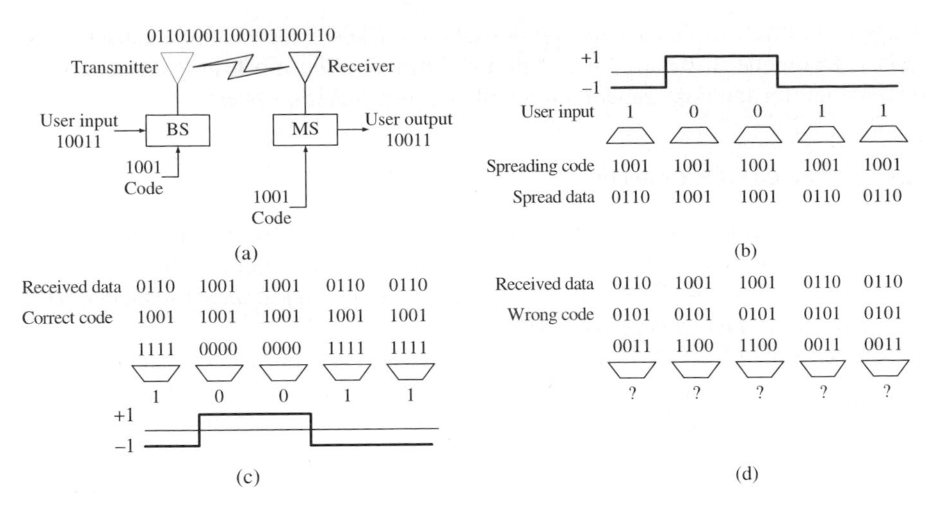


*Figure 9.* CDMA working principle, (a) general scheme, (b) spreading process, (c) despreading using correct code, (d) despreading using wrong code

**Figure 9** gives more detail explanation of spreading and dispreading process. In **Figure 9a** general spreading-despreading process is shown, user input data 10011 is spread by sequence 1001 which results in spread data 0110 1001 1001 0110 0110, this process is given by **Figure 9b**. Next, if encoded sequence is received by correct, destined user original data can be recovered easily via using of correct, spreading code (1001), which is shown in **Figure 9c** As shown in **Figure 9d** in case when spread data is received by wrong user, original data cannot be recovered without knowledge about correct spreading sequence.

Nowadays CDMA is most widely used multiple access scheme and integrated into almost all 3G mobile phone systems like WCDMA (UMTS) and cdma2000. Compared to other schemes CDMA has the advantage of a soft handover and soft capacity. Soft capacity in CDMA systems describes the fact that CDMA systems can add more and more users to a cell with tradeoff the QoS (Quality of Service) in system, i.e. there is no hard limit. Also CDMA can be used in combination with FDMA/TDMA access schemes to increase the capacity of a cell and frequency planning in CDMA systems is much easier compared to the TDMA/FDMA schemes. However there are several disadvantages within the CDMA system, such as, complexity of receivers compared to FDMA/TDMA based networks, needs more complicated power control for senders due to the *near-far problem* (see Section 2.3).

## 2. CDMA BASED SINGLE CARRIER RADIO ACCESS

Code division multiple access communication is rapidly replacing time- and frequency-division methods as the cornerstone of wireless communication and mobile radio. Above we have discussed multiple access techniques, including

CDMA, in brief. In this section we describe the CDMA based single carrier radio access technique in detail, since it is now attracting the most attention as a core technology for the next-generation mobile communication system.

## 2.1 The CDMA Concept

In CDMA systems all users transmit in the same bandwidth simultaneously, but are separated from each other via the use of a set of spreading sequences (codes). There are two basic types of CDMA implementation methodologies: *direct sequence (DS)* systems and *frequency hopping (FH)*.

The concept of DS CDMA is shown in **Figure 10**. As illustrated in **Figure 10** each user's narrowband signal is spread over a wider bandwidth. This wider bandwidth is greater than the minimum bandwidth required to transmit the information. Each user's narrowband signal is spread by a different wideband code. Each of the codes is orthogonal to one another, and channelization of simultaneous users is achieved by the use of this set of spreading codes. All the spread wideband signals of different users are added together to form a composite signal, which is transmitted over the radio channel in the same frequency band. The resulting composite signal has a noiselike spectrum, and in fact can be intentionally made to look like noise to all but the intended radio receiver. The receiver is able to distinguish among the different users by using a copy of the original code. The receiver separates the desired user out of the composite signal by correlating the composite signal with original code. All the users with codes that do not match the code of the desired users are rejected.

**Figure 11** shows concept of FH CDMA system. In a FH method, a pseudo-random sequence is used to change the radio signal frequency, across a broad frequency band in a random fashion. A spread-spectrum modulation implies that the radio transmitter frequency hops from channel to channel in a predetermined but pseudorandom manner. The RF signal is dehopped at the receiver end via frequency synthesizer controlled by a pseudorandom sequence generator synchronized to the transmitter's pseudorandom sequence generator. If the rate change of the carrier frequency is greater than the symbol rate, then the system is referred to as a *fast frequency hopping system*. In **Figure 12a** there are several frequency hops during the one data symbol, hence this system refers to *fast frequency hopping*. Fast FH

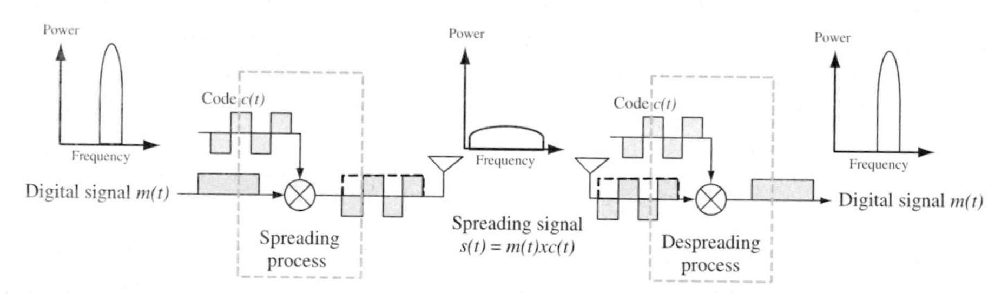


*Figure 10.* Concept of DS-CDMA

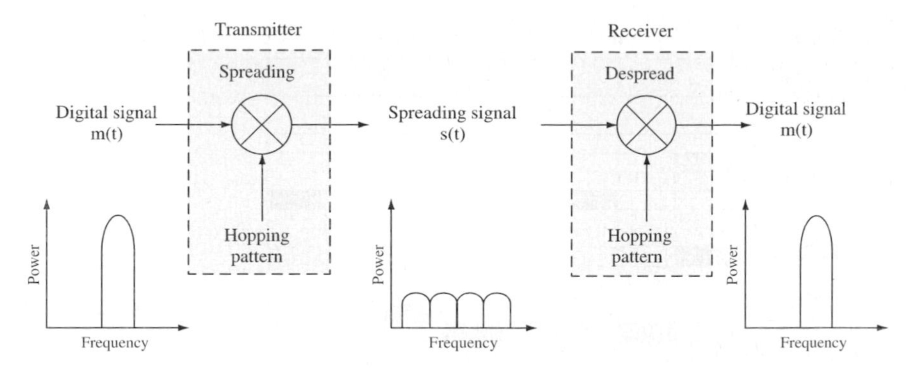


*Figure 11.* Concept of FH-CDMA

systems are very expensive with current technologies and are not at all common. If the channel changes at a rate less than or equal to the symbol rate, it is called *slow frequency hopping*. As demonstrated in **Figure 12b**, *slow FH* there are several data symbols during the one frequency hop duration, in other words symbol duration is smaller than the hop duration.

A frequency hopped system provides a level of security, especially when a large number of channels are used, since an unintended receiver that does not know the pseudorandom sequence of frequency slots must return rapidly to search for the signal it wishes to intercept. Multiple simultaneous transmission from several users is possible using FH, as long as each uses different frequency hopping sequences and none of them "collide" at any given instant of time.

**Figure 13** shows the working principle of a DS CDMA scheme by the example of three users simultaneously transmitting three separate messages $m_1(t)$=[+1,+1, -1], $m_2(t)$,=[1, -1, +1] and $m_3(t)$=[-1, +1, -1]. Users are separated from each other by the multiplication of orthogonal codes $c_1(t)$, $c_2(t)$, and $c_3(t)$, which are equal to [+1, -1, +1, -1], [+1, +1, -1, -1], and [+1, -1, -1, +1], respectively. Messages $m_1(t)$, $m_2(t)$ and $m_3(t)$ are multiplied by the code $c_1(t)$, $c_2(t)$ and $c_3(t)$, respectively. The resulting products are added together and form the signal $s(t) = m_1(t) \times c_1(t) + m_2(t) \times c_2(t) + m_3(t) \times c_3(t)$ which is transmitted through the radio channel. At receiver side to recover the each user's data received signal $r(t)$ is first multiplied by the code sequences $c_1(t)$, $c_2(t)$ and $c_3(t)$, assigned to users 1, 2, and 3, respectively. Next, the resulting products $r(t) \times c_1(t)$, $r(t) \times c_2(t)$, *and* $r(t) \times c_3(t)$, are integrated over one symbol period. After decision (or detection) and digital-to-analog conversion procedures we get the desired stream of information. In our example we assume the ideal channel with no fading and perfect synchronization of the codes at the receiver, hence the received signal $r(t)$is equal to$s(t)$, and the recovered messages $\tilde{m}_1(t), \tilde{m}_2(t)$ *and* $\tilde{m}_3(t)$ match the original messages perfectly.

**Figure 14** shows waveforms for the three user messages $m_1(t)$, $m_2(t)$ and $m_3(t)$, orthogonal codes $c_1(t)$, $c_2(t)$, $c_3(t)$, and corresponding spread messages $m_1(t) \times c_1(t)$, $m_2(t) \times c_2(t)$ and $m_3(t) \times c_3(t)$. In this example the chip rate ($1/T_c$) of the

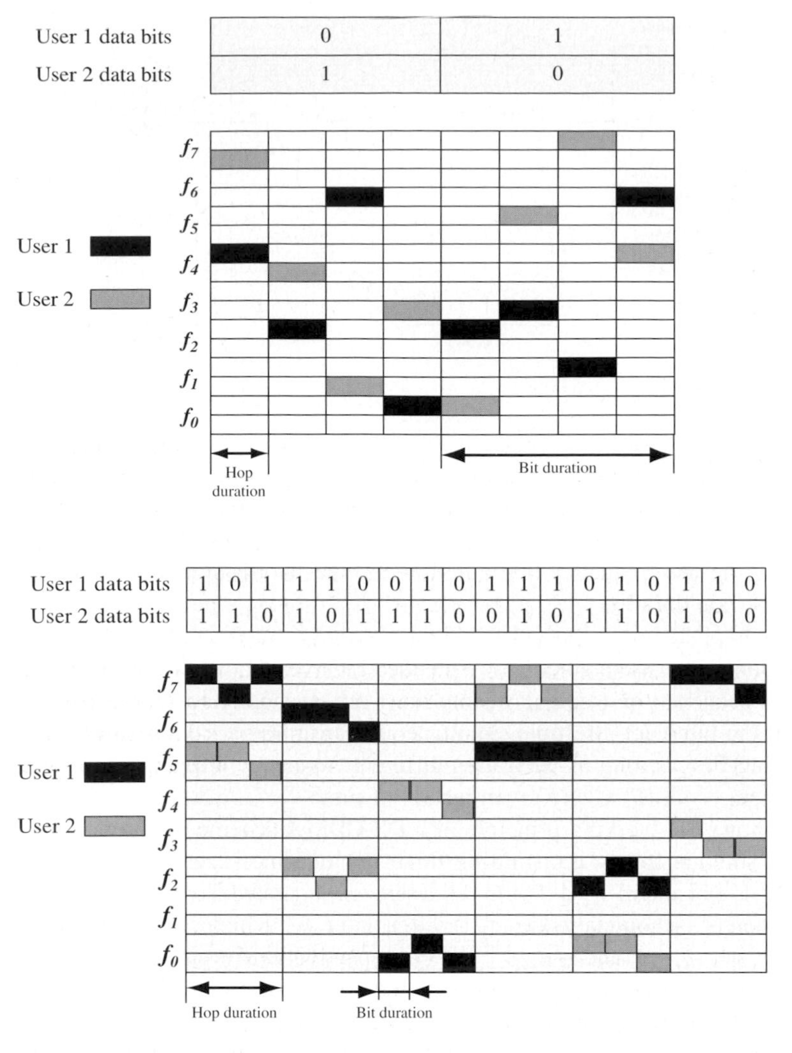


*Figure 12.* An example of frequency hopping patter. (a) fast frequency hopping, (b) slow frequency hopping

orthogonal code is running at four times the bit rate ($1/T_b$). Therefore, we have a processing gain or an effective bandwidth expansion factor of four.

**Figure 15** shows the waveforms at the different points of the receiver. As we mentioned above the received signal $r(t)$ is equal to $s(t)$ and is the summation of the three signals, in other words

$$r(t) = \sum_{n=1}^{n} m_n(t)c_n(t), \tag{1}$$

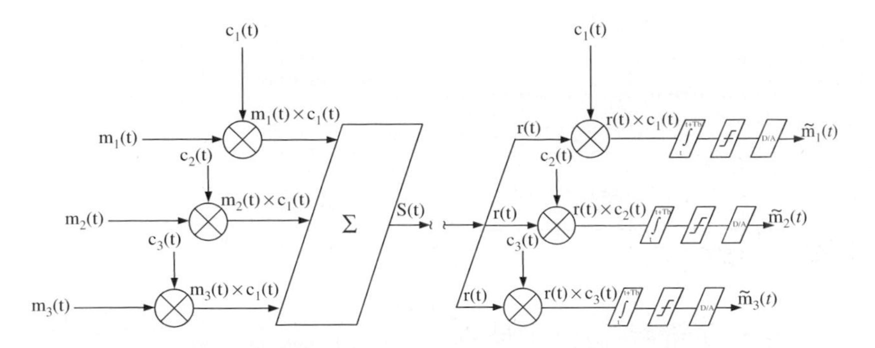


*Figure 13.* CDMA system example

where $N$ is the total number of users (in this example $N = 3$). Multiplying the $r(t)$ by the three corresponding orthogonal codes $c_1(t)$, $c_2(t)$ and $c_3(t)$ we obtain waveforms represent by $r(t) \times c_1(t)$, $r(t) \times c_2(t)$, and $r(t) \times c_3(t)$, respectively.

**Figure 16** shows the signals at the output of the integrators (bold line) and recovered messages (dash line). The integrator adds up the signal power over one bit interval of the baseband message, and the decision threshold decides, based on the output of the integrator, whether or not the particular bit is a +1 or -1. If the output of the integrator greater than 0, then the decision is a +1, otherwise +1. The digital-to-analog (D/A) converter transforms the decision into the recovered waveforms $\tilde{m}_1(t)$, $\tilde{m}_2(t)$ and $\tilde{m}_3(t)$. Since in our example the received signal $r(t)$ is equal to $s(t)$, the recovered messages $\tilde{m}_1(t)$, $\tilde{m}_2(t)$ and $\tilde{m}_3(t)$ match the original messages perfectly.

## 2.2 Spreading Codes

From previous sections we know that in CDMA system all users share the same RF band. In order to avoid mutual interference, spreading codes are used to separate single users. In the IS-95 CDMA system downlink and uplink channels use different codes to channelize individual users. The downlink uses orthogonal *Walsh codes*, whereas uplink uses *PN (pseudorandom noise) codes* for channelization. In cdma2000 new type of code, the *Quasi-orthogonal code* is introduced.

For CDMA there are three conditions that must be met by a set of spreading sequences:

1. The cross-correlation should be zero or very small.
2. Each sequence in the set has an equal number of 1s and -1s, or the number of 1s differs from the number of -1s by at most one.
3. The scaled dot product of each code should equal to 1.

*Walsh codes* used in IS-95 are a set of 64 binary orthogonal sequences. In the forward CDMA link, Walsh codes are used to separate users. In any given sector, each forward channel is assigned a distinct Walsh code. In the reverse CDMA link, the 64 Walsh sequences are used for waveform encoding. One Walsh function is

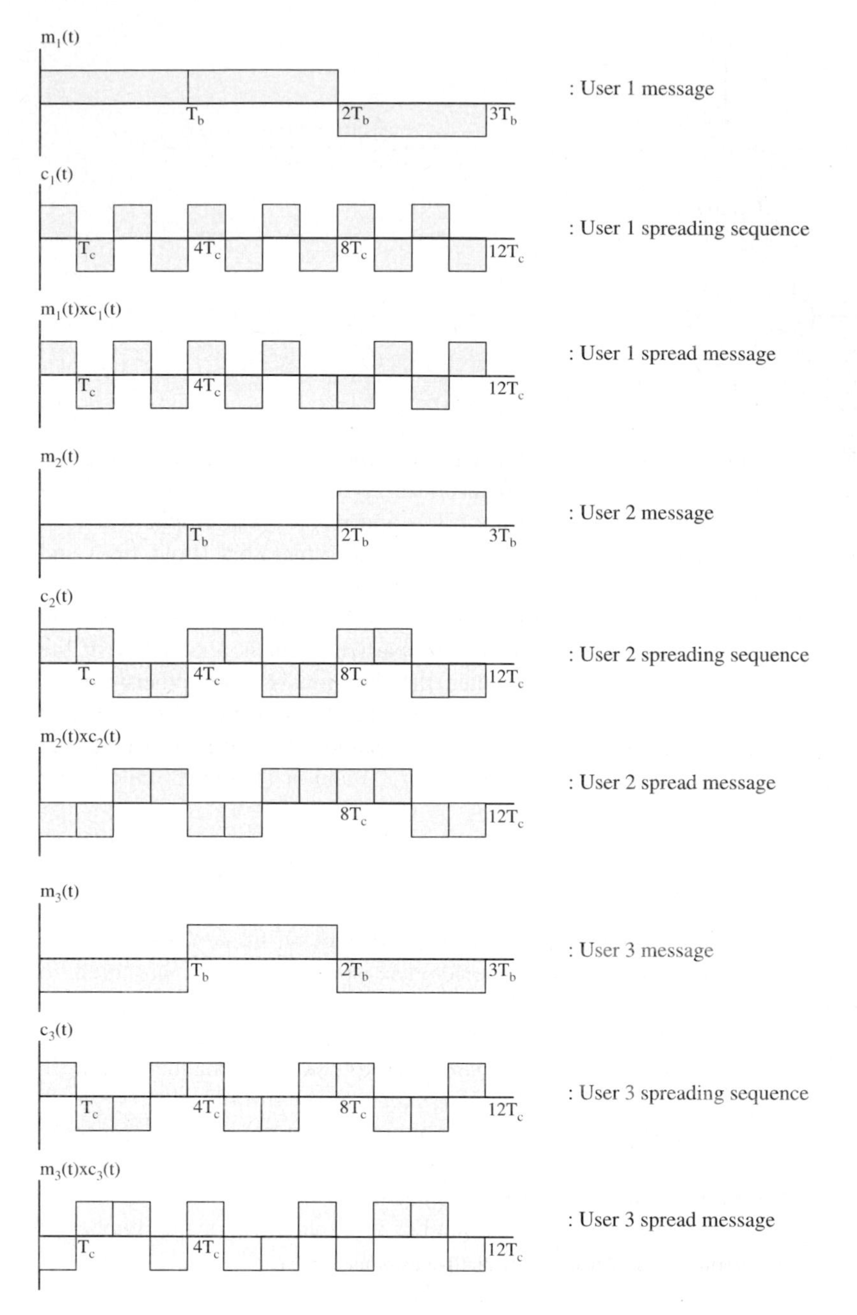


*Figure 14.* The waveforms for the baseband messages $m_1(t), m_2(t), and\ m_3(t)$, orthogonal codes $c_1(t), c_2(t), andc_3(t)$, and spread messages $m_1(t)c_1(t), m_2(t)c_2(t), and\ m_3(t)c_3(t)$

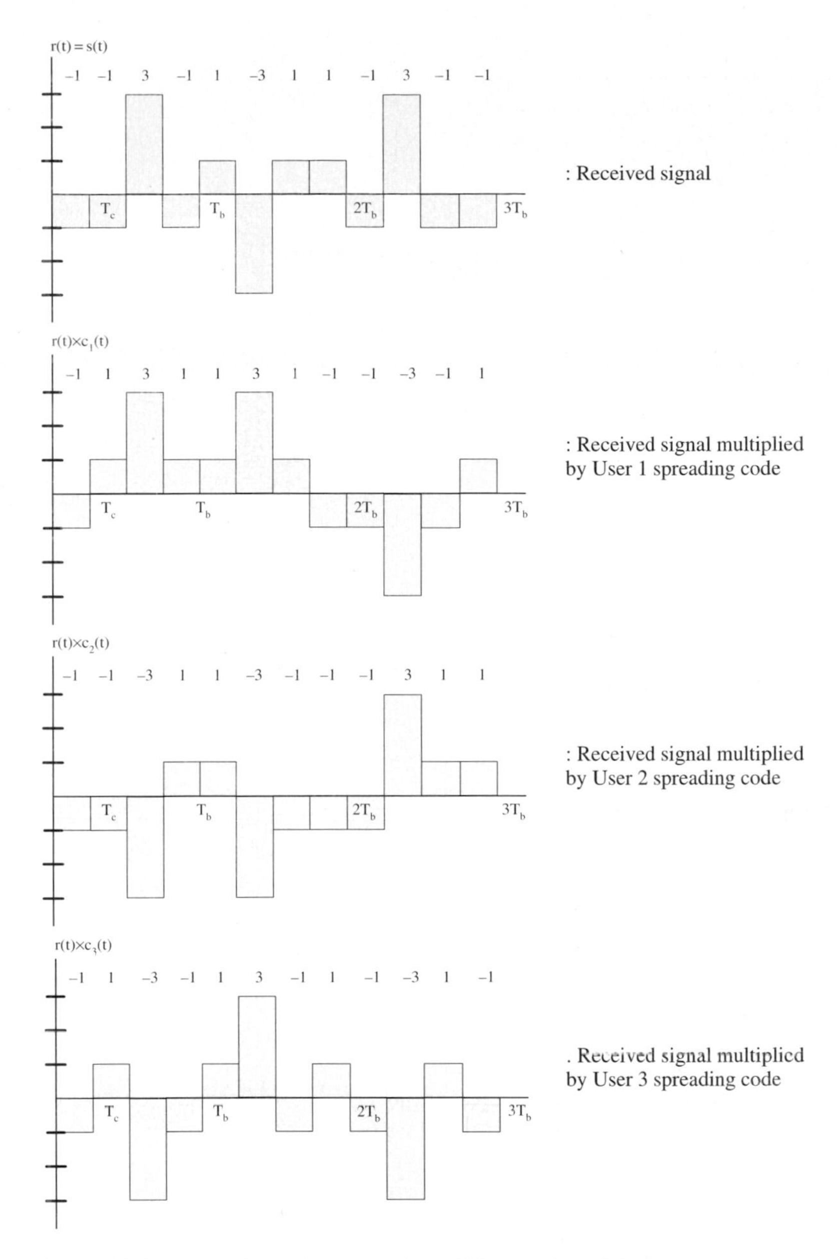


*Figure 15.* Time waveforms for the signals at different points of the receiver

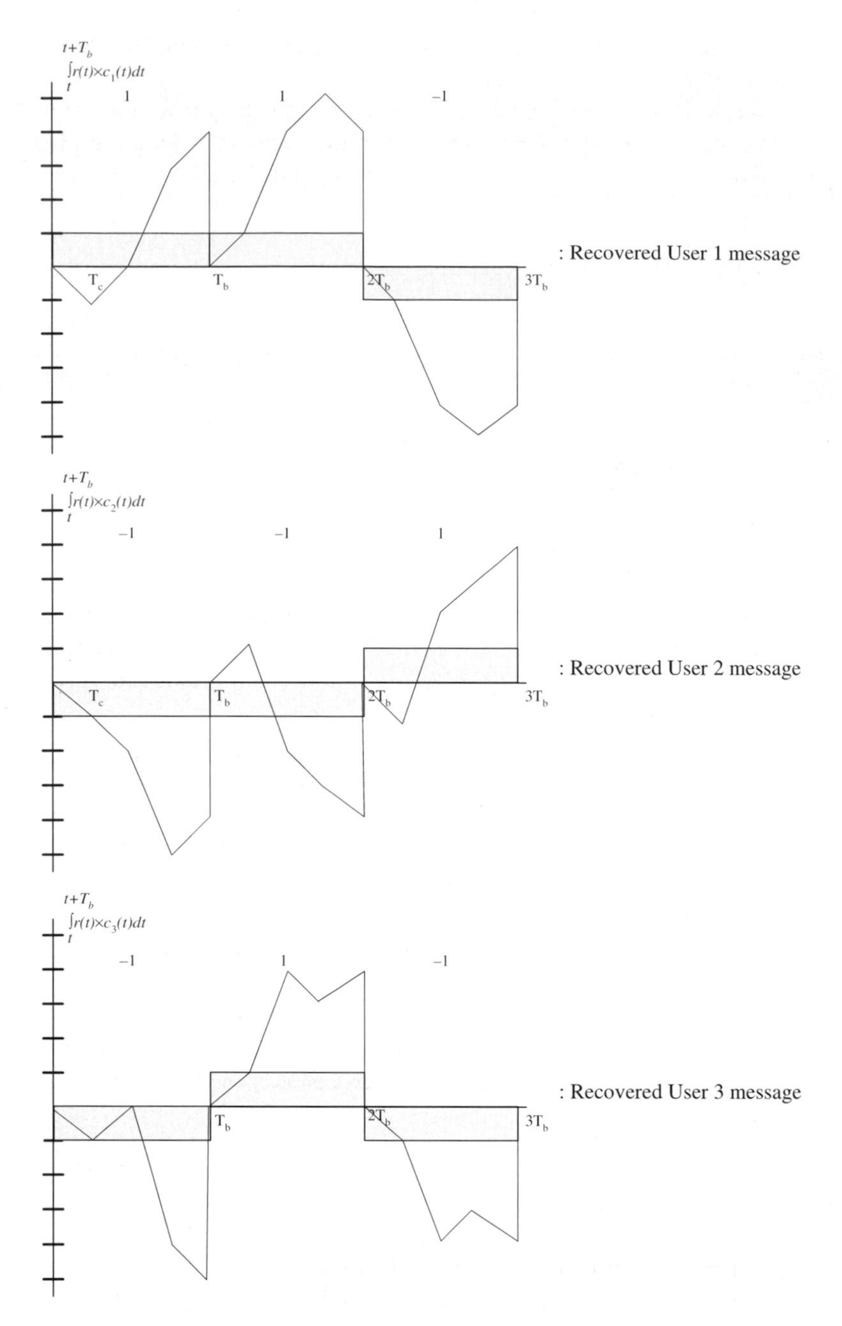


*Figure 16.* Time waveforms at the output of the integrators (bold line) and decision threshold (dash line)

transmitted for six coded bits. In cdma2000, the Walsh code length limit is 128 for 1x rates and, 256 for 3x rates.

Main parameter of Walsh code sequences is that they are orthogonal. Two real-valued waveforms $x$ and $y$ are said to be orthogonal if their cross-correlation $R_{xy}(0)$ over T is zero, where

$$(2) \qquad R_{xy}(0) = \int_0^T x(t)y(t)dt$$

In discrete time, the two sequences $x$ and $y$ are orthogonal if their cross-product $R_{xy}(0)$ is zero, where

$$(3) \qquad R_{xy}(0) = x^T y = \sum_{i=1}^{I} x_i y_i$$

T in (3) denotes the transpose of the column vector.

For the special case of binary sequences, the values of 0 and 1 may be viewed as having opposite polarity (-1 and 1), and when the product of two binary sequences results in an equal number of 1's and 0's, the cross-correlation is zero. For example, the following two codes, $x$ = [1 -1 1 -1] and $y$ = [1 1 -1 -1], are orthogonal, since their cross-product is zero:

$$R_{xy}(0) = (1)(1) + (-1)(1) + (1)(-1) + (-1)(-1) = 0$$

Walsh codes generated using an iterative process of constructing Hadamard matrix starting with $H_1$=[1]. The Hadamard matrix is built by using the function

$$(4) \qquad H_{2n} = \begin{bmatrix} H_n & H_n \\ H_n & \overline{H}_n \end{bmatrix}$$

where $\bar{H}_n$ contains the same but inverted elements of $H_n$. Therefore, to obtain a set of eight orthogonal Walsh codes, we need to generate a Hadamard matrix of order 8, or

$$H_8 = \begin{bmatrix} +1 & +1 & +1 & +1 & +1 & +1 & +1 & +1 \\ +1 & -1 & +1 & -1 & +1 & -1 & +1 & -1 \\ +1 & +1 & -1 & -1 & +1 & +1 & -1 & -1 \\ +1 & -1 & -1 & +1 & +1 & -1 & -1 & +1 \\ +1 & +1 & +1 & +1 & -1 & -1 & -1 & -1 \\ +1 & -1 & +1 & -1 & +1 & -1 & +1 & -1 \\ +1 & +1 & -1 & -1 & -1 & -1 & +1 & +1 \\ +1 & -1 & -1 & +1 & -1 & +1 & +1 & -1 \end{bmatrix}$$

where, each row of the matrix above except the first, can be used as orthogonal spreading sequence. The 1st sequence of Hadamard matrix consists of all 1s and thus cannot be used for channelization.

Earlier, in **Section 2.1**, we have illustrated orthogonal Walsh codes ability to provide channelization of different users. However, this ability heavily depends on the orthogonality of the codes during the all stages of the transmission. In practice, the IS-95 CDMA system uses a pilot channel and sync channel to synchronize the downlink and to ensure that the link is coherent. In the uplink, which does not have sync and pilot channels, another type of codes, ***PN codes*** are used for channelization, due to the noncoherent nature of the uplink

*PN sequences* have an important property: time-shifted versions of the same PN sequence have very little correlation with each other, in other words low *autocorrelation* property. We define the discrete-time autocorrelation of a real valued sequence $\boldsymbol{x}$ to be

$$(5) \qquad R_x(i) = \sum_{j=0}^{J-1} x_j x_{j-1}$$

In other words, for each successive shift $i$, we calculate the summation of the product of $x_j$ and its shifted version $x_{j-i}$.

PN code sets can be generated from linear feedback shift registers, as shown in **Figure 17**. The register starts with an initial sequence of bits. In each step, the content of the register is shifted one place to the right and it is also fed back to the leftmost place, the output of the last stage and the output of the one intermediate stage are combined and fed as input to the first stage. The output bits of the last stage form the PN code.

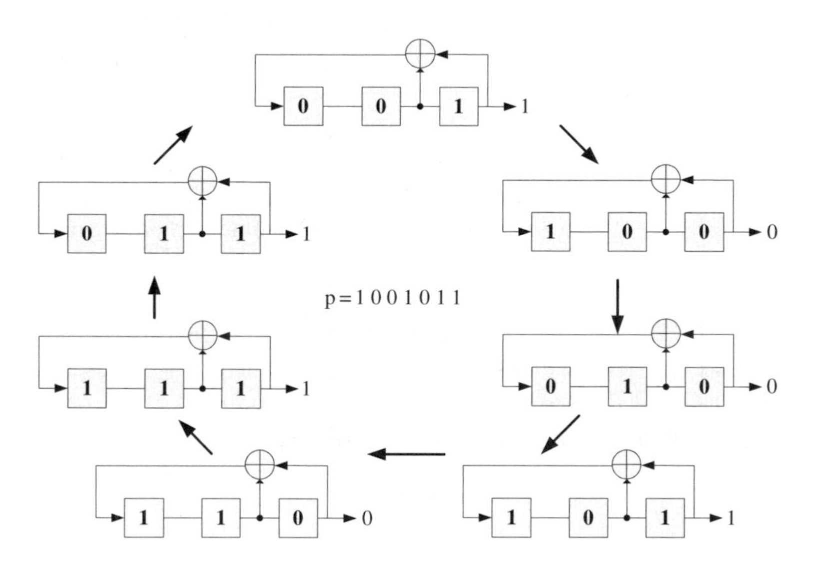


*Figure 17.* Example for a PN sequence generated by a linear feedback shift register of three stages

The code generated in this manner is called a maximal-length shift register code, and the length L of this code is

(6) $L = 2^m - 1$

where $m$ is the number of stages of the register. In example given by **Figure 17** the linear feedback shift register with three stages is shown. An initial state of [0 0 1] is used for the register. After clocking the bits through the register, we obtain the required PN sequence, which is $p = [1001011]$.

Note that at shift L=$2^3$–1=7, the state of the register returns to that of the initial state, and further shifting of the bits yields another identical sequence of outputs. A PN code set of 7 codes can be generated by successively shifting **p**, and by changing 0s to -1s we obtain

$$p_1 = [+1\ \ -1\ \ -1\ \ +1\ \ -1\ \ +1\ \ +1]$$
$$p_2 = [+1\ \ +1\ \ -1\ \ -1\ \ +1\ \ -1\ \ +1]$$
$$p_3 = [+1\ \ +1\ \ +1\ \ -1\ \ -1\ \ +1\ \ -1]$$
$$p_4 = [-1\ \ +1\ \ +1\ \ +1\ \ -1\ \ -1\ \ +1]$$
$$p_5 = [+1\ \ -1\ \ +1\ \ +1\ \ +1\ \ -1\ \ -1]$$
$$p_6 = [-1\ \ +1\ \ -1\ \ +1\ \ +1\ \ +1\ \ -1]$$
$$p_7 = [-1\ \ -1\ \ +1\ \ -1\ \ +1\ \ +1\ \ +1]$$

We can easily verify that these codes satisfy the three conditions outlined earlier.

**Figure 18** shows the *channelization using PN codes.* Suppose the same two users A, and B wish to send two separate messages:

- User A signal $m_1(t)=[+1\ -1]$, spreading code $p_1(t) = [+1 - 1 - 1 + 1 - 1 + 1 + 1]$
- User B signal $m_2(t)=[-1\ +1]$, spreading code $p_2(t) = [-1 + 1 - 1 + 1 + 1 + 1 - 1]$.

Each message is spread by its assigned PN code:

- For message one: $m_1(t)p_1(t) = [+1 - 1 - 1 + 1 - 1 + 1 + 1 - 1 + 1 + 1 - 1 + 1 - 1 - 1]$;
- For message two: $m_2(t)p_2(t) = [+1 - 1 + 1 - 1 - 1 - 1 + 1 - 1 + 1 - 1 + 1 + 1 + 1 - 1]$;

The spread spectrum signals for two messages are combined to form a composite signal s(t):

$$s(t) = m_1 p_1(t) + m_2 p_2(t) =$$
$$= [2\ \ -2\ \ 0\ \ 0\ \ -2\ \ 0\ \ 2\ \ -2\ \ 2\ \ 0\ \ 0\ \ 2\ \ 0\ \ -2]$$

At the receiver of user B, the composite signal is multiplied by the PN code corresponding to the user B:

$$s(t)p_2(t) = [-2\ \ -2\ \ 0\ \ 0\ \ -2\ \ 0\ \ -2\ \ 2\ \ 2\ \ 0\ \ 0\ \ 2\ \ 0\ \ 2]$$

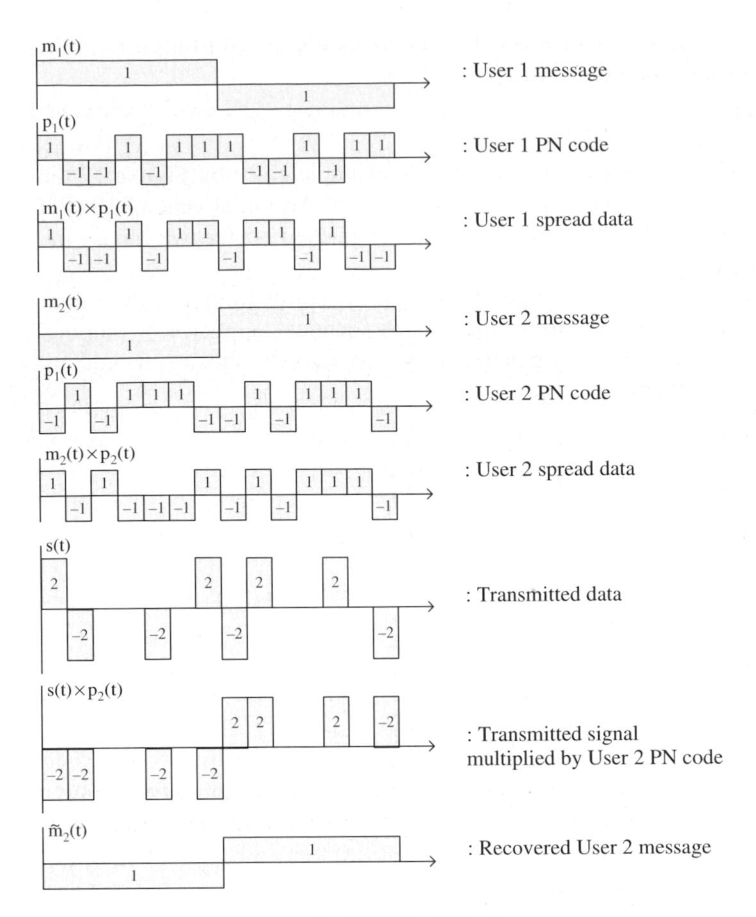


*Figure 18.* Example of channelization using PN code sequences

Then the receiver integrates all the values over each bit period, which results in $M_2(t) = [-8 \;\; 8]$ function for user *B*. After the decision threshold we obtain the result $\tilde{m}_2(t) = [-1 \;\; +1]$ for user *B*. may try to decode the symbols for user A in the same manner.

The two short codes of length $2^{15}–1$ and one long code length of $2^{42}–1$ used in IS-95 CDMA system. For cdma2000 Spreading Rate 3, the short code length is 3 times the short code length given above or $3x2^{15}$ in length.

All base stations and all mobiles use the same three PN sequences. In uplink direction long PN code used for channelization, by assigning different time shifted versions of the long code to different users, whereas short PN codes used for scrambling users data.

In downlink channel each base station is also assigned a unique, time shifted version of the short PN code that is superimposed on top of the Walsh code. This is done to provide isolation among the different base stations or sectors, which is necessary because each base station uses the same 64 Walsh code set. Scrambling user data in downlink done via using of long PN code.

**Table 1** summarizes the Section 2.2 and gives main parameters of spreading codes

## 2.3 Key Features of CDMA

As discussed earlier, CDMA offers many advantages over TDMA and FDMA. CDMA is a scheme by which multiple users are assigned radio resources using DS-SS techniques. Nowadays, the most prominent CDMA applications are mobile communication systems like IS-95, cdma2000 or WCDMA. To apply CDMA in a mobile communications systems there are specific additional methods which are required to be implemented in all these systems. Methods such as power control and soft handover have to be applied to control the inter-user interference and to be able to separate the users by their respective codes. In this section we describe some basic CDMA principles, such as frequency allocation, power control, handover, and etc.

**Power control** is one of the most necessary mechanisms exploited in cellular communication systems. Performance limiting factors, such as, varying path loss and fading result in the need to control the mobile's transmission power. Power control is where the transmit power from each user is controlled such that the received power of each user at the BS is equal to one other.

Especially power control is essential in CDMA based cellular networks since in CDMA all users share the same frequency separated via using of different spreading codes and each user's signals acts as random interference to other users. This issue is also known as the *near-far problem* in a spread-spectrum multiple access systems, and arises when a mobile user near a cell jams a user that is distant from the cell (assuming both are transmitting at the same power). The problem is this: consider a receiver and two transmitters (one close to the receiver; the

*Table 1.* Spreading codes parameters

| | Length | Downlink | Uplink |
|---|---|---|---|
| Walsh codes | 64 in IS-95<br>128 in cdma2000 Rate 1<br>256 in cdma2000 Rate 3 | Used for channelization, except 1st sequence that consists all 1s | Used for waveform encoding (orthogonal modulation) |
| Long PN code | $2^{42}$-1 | Used for scrambling | Used for channelization |
| Short PN code | $2^{15}$-1 in IS-95 and cdma2000 Rate 1<br>$3x2^{15}$ in cdma2000 Rate 3 | Used to separate individual cells or sectors | Used for scrambling |

other far away). If both transmitters transmit simultaneously and at equal powers, then the receiver will receive more power from the nearer transmitter. This makes the farther transmitter more difficult, if not impossible, to "understand." Since one transmission's signal is the other's noise the signal-to-noise ratio (SNR) for the farther transmitter is much lower. If the nearer transmitter transmits a signal that is orders of magnitude higher than the farther transmitter then the SNR for the farther transmitter may be below detectability and the farther transmitter may just as well not transmit. This effectively jams the communication channel. In CDMA systems this is commonly solved by *power control.* **Figure 19** demonstrates power control mechanism working principle. There are four MSs located at different distances from BS; if there is no power control mechanism user D signal reaches the BS with too low power since this user is located too far from BS and signals from other MSs reject the user D signal. Using power control mechanism we can achieve the equal power signals from different MSs at the receiver.

There two kinds of power control mechanisms:

- *Open-loop power control* where an original estimate is made by the mobile.
- *Closed-loop power control* where a faster correction is made to this original estimate, based on instruction provided to the mobile by the BS

In the open loop power control, the MS adjusts its own transmit power on the basis of the received downlink signal, whereas in a closed loop the BS measures the received signal strength and transmits a power control command to the MS. In consequence, the MS adjusts it's transmit power on the basis of the received uplink signal.

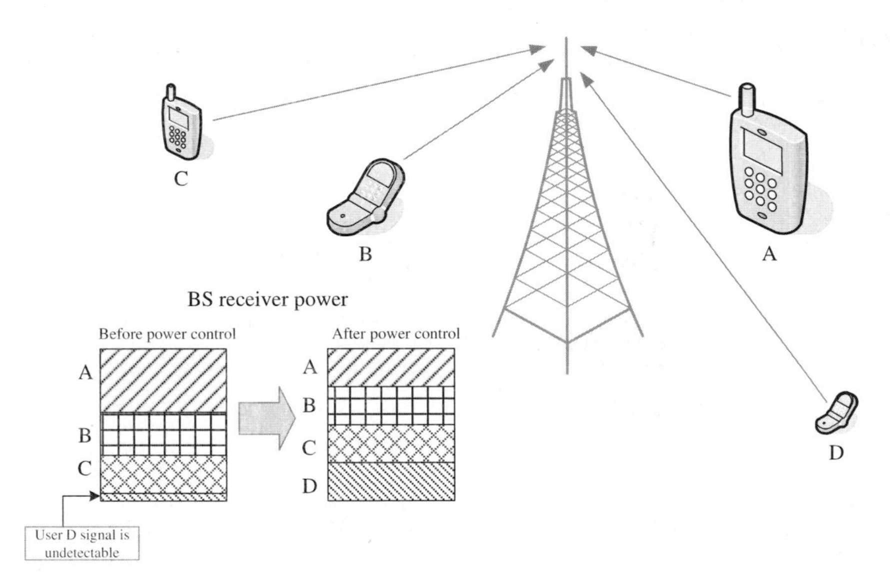


*Figure 19.* Near-far problem example

First CDMA standard, IS-95, utilized the both mechanisms, whereas current CDMA systems like cdma2000 and WCDMA (UMTS) exploit only closed-loop power control. Thus, in this section much attention is paid to closed-loop power control mechanism.

In ***open-loop power control*** each MS measures the received signal strength of the pilot signal, and depending on this measurement and information from the link power budget that is transmitted during initial synchronization, the downlink path loss is estimated. Assuming a similar path loss for the uplink, the MS uses this information to determine its transmitter power. Leaving out the calculation process we can say that MS power can be achieved as:

(7) *Mobile_power(dBm)=target_SNR(dB)+BS_power(dBm)*
*+total_uplink_noise_and_interference(dBm)-received_power(dBm)*
*=constant(dB)-received_power(dBm)*

In IS-95, the nominal value of the *constant* in (7) is specified to be -73 dB. This value can be attributed to the nominal values -13 dB for the target SNR, -100 dBm for the uplink noise and interference, and 40 dBm (10 W) for the BS power. The actual values of these parameters may be different and data for calibrating the *constant* in (7) are broadcast to the MSs on the sync channel.

*Open loop power control* is used to compensate for slow-varying and log-normal shadowing effects where there is a correlation between forward and uplinks are on different frequencies, the open loop power control is inadequate and too slow compensate for fast Rayleigh fading. To compensate for power fluctuations due to fast Rayleigh fading the ***closed loop power control*** is used.

Once mobile gets on a traffic channel and starts to communicate with the base station, the closed-loop power control process operates along with the open-loop power control. The calculation of downlink path loss through the measurement of the BS received signal strength can be used as a rough estimate of the path loss on the uplink. The true value, however, must be measured at the BS upon reception of the MS's signals. At the BS, the measured signal strength is compared with the desired strength, and a power adjustment command is generated. If the average power level is greater than the threshold, the power command generator generates a "1" to instruct the MS to decrease power. If the average power is less than the desired level, a "0" is generated to instruct the mobile to increase power. These commands instruct the MS to adjust transmitter power by a predetermined amount, usually 1 dB. Ideally, frame error rate (FER) is good indicator of link quality. But because it takes a long time for the BS to accumulate enough bits to calculate FER, $E_b/N_0$ is used as an indicator of uplink quality.

**Figure 20** shows closed loop power control working principle on a fading channel at low speed. Closed loop power control commands the mobile station to use a transmit power proportional to the inverse of the received power (or SNR). Provided the mobile station has enough headroom to ramp the power up, only very little residual fading is left and the channel becomes an essentially non-fading channel as seen from the BS receiver. Although, this fading removal is highly desirable from

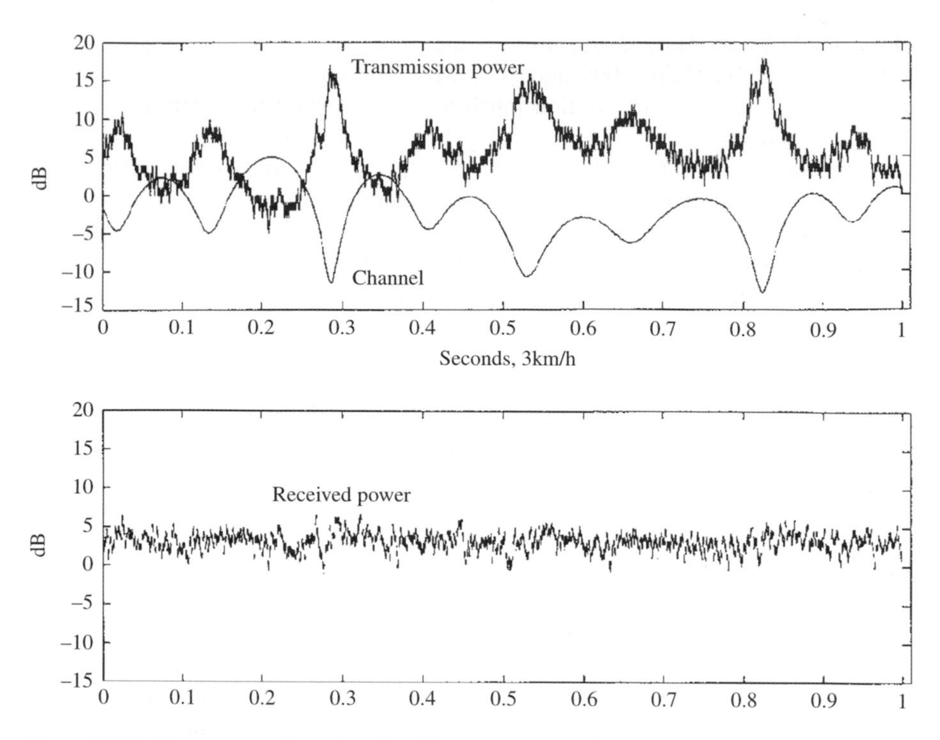


*Figure 20.* Fading compensation using closed loop power control

the receiver point of view, however it comes at the expense of increased average transmit power at the transmitting side. This means that a mobile station in a deep fade, i.e. using a large transmission power, will cause increased interference to other cells. **Figure 20** illustrates this point.

Closed-loop power control has an inner and an outer loop. Thus far we only have described the *inner-loop* of the closed-loop power control process. The premise of the inner loop is that there exists a predetermined SNR threshold by which power-up and power-down decisions are made. The closed-loop power control also employs what is called an *outer-loop power control.* This mechanism ensures that the power control strategy is operating correctly. The FER at the BS is measured and compared with the desired error rate, and if the difference between error rates is large, then the power command threshold is adjusted to yield the desired FER. Both, inner-loop and outer-loop power control mechanisms are illustrated in **Figure 21**.

Ideally power control is not needed in the downlink. Though in downlink direction the near-far problem does not exist and downlink power control is not necessary as uplink power control. However, in real life, one particular mobile may be nearby a significant jammer and experience a large background interference, or a mobile may suffer a large path loss such that arriving composite signal is on the order of the thermal noise. Thus, downlink power control is still needed. When downlink

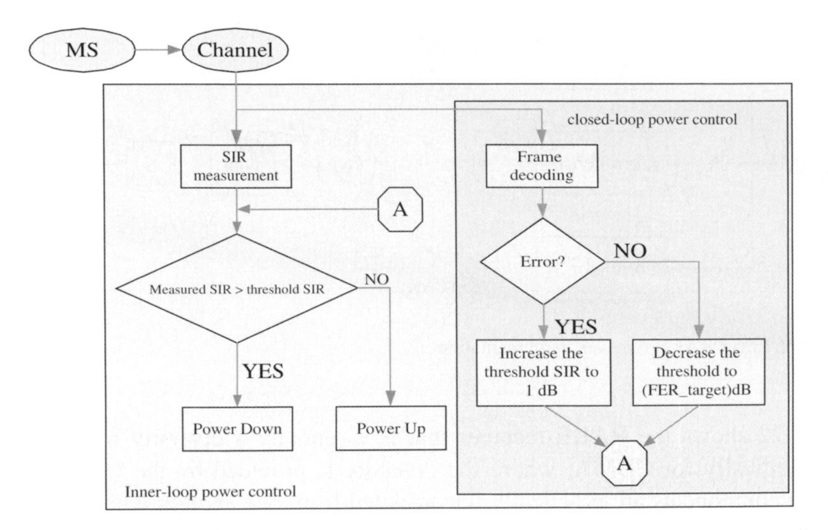


*Figure 21.* Inner-and outer-closed loop power control mechanism working principle

power control is enabled, the BS periodically reduces the power transmitted to an individual MS. This process continues until the MS senses an increase in the downlink FER. The MS reports the number of FER to BS, and the BS depending on this information can decide whether to increase power by a small amount, nominally 0.5 dB. Before the BS complies with the request, it must consider other requests, loading, and the current transmitted power.

The IS-95 system uses a combination of open-loop and closed-loop power control with rate of 800 Hz or 1.25 ms. Unlike IS-95 where closed loop power control was applied only to the reverse link, both CDMA2000 and WCDMA employ power control in the uplink and downlink directions. The only difference between the two technologies is the rate of the power control. CDMA2000 operates at a rate of 800 Hz, while WCDMA operates power control at a rate of 1600 Hz

**Rake receiver.** One of the main advantages of CDMA systems is their ability to use signals that arrive in the receivers with different time delays, due to *multipath propagation*. FDMA and TDMA, which are narrow band systems, cannot distinguish between the multipath arrivals, and resort to equalization to mitigate the negative effects of multipath. Due to its wide bandwidth and *rake receivers*, CDMA uses the multipath signals and combines them to make a more reliable signal at the receivers.

A rake receiver is a radio receiver designed to counter the effects of multipath fading. It does this by using several "sub-receivers" or “fingers” each delayed slightly in order to tune in to the individual multipath components. Each component is decoded independently, but at a later stage combined in order to make the most use of the different transmission characteristics of each transmission path. This could very well result in higher SNR ratio (or $E_b/N_o$) in a multipath environment than in a "clean" environment.

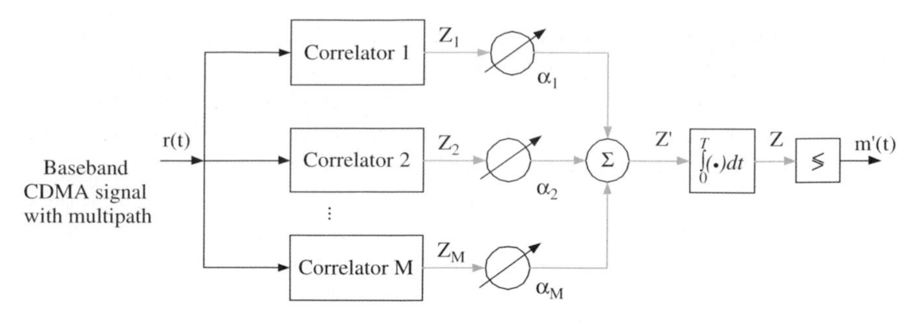


*Figure 22.* An M-finger RAKE-receiver implementation

In **Figure 22** shows the RAKE-receiver that is essentially a diversity receiver designed specifically for CDMA, where the diversity is provided by the fact that the multipath components are practically uncorrelated from one another when their relative propagation delay exceeds a chip period. As shown in **Figure 22**, a RAKE-receiver utilizes multiple correlators to separately detect the M strongest multipath components. The outputs of each correlator are then weighted to provide a better estimate of the transmitted signal than is provided by a single component. Demodulation and bit decision are then based on the weighted outputs of the M correlators.

To explore the performance of a RAKE-receiver, assume M correlators are used in a CDMA receiver to capture the M strongest multipath components. A weighted network is used to provide a linear combination of the correlator output for bit detection. Correlator 1 is synchronized to the strongest multipath $m_1$. Multipath component $m_2$ arrives $\tau_1$ later than $m_1$ where $\tau_2 - \tau_1$ is assumed to be greater than a chip duration. The second correlator is synchronized $m_2$. It correlates strongly with $m_2$, but has low correlation with m1. The M decision statistics are weighted to form an overall decision statistics as shown in **Figure 22**. The outputs of the M correlators are denoted as $Z_1, Z_2, \ldots, Z_M$. They are weighted by $\alpha_1, \alpha_2, \ldots$ and $\alpha_M$, respectively. The weighting coefficients are based on the power or the SNR from each correlator output. If the power or SNR is small out of particular correlator, it will be assigned a small weighting factor. Just as in the case of a *maximal ration combining* diversity scheme, the overall signal Z' is given by

$$Z' = \sum_{m=1}^{M} \alpha_m Z_m \tag{8}$$

The weighting coefficients $\alpha_m$, are normalized to the output signal power of the correlator in such a way that the coefficients sum to unity, as shown below:

$$\alpha_m = \frac{Z_m^2}{\sum_{m=1}^{M} Z_m^2} \tag{9}$$

In CDMA, both the base station and mobile receivers use RAKE receiver techniques, e.g. IS-95 and WCDMA. Although there are several differences between the RAKE receiver in the MS and BS, all the basic principles presented here are the same. Each correlator in a RAKE receiver is called a RAKE-receiver finger. The base station combines the outputs of its RAKE-receiver fingers noncoherently. i.e., the outputs are added in power. The mobile receiver combines its RAKE-receiver finger outputs coherently, i.e., the outputs are added in voltage. Typically, mobile receivers have 3 RAKE-receiver fingers and base station receivers have 4 or 5 depending on the equipment manufacturer.

The reason is why it is called a "RAKE" receiver is that most block diagrams of the device resemble a garden rake, which can illustrate the RAKE receiver's operation. The manner in which a garden rake eventually picks up debris off a patch of grass resembles the way the RAKE's fingers work together to recover multiple versions of a transmitter's signal.

**Handover.** In a mobile communications environment, as a user moves from the coverage area of one base station to the coverage area of another BS, a handover must occur to transition the communication link from one BS to the next. Handovers in CDMA are fundamentally different from handovers in TDMA systems. While in a TDMA system handover is a short procedure, and the normal state of affairs is a non-handover situation, the situation in a CDMA system is dramatically different. A MS communicating with its serving BS can spend a large part of the connection time in a soft handover state.

*Soft handover* refers to the state where the mobile is in communication with multiple Base Stations at the same time. Soft handover is a *make-before-break* type of handover, whereby a mobile acquires a target code channel before breaking an existing one. Soft handover is a special attribute of CDMA that is enabled by universal frequency reuse. **Figure 23** shows the soft handover process, when MS moves from cell A to cell B.

During the soft handover process MS has to employ one of its RAKE receiver fingers for each received BS. Note that each received multipath component requires a RAKE finger of its own. Each separate link from a BS is called a soft handover branch. Since, all BSs use the same frequency in a soft handover, a MS can consider their signals as just additional multipath components. An important difference between a multipath component and a soft handover branch is that each branch is coded with a different spreading code, whereas multipath components are just time delayed versions of the same signal.

Note that during the soft handover process two power control loops per connection are active, one for each base station.

**Figure 24** shows the soft handover process example when mobile MS moves from the coverage area of BS1 to the BS2 serving area. The soft handover typically uses pilot channel $E_c/N_0$as the handover measurement quantity. The following definitions are used to describe the handover process:

*Active set*: The active set contains the pilots of those sectors that are actively exchanging traffic channel information with the mobile.

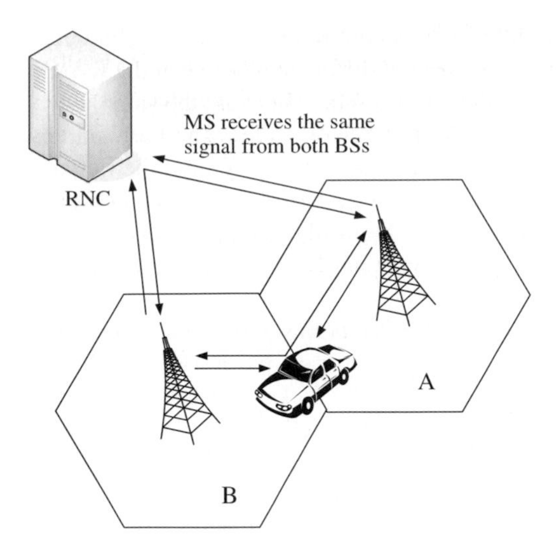


*Figure 23.* Soft handover

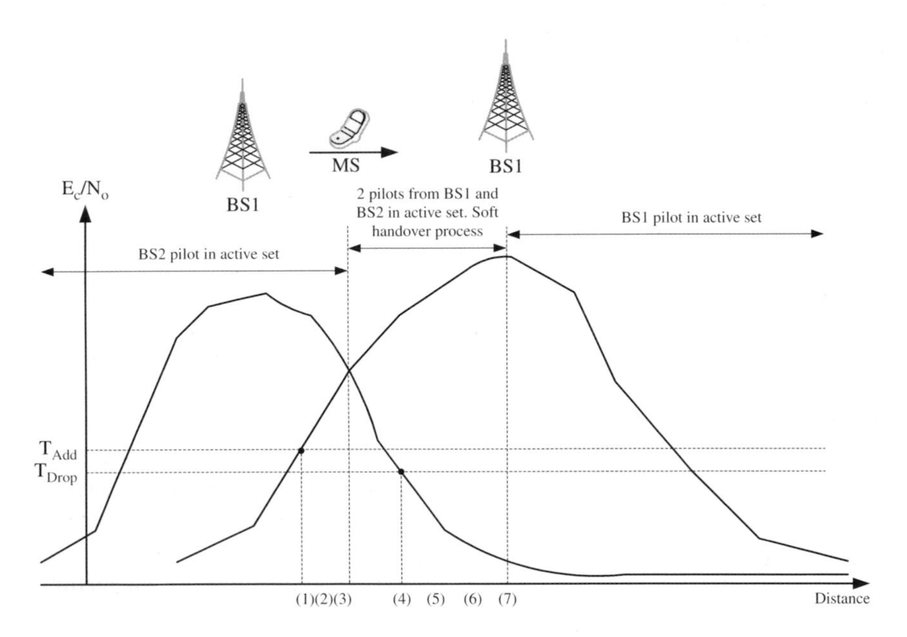


*Figure 24.* Soft handover process example

*Neighbor set*: The neighbor set or monitored set is the list of cells that MS continuously measures, but whose pilot $E_c/N_0$ are not strong enough to be added to the active set.

The following are the steps during the handover process:

1. MS is being served by BS1 only, and its active set contains only BS1 pilot. The MS measures the $E_c/N_0$ of BS2 pilot and finds it to be greater than *pilot detection threshold* $T_{Add}$. The MS sends a pilot measurement message and moves BS2 pilot from the neighbor set to the candidate set.
2. The MS receives a handover direction message from BS1 and starts communicating with BS2 on a new traffic channel. Handover direction message contains the PN offset of BS2 and the Walsh code of the newly assigned traffic channel.
3. The MS moves BS2 pilot from the candidate set to the active set. After acquiring the forward traffic channel specified in *the handover direction message*, the MS sends a *handover completion message*. Now active set contains two pilots.
4. The mobile detects that BS1 pilot has now dropped below the *pilot drop threshold* $T_{Drop}$, and starts the drop timer.
5. The drop timer reaches the *handover drop timer expiration value* $T_{TDrop}$ and the MS sends a pilot strength measurement message.
6. The MS receives a *handover direction message* which contains only the PN offset of BS2.
7. The mobile moves BS1 pilot from the active set to the neighbor set, and it sends a *handover completion message*.

Soft handover is typically employed in cell boundary areas, where cells overlap. When MS moves from one sector to another within the same cell ***softer handover*** process is occurs. From a MS's point of view it is a just another soft handover. The difference is only meaningful to the network, since a softer handover is an internal procedure for a BS, which saves the transmission capacity between BSs and the BS controller (RNC). The uplink softer handover branches can be combined within the BS, which is a faster procedure, and uses less of the fixed infrastructure's transport resources than most other types of handover procedures in CDMA system.

Placing in soft handover any additional BSs that can be detected by the mobile station, as soon as possible, results in reduced call dropping probabilities, increased capacity and coverage, and improved voice quality in cell boundaries, which usually has poor coverage coupled with increased interference from other cells.

In this section main attention is paid to soft handover. However, note that hard handover process is also important in CDMA systems, e.g. in WCDMA hard handover procedure can be used to change the radio frequency band of the connection between MS and BS or to change the cell on the same frequency when no network support of macro diversity exist. It can be also used to change the mode between FDD and TDD.

**Capacity.** The capacity of a CDMA system is proportional to the *processing gain* of the system, which is the ratio of the spread bandwidth to the data rate. A general expression for the signal-to-noise (SNR) power ratio for a particular mobile user at the base station given by

$$(10) \qquad \frac{S}{N} = \frac{R \cdot E_b}{B \cdot N_0} = \frac{E_b/N_0}{B/R}$$

where, $S = E_b/T_b = RE_b$is the carrier power and $N = BN_0$ is the interference power at the base station receiver. The quantity $E_b/N_0$ is the bit energy to noise power spectral density ratio, and $B/R$ is the processing gain of the system. Let $K$ denote the number of mobile users. If power control is used to ensure that every mobile has the same received power, the SNR of one user can be written as

$$(11) \qquad \frac{S}{N} = \frac{1}{K-1}$$

This is so because the total interference power in the band is equal to the sum of powers from individual users. Substituting S/N from (7) into (8), the capacity for a CDMA system is found to be

$$(12) \qquad K \approx K-1 = \frac{B/R}{E_b/N_0}$$

The capacity of a CDMA system is limited by the interference caused by other users simultaneously occupying the same bandwidth; this interference is reduced by the processing gain of the system.

The IS-95 CDMA standard specifies that each user conveys baseband information at 9.6 kbps, which is the rate of the vocoder output. The rate of the spread signal is 1.2288 Mcps, resulting in processing gain equal to

$$(13) \qquad P_G = \frac{B}{R} = \frac{1.2288 \cdot 10^6}{9.6 \cdot 10^3} = 128$$

Assuming the required $E_b/N_0$=6dB=4 we can derive the single cell CDMA system capacity using (8)

$$(14) \qquad K \approx \frac{B/R}{E_b/N_0} = \frac{128}{4} = 32 \; users$$

Equation (12) is effectively a model that describes the number of users a single CDMA cell can support. In reality particular cell is bordered by other CDMA cells that are serving other users. **Figure 25** shows an example when the signal powers from users located in different cells constitute interference each other. This effect calls ***effect of loading***, cell $A$ in example given above is said to be loaded by users from other cells.

Equation to account for the effect of loading given as:

$$(15) \qquad \frac{E_b}{N_0} = \frac{1}{K-1} \frac{B}{R} \left( \frac{1}{1+\eta} \right)$$

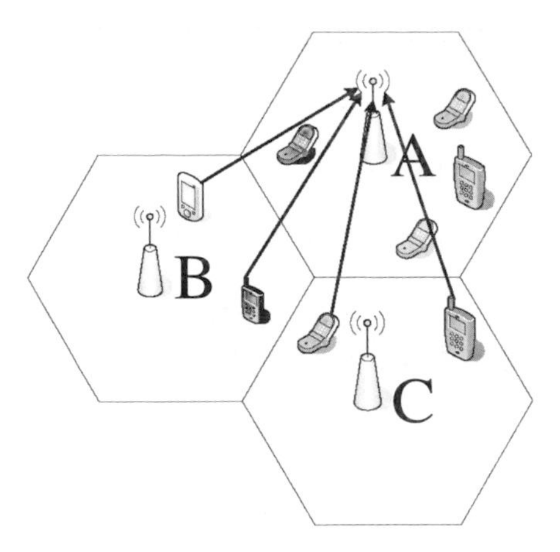


*Figure 25.* Interference introduced by users in the neighboring cell

where $\eta$ is the loading factor, between 0% and 100%. The inverse of the factor $(1+\eta)$ is sometimes known as the ***frequency reuse factor*** F; that is

$$F = \frac{1}{1+\eta} \quad (16)$$

In the single cell case the frequency reuse factor is ideally 1, however in real environment with multicell case, as the loading $\eta$ increases, the frequency reuse factor correspondingly decreases.

Instead of an omnidirectional antenna, which has an antenna pattern over 360 degrees, cells can be sectorized to several sectors, e.g. in example above cell A can be sectorized to six sectors so that each sector is only receiving signals over 60 degrees (**Figure 26**). In effect, a sectorized antenna rejects interference from users that are now within its antenna pattern. This arrangement decreases the effect of loading by a factor of approximately 6. This factor is called ***sectorization gain*** $G_s$. In reality, $G_s$ is typically around 2.5 and 5 for three- and six-sector configured systems, respectively.

Equation (15) is thus modified to account for the effect of sectorization:

$$\frac{E_b}{N_0} = \frac{1}{K-1}\frac{B}{R} \cdot F \cdot G_s \quad (17)$$

Equation (15) assumes that all users are transmitting 100% of the time. In practice CDMA systems uses variable rate vocoders, which means that the output rate of the vocoder is adjusted according to a user's actual speech pattern. The effect of this variable-rate vocoding is the reduction of overall transmitted power and hence interference. By employing variable-rate vocoding, the system reduces the total interference power by this ***voice activity power.***

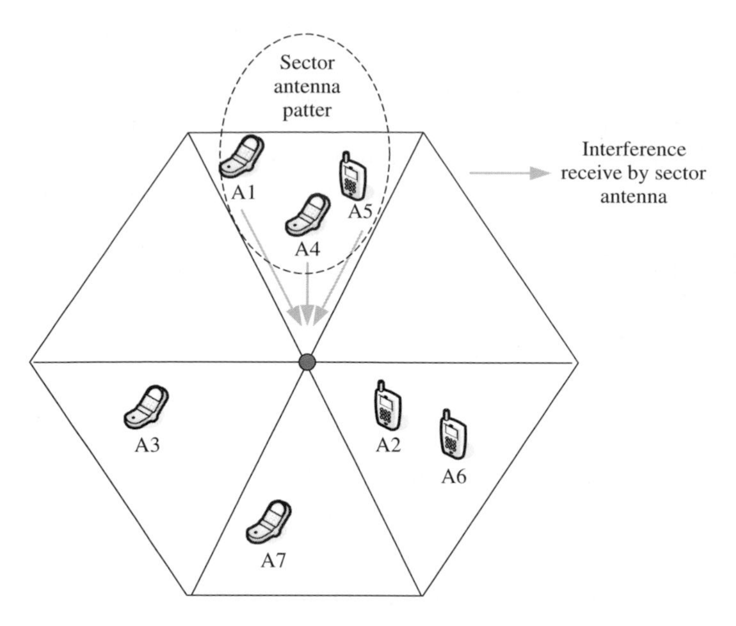


*Figure 26.* Cell sectorization using 60° directional antenna (6 sectors per cell)

Thus, (17) is again modified to account for the effect of voice activity:

$$\frac{E_b}{N_0} = \frac{1}{K-1}\frac{B}{R}\frac{1}{D_v} \cdot F \cdot G_s \tag{18}$$

where $D_v$ is the voice activity factor. Solving (18) for $M$ gives:

$$K = 1 + \frac{(B/R)}{(E_b/N_0)}\frac{1}{D_v} \cdot F \cdot G_s \tag{19}$$

If M is large, then

$$K \approx \frac{(B/R)}{(E_b/N_0)}\frac{1}{D_v} \cdot F \cdot G_s \tag{20}$$

In real systems voice activity power $D_v$ is typically around $0.35 \sim 5$.

Taking into account the all parameters above, we can update the equation (14) for multicell environment. Assuming the voice activity power $D_v = 0.4$, sectorization gain $G_s = 2.5$ (3 sectors per cell) and frequency reuse factor $F = 0.6$ we get

$$K \approx \frac{(B/R)}{(E_b/N_0)}\frac{1}{D_v} \cdot F \cdot G_s = \frac{128}{4} \cdot \frac{1}{0.4} \cdot 0.6 \cdot 2.5 = 120 \tag{21}$$

$K = 120$ channels per cell or *40* channels per sector.

Resulting from statements above, we can draw several conclusions regarding CDMA capacity:

1. Capacity is directly proportional to the processing gain of the system.
2. Capacity is inversely proportional to the required $E_b/N_0$ of the link. The lower the required threshold $E_b/N_0$, the higher the system capacity.
3. Capacity can be increased if one can decrease the amount of loading from users in adjacent cells.
4. Spatial filtering, such as sectorization, increases system capacity. For example, a six-sector cell would have more capacity than a three sector cell.

## 3. MULTI-CARRIER TRANSMISSION

The principle of multi-carrier transmission is to convert a serial high-rate data stream onto multiple parallel low rate sub-streams. Each sub-stream is modulated on another sub-carrier. Since the symbol rate on each sub-carrier is much less than the initial serial data symbol rate, the effects of delay spread, i.e., ISI, significantly decrease, reducing the complexity of the equalizer. **Figure 27** illustrates an example of multi-carrier modulation with 4 sub-carriers.

One of the efficient multi-carrier modulation techniques is an Orthogonal frequency-division multiplexing (OFDM). In OFDM the frequencies and modulation of frequency-division multiplexing are arranged to be orthogonal with each other which almost eliminates the interference between channels. Although the principles and some of the benefits have been known for 40 years, it is made popular today by the lower cost and availability of digital signal processing components.

### 3.1 Orthogonal Frequency Division Multiplexing.

OFDM can be simply defined as a form of multicarrier modulation where its carrier spacing is carefully selected so that each subcarrier is orthogonal to the other subcarriers. As is well known, orthogonal signals can be separated at the receiver

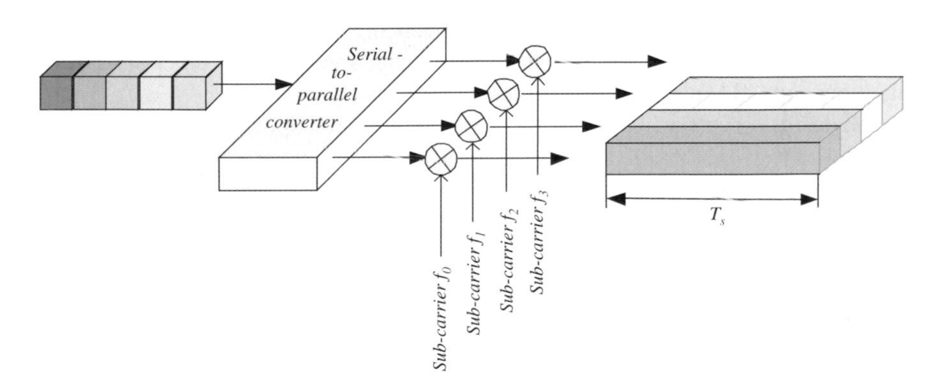


*Figure 27.* Multi-carrier modulation with 4 sub-carriers

by correlation techniques; hence, intersymbol interference among channels can be eliminated. Orthogonality can be achieved by carefully selecting carrier spacing, such as letting the carrier spacing be equal to the reciprocal of the useful symbol period. In order to occupy sufficient bandwidth to gain advantages of the OFDM system, it would be good to group a number of users together to form a wideband system, in order to interleave data in time and frequency (depends how broad one user signal is).

A communication system with multi-carrier modulation transmits $N_c$ complex valued source symbols $S_n$, $n=0,\ldots,N_c-1$, in parallel on $N_c$ sub-carriers. The source symbol duration $T_d$ of the serial data symbols results after serial-to-parallel conversion in the OFDM symbol duration

$$T_S = N_c T_d \quad (22)$$

In order to achieve orthogonality each of $N_c$ sub-streams modulated on sub-carriers with a spacing of

$$F_S = \frac{1}{T_S} \quad (23)$$

presuming a rectangular pulse shaping. The $N_c$ parallel modulated source symbols $S_n$ are referred to as an OFDM symbol.

The $N_c$ sub-carrier frequencies are located at

$$f_n = \frac{n}{T_S}, \quad n = 0, \ldots, N_c-1 \quad (24)$$

As an example, **Figure 28** shows four subcarriers from one OFDM signal. In this example, all subcarriers have the same phase and amplitude, but in practice the amplitudes and phases may be modulated differently for each subcarrier. Note that each subcarrier has exactly an integer number of cycles in the interval T, and the number of cycles between adjacent subcarriers differs by exactly one. This property accounts for the orthogonality between the subcarriers. As it is shown from **Figure 29** at the maximum of each sub-carrier spectrum, all other subcarrier spectra are zero. Because an OFDM receiver essentially.

calculates the spectrum values at those points that correspond to the maxima of individual subcarriers, it can demodulate each subcarrier free from any interference from other subcarriers

A key advantage of using OFDM is that multi-carrier modulation can be implemented in the discrete domain by using and IDFT, or a more computationally efficient IFFT. The block diagram of a multi-carrier modulator employing OFDM based on IDFT and a multicarrier demodulator employing inverse OFDM based on a DFT is illustrated in **Figure 30**

When the number of sub-carriers increases, the OFDM symbol duration $T_s$ becomes large compared to the duration of the impulse response $\tau_{\max}$ of the channel,

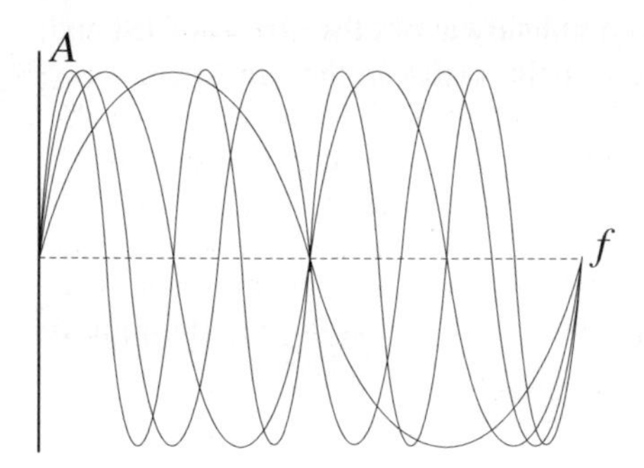


*Figure 28.* Example of four subcarriers within one OFDM symbol

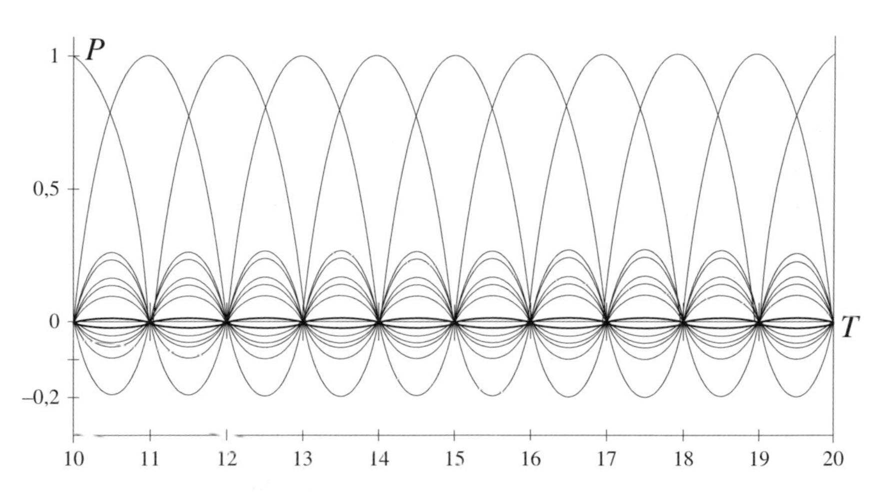


*Figure 29.* Spectra of individual sub-carriers

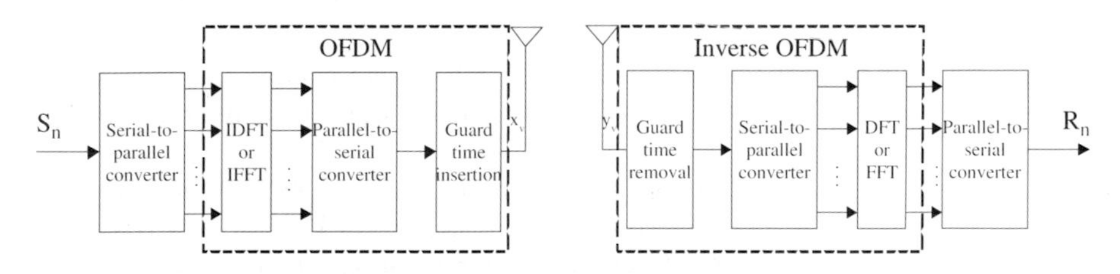


*Figure 30.* Digital multi-carrier transmission system applying OFDM

and the amount of ISI reduces. However, to completely avoid the effects of ISI and, thus, maintain the orthogonality between the signals on the on the sub-carriers, i.e., to also avoid ICI, a guard interval duration

$$T_g \geq \tau_{\max} \tag{25}$$

has to be inserted between adjacent OFDM symbols. The guard interval is a cyclic extension of each OFDM symbol which is obtained by extending the duration of an OFDM symbol to

$$T_s^{'} = T_g + T_s \tag{26}$$

A block of subsequent OFDM symbols, where the information transmitted within these OFDM symbols belongs together, e.g., due to coding and/or spreading in time and frequency direction, is referred to as an OFDM frame.

The benefits of using OFDM are many, including high spectrum efficiency, resistance against multipath interference (particularly in wireless communications), simple digital realization by using FFT operation, flexible spectrum allocation and low complex receivers due to the avoidance of ISI and ICI with sufficiently long guard interval.

An extremely important benefit from using multiple sub-carriers is that because each carrier operates at a relatively low symbol rate, the duration of each symbol is relatively long. If one sends, say, a million bits per second over a single baseband channel, then the duration of each bit must be under a microsecond. This imposes severe constraints on synchronization and removal of multipath interference. If the same million bits per second are spread among $N_c$ subcarriers, the duration of each bit can be longer by a factor of $N_c$, and the constraints of timing and multipath sensitivity are greatly relaxed. For moving vehicles, the Doppler Effect on signal timing is another constraint that causes difficulties for some other modulation schemes.

However, OFDM suffers from time-variations in the channel, or presence of a carrier frequency offset. This is due to the fact that the OFDM subcarriers are spaced closely in frequency. Imperfect frequency synchronization causes a loss in subcarrier orthogonality which severely degrades performance.

Because the signal is the sum of a large number of subcarriers, it tends to have a high peak-to-average power ratio (PAPR). Also, it is necessary to minimize intermodulation between the subcarriers, which would effectively raise the noise floor both in-channel and out of channel. For this reason circuitry must be very linear. This is demanding, especially in relation to high power RF circuitry, which also needs to be efficient in order to minimize power consumption.

Radio access techniques are often combined to hybrid schemes in communication systems like GSM where TDMA and FDMA are applied, or UMTS where CDMA, TDMA and FDMA are used. These hybrid combinations additionally increase the user capacity and flexibility of the system. Nowadays much attention paid to the systems combined with OFDM. For example the combination of OFDM with DS-CDMA or FDMA offers the possibility to overload an otherwise limited systems. Next in this section we describe the different hybrid multiple radio access schemes.

## 3.2 Multi-carrier FDMA (OFDMA)

OFDMA is a combination of modulation scheme that resembles OFDM and a multiple access scheme that combines TDMA and FDMA. OFDMA typically uses a FFT size much higher than OFDM, and divides the available sub-carriers into logical groups called sub-channels. Unlike OFDM that transmits single user information on all subcarriers at any given time, OFDMA allows multiple users to transmit simultaneously on the different subcarriers per OFDM symbol. Therefore, an OFDMA system with, e.g., $N_c = 1024$ sub-carriers and adaptive sub-carrier allocation is able to handle thousand of users. Another approach is that OFDMA may transmit different amounts of energy in each sub-channel (**Figure 31**).

**Figure 32** illustrates the simplest OFDMA scheme with one sub-carrier per user. At the base station the received signal, being the sum of K users' signals, acts as an OFDM signal due to its multipoint to point nature. Unlike conventional FDMA, which requires $K$ demodulators to handle simultaneous $K$ users, OFDMA requires only a single demodulator, followed by an $N_c$-point DFT.

The basic components of an OFDMA transmitter at the terminal station are FEC channel coding, mapping, sub-carrier assignment, and single carrier modulator (multi-carrier modulator in the case that several sub-carriers assigned per user).

A very accurate clock and carrier synchronization is essential for an OFDMA system, to ensure orthogonality between the $K$modulated signals originating from different terminal stations.

Nowadays, OFDMA is being considered as a modulation and multiple access method for 4th generation wireless networks, and currently the modulation of choice for high speed data access systems such as IEEE 802.11a/g wireless LAN (Wi-Fi) and IEEE 802.16a/d/e wireless broadband access systems (WiBro, WiMAX).

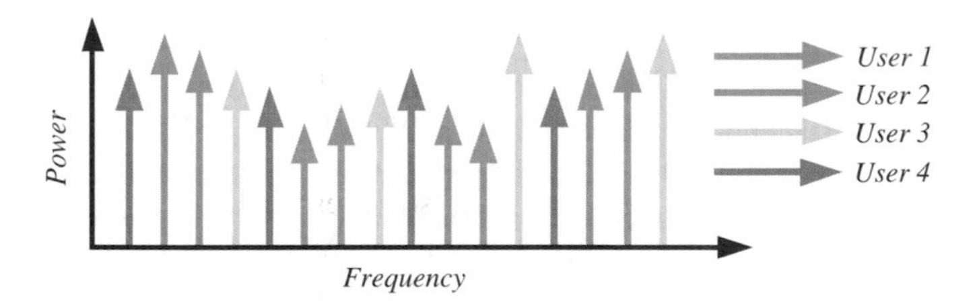


*Figure 31.* Example of four users sharing the same OFDM symbol

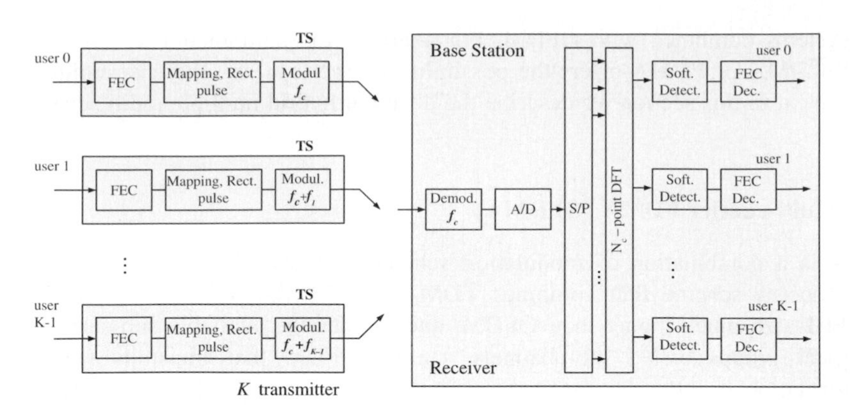


*Figure 32.* Basic principle of OFDMA

## 3.3 Multi-carrier Spread Spectrum

There are various combinations of multi-carrier modulation with the spread spectrum technique as multiple access schemes have been introduced. It has been shown that multi-carrier spread spectrum (MC-SS) offers high spectral efficiency, robustness and flexibility.

There are two general schemes of multi-carrier spread spectrum, namely MC-CDMA (OFDM-CDMA) and MC DS-CDMA.

In both schemes, the different users share the same bandwidth at the same time and separate the data by applying different user specific spreading codes, i.e., the separation of users signals is carried out in the code domain. Moreover, both schemes apply multi-carrier modulation to reduce the symbol rate and, thus, the amount of ISI per sub-channel. This ISI reduction is significant in spread spectrum systems where high chip rates occur.

The MC-CDMA transmitter spreads the original signal using a given spreading code in the frequency domain. In other words, a fraction of the symbol corresponding to a chip of the spreading code is transmitted through a different subcarrier. Multi-carrier modulation is realized by using low-complex OFDM operation. **Figure 33** demonstrates the general principle of MC-CDMA. Each symbols of the serial data stream is copied on the sub-streams before multiplying it with a chip of the spreading code assigned to the specific user. Mapping of the chips in the frequency domain allows for simple methods of signal detection.

This concept was proposed with OFDM for optimum use of the available bandwidth. For multi-carrier transmission, it is essential to have frequency nonselective fading over each subcarrier, hence in MC-CDMA the number of subcarriers $N_c$ has to be chosen sufficiently large to guarantee frequency nonselective fading on each subchannel.

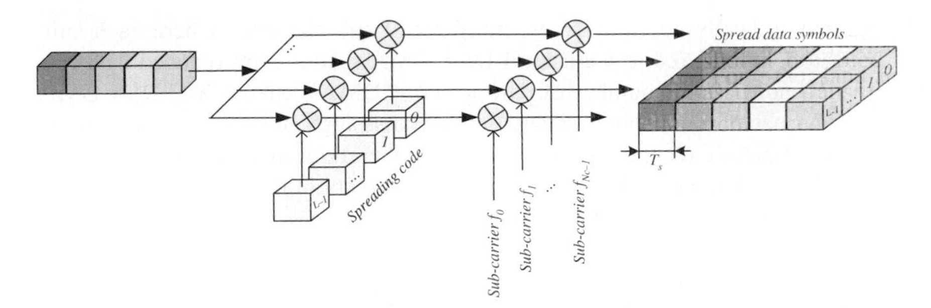


*Figure 33.* MC-CDMA signal generation for one user

Note that one of the IMT-2000 family of protocols is based on MC-CDMA technology. The IMT-MC (multicarrier) protocol (cdma2000) uses MC-CDMA spreading in the downlink, although in the uplink direction, the IMT-MC uses DS-CDMA.

Another success combination of multi-carrier modulation technique with spread spectrum is MC DS-CDMA. Unlike MC-CDMA, that maps the chips of a spread data symbol in frequency direction over a several parallel sub-channels, MC DS-CDMA maps the chips of spread data symbol in the time direction over several multi-carrier symbols. The principle of MC DS-CDMA is illustrated in **Figure 34**. MC DS-CDMA serial-to-parallel converts the high-rate data symbols into parallel low-rate sub-streams before spreading the data symbols on each sub-channel with a user-specific spreading code in time direction, which corresponds to direct sequence spreading on each sub-channel. The same spreading codes can be applied on the different sub-channels.

MC DS-CDMA systems have been proposed with different multi-carrier modulation schemes, also without OFDM, such that within the description of MC DS-CDMA the general term multi-carrier symbol instead of OFDM symbol is used. The MC DS-CDMA schemes can be subdivided in schemes with broadband sub-channels and schemes with narrowband sub-channels. Systems with broadband

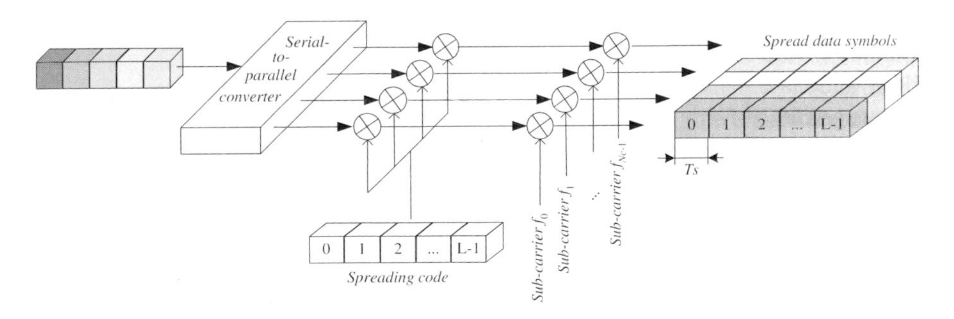


*Figure 34.* MC DS-CDMA signal generation for one user

sub-channels typically apply only few numbers of sub-channels, where each sub-channel can be considered as a classical DS-CDMA system with reduced data rate and ISI, depending in the number of parallel DS-CDMA systems. MC DS-CDMA systems with narrowband sub-channels typically use high numbers of sub-carriers and can be efficiently realized using the OFDM operation. Since each sub-channel is narrowband and spreading is performed in time direction, these schemes can only achieve a time diversity gain if no additional measures as coding or interleaving are applied.

It can be noted that both schemes have a generic architecture. In the case where the number of sub-carriers $N_c = 1$, the classical DS-CDMA transmission scheme is obtained, whereas without spreading (spreading gain $P_G = 1$) it results in a pure OFDM system.

**Table 2** below gives the main characteristics of different MC-SS concepts and summarize the main advantages and drawbacks of different schemes.

### 3.4 Multi-carrier Code-select CDMA

As we mentioned above a main disadvantage of multi-carrier systems is the high PAPR of the output signal, which may take values within a range that is the proportional to the number of carriers in the system. High peak power in transmitted signal will occasionally reach the amplifier saturation region and cause signal distortion, which results in performance degradation. To reduce PAPR in OFDM

*Table 2.* Comparison table between MC-CDMA and MC DS-CDMA

| Parameter | MC-CDMA | MC DS-CDMA |
|---|---|---|
| *Spreading* | Frequency direction | Time direction |
| *Subcarrier spacing* | $F_s = \frac{P_G}{N_c T_d}$ | $F_s \geq \frac{P_G}{N_c T_d}$ |
| *Detection algorithm* | MRC, EGC, ZF, MLD, equalization IC. | Correlation detector (coherent RAKE) |
| *Specific characteristics* | Very efficient for the synchronous downlink via using orthogonal codes | Designed especially for an asynchronous uplink |
| *Applications* | Synchronous uplink and downlink | Asynchronous uplink and downlink |
| *Advantages* | – Simple implementation<br>– Low complex receivers<br>– High spectral efficiency<br>– High frequency diversity | – Low PAPR in the uplink<br>– High time diversity |
| *Disadvantages* | – High PAPR especially in the uplink<br>– Synchronous transmission | – ISI and/or ICI can occur<br>– More complex receivers<br>– Less spectral efficient if other multi-carrier modulation than OFDM is used. |

based systems several proposals have been suggested and studied. Although, most of the PAPR reduction schemes are at the expense of BER performance, there are several interesting schemes to avoid the large amplitude fluctuations in Multi-carrier CDMA systems. One of such schemes is the Multi-carrier Code-select CDMA, namely MC CS-CDMA.

Main difference between conventional MC DS-CDMA and MC CS-CDMA is the so called code selection process added at transmitter side. **Figure 35** shows the simple single carrier code selection scheme with M=3 code selecting (CS) bits. As illustrated in **Figure 35**, stream of serial bits are first parallelized into $M+1$ substreams, where $M$ code selecting bits are going into *Spreading Code Block (SCB)*. Depending on the CS bits combination SCB chooses one of the L=$2^M$ spreading codes and $(M+1)^{th}$ bit is spread by this spreading sequence. At the receiver side we have the $L$ parallel fingers, one per each spreading code, and by simple correlation process we determine the spreading code and depending on this code Decision Block recovers the bits were transmitted.

Combining the CS CDMA scheme with conventional MC DS-CDMA we achieve the following advantages:

- Decreased number of subcarriers
- Low PAPR
- Bandwidth efficiency

All of this advantages only at the cost of receiver complexity and does not affect on the data rate and performance of the system. On the contrary due to the decreased number of subcarriers we can use the remain bandwidth for the purpose of performance improvement (*e.g. frequency repetition*).

**Figure 36** shows transmitter schematic for MC CS-CDMA system. At the transmitter side, the binary bit stream is first serial-to-parallel converted into U parallel substream. Next, $M$ bits of each group select a spreading code of SCB (spreading code block). The spreading code which is selected is then spreads the each $M+1$ bit of parallel substreams. Each SCB in MC CS-CDMA consists of code sequence sets with $L=2^M$ code sequences.

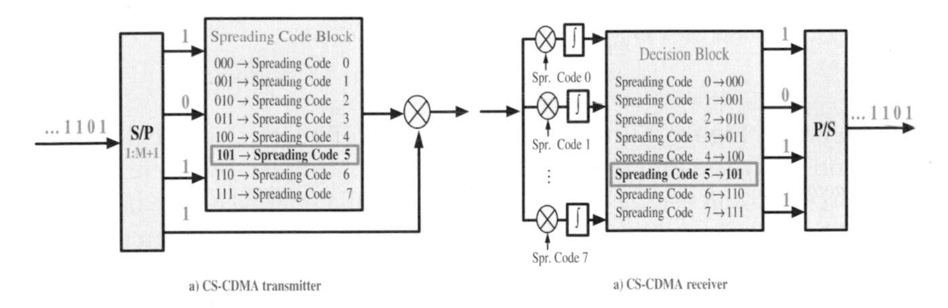


*Figure 35.* CS-CDMA concept

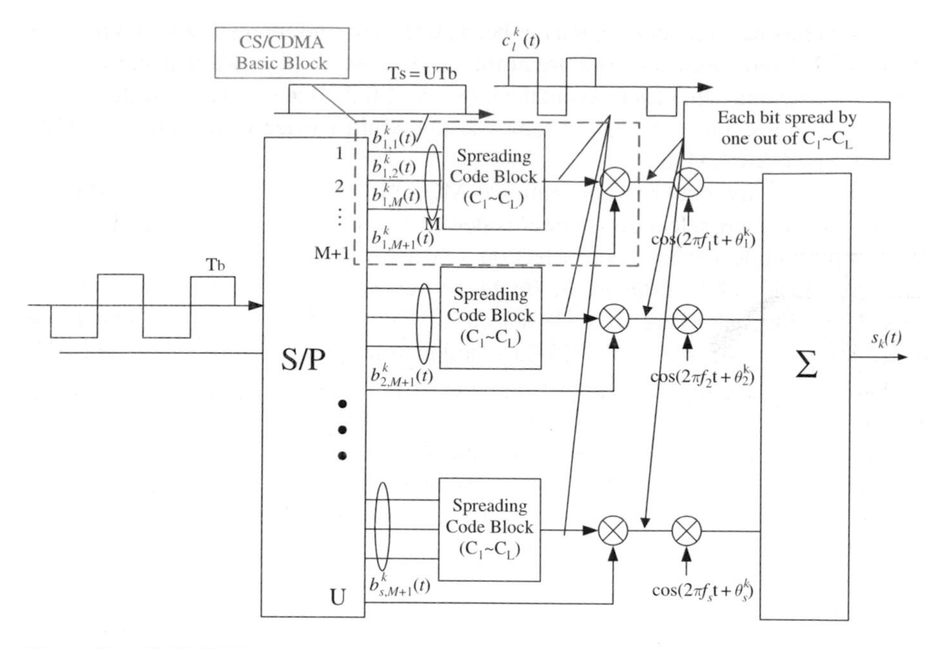


*Figure 36.* MC CS-CDMA system transmitter scheme

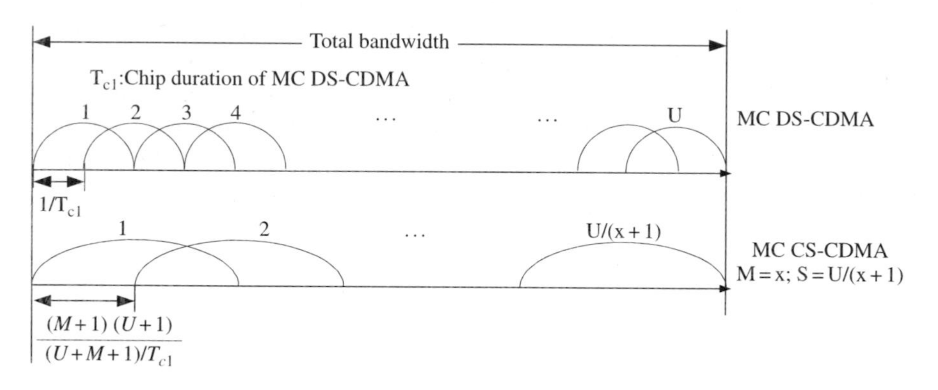


*Figure 37.* Frequency domain view of MC DS-CDMA and MC CS-CDMA

As it is shown in **Figure 37** increasing the number of code select bits MC CS-CDMA system can decrease the number of parallel subcarriers, but increases the subcarrier spacing distance for each carrier, achieving the improved time diversity.

MC CS-CDMA system is robust with respect to multipath interference and multiuser interference due to increasing spreading gain and diversity gain. Also note that MC CS-CDMA achieves lower PAPR than conventional MC DS-CDMA due to the reduced number of subcarriers. However, the advantages of MC CS-CDMA is at the expense of higher system complexity of receiver when increasing of various parallel input bits M of SCB and the number of fingers of Rake receiver in each subcarrier.

## REFERENCES


1. Agrawal D.P., Zeng Q., 2003, *Introduction to Wireless and Mobile Systems,* Brook/Cole, Pacific Grove.
2. Etoh M., 2005, *Next Generation Mobile Systems 3G and Beyond*, Wiley, Wiltshire.
3. Fazel K., Kaiser S., 2003, *Multi-Carrier and Spread Spectrum Systems,* Wiley, Wiltshire.
4. Glisik S.G., 2006, *Advanced Wireless Networks. 4G Technologies,* Wiley, Wiltshire.
5. Groe J.B., Larson L.E., 2000, *CDMA Mobile Radio Design,* Artech House, London.
6. Harada H., Prasad R., 2002, *Simulation and Software Radio for Mobile Communications,* Artech House, London.
7. Holma H., Toskala A., 2004, *WCDMA for UMTS, 3rd* ed., Wiley, Cornwall.
8. Korhonen J., 2003, *Introduction to 3G Mobile Communications,* 2nd ed., Artech House, Norwood.
9. Lee J.S., Miller L.R., 1998, *CDMA Systems Engineering Handbook,* Artech House, London.
10. Mishra A.R., 2004, *Fundamentals of Cellular Network Planning and Optimisation. 2G/2.5G/3G... Evolution to 4G,* Wiley, Wiltshire.
11. Nee R., Prasad R., 2000, *OFDM Wireless Multimedia Communications,* Artech House, London.
12. Rappaport T.S., 2002, *Wireless Communications. Principles and Practice,* 2nd ed., Prentice Hall, Upper Saddle River.
13. Ryu K.W., Park Y.W., *et al.*, 2004, *Performance of multicarrier code select CDMA for high data transmission*, IEEE 60th VTC2004-Fall, pp. 5054–5058, Vol. 7., Los Angeles.
14. Schiller J., 2003, *Mobile Communications,* 2nd ed., Addison-Wesley, Kent.
15. Schulze N, Luders C., 2005, *Theory and Applications of OFDM and CDMA Wideband Wireless Communications,* Wiley, Wiltshire.
16. Tachikawa K., 2002, *W-CDMA Mobile Communication System,* Wiley, Comwall.
17. Webb W., 1998, *Understanding Cellular Radio,* Artech House, London.
18. Wikipedia The Free Encyclopedia , http://www.wikipedia.org
19. Wilkinson N., 2002, *Next Generation Network Services,* Wiley, Guildford.
20. Yang S.C., 2004, *3G CDMA2000 Wireless System Engineering*, Artech House, Norwood.
21. Yang S.C., 1998, *CDMA RF System Engineering,* Artech House, London.
22. Zigangirov K.Sh., 2004, *Theory of Code Division Multiple Access Communication,* IEEE press and Wiley-Interscience, Piscataway.

CHAPTER 3

# FUNDAMENTALS OF SINGLE-CARRIER CDMA TECHNOLOGIES

F. ADACHI, D. GARG, A. NAKAJIMA, K. TAKEDA, L. LIU,
AND H. TOMEBA
*Tohoku University*

**Abstract:** A broad range of wireless services of e.g., 100Mbps-to-1Gbps are demanded for the beyond 3rd generation (3G) wireless mobile communications systems. Wireless channels for such high-speed data transmissions are characterized by severely frequency-selective channel, which is caused by many interfering paths with different time delays. Promising wireless access technique that can overcome the channel frequency-selectivity and even improve the transmission performance is code division multiple access (CDMA). There are two approaches in CDMA: multi-carrier (MC)-CDMA and single-carrier (SC)-CDMA (direct-sequence CDMA or DS-CDMA is another popular terminology, but in this Chapter, the terminology "SC-CDMA" is used). Both MC- and SC-CDMA techniques have flexibility for providing variable rate transmissions, yet retaining multiple access capability. Their special case is orthogonal frequency division multiplexing (OFDM) and non-spread single carrier (SC) transmission, respectively. A lot of attention has been paid to MC-CDMA. However, it was recently shown that SC-CDMA can achieve a good performance comparable to MC-CDMA if proper frequency-domain equalization (FDE) is adopted. In this chapter, various techniques for improving SC-CDMA transmission performance are presented

**Keywords:** CDMA, frequency-domain equalization, space-time block coded transmit diversity, MIMO, space division multiplexing, hybrid ARQ

## 1. INTRODUCTION

During the last 25 years, wireless technologies have enhanced our communications networks by providing an important capability, i.e., mobility. Before the introduction of wireless mobile networks, communication was possible only from/to fixed places (i.e., houses, offices and public phones). In fixed and wireless networks, voice conversation was a long-time dominant service, but the introduction of Internet communications services in the fixed networks has been changing our society at a very rapid pace. In line with the recent explosive expansion of Internet traffic in the fixed networks, demands for broad ranges of services are becoming stronger even in wireless networks. People



*Y. Park and F. Adachi (eds.), Enhanced Radio Access Technologies for Next Generation Mobile Communication*, 81–120.

want to be connected anytime, anywhere with the networks for not only making voice conversations with people but also data conversation (i.e., downloading/uploading their information). A variety of services are now available over the 2nd and 3rd generation (2G and 3G) wireless networks, including e-mailing, Web access, and on-line services ranging from bank transactions to entertainment.

The 3G wireless networks based on single-carrier code division multiple access (SC-CDMA) technique, with much higher data rates of up to 384kbps (around 14Mbps in the later stage) and better representation than the present 2G wireless networks, were put into services in some countries and their deployment speed has since accelerated. However, the capabilities of 3G wireless networks will sooner or later be insufficient to cope with the increasing demands for broadband services, which is now in full force in fixed networks. Demands for downloading of ever increasing volume of information will become higher and higher even in mobile communications networks. Most of the services may contain high resolution and short delay streaming video combined with high fidelity audio. 3G wireless networks will be continuously evolving with high speed downlink packet access (HSDPA) technique, multiple-input/multiple-output (MIMO) antenna technique, etc, for providing packet data services of around 14Mbps as the mid-term evolution and of 30~100Mbps as the long-term evolution. The evolution of 3G wireless networks will be followed by the development of next generation wireless networks, called 4th generation (4G) wireless networks, that support extremely higher-speed packet data services, of e.g., 100M~1Gbps, than 3G wireless networks. Most of the wireless techniques to be developed for the 3.9G wireless networks will be used for the 4G wireless networks.

The broadband SC-CDMA technologies for the 4G wireless networks will be introduced in this chapter. Another promising access technology is multi-carrier CDMA (MC-CDMA), which has recently been attracting much attention. Before briefly compare SC- and MC-CDMA techniques in **Section 3**, the broadband propagation channel is discussed in **Section 2**. Frequency-domain equalization is a key technique for improving the transmission performance of SC-CDMA in a broadband channel. This will be introduced in **Section 4**. MIMO antenna techniques can be used for achieving better transmission performance and higher transmission rate with a limited bandwidth. First, space-time block coding for achieving better transmission performance is presented in **Section 5** and then space-division multiplexing (SDM) for increasing the transmission rate in **Section 6**. Multi-access interference (MAI) limits the transmission performance. Block spreading can be used to avoid MAI. This will be discussed in **Section 7**. Since a major transmission mode in beyond 3G systems will be packet based, also introduced in **Section 8** is the hybrid automatic repeat request (HARQ).

## 2. BROADBAND CHANNEL

A simplified propagation model is illustrated in **Figure 1**. There are several large obstacles between a base station (BS) and a mobile station (MS) and also many local scatterers in the vicinity of the MS. The reflection of the transmitted signal

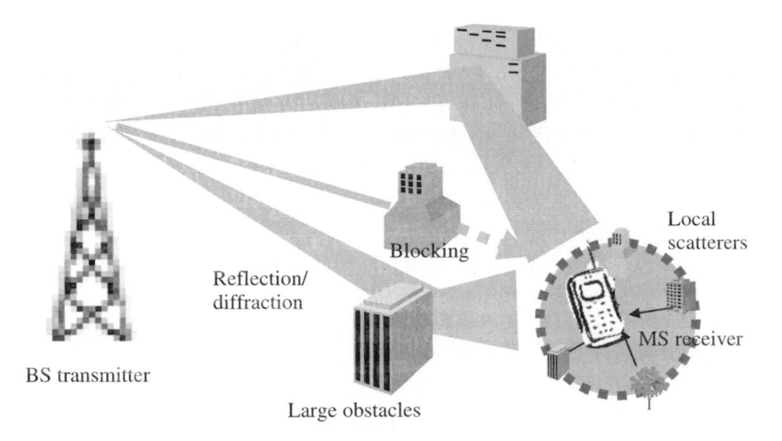


*Figure 1.* Propagation model

by large obstacles creates the propagation paths with different time delays, where time delay difference is more than the inverse of signal bandwidth $W$. Each path is a cluster of irresolvable multipaths, having a time delay difference of shorter than $1/W$, created by reflection or diffraction of the transmitted signal reaching the surroundings of a MS by local scatterers (such as neighboring buildings). They interfere with each other and the received signal power changes rapidly in a random manner with a period of about half carrier wavelength as the MS moves.

When multiple paths with different time delays are present, many impulses with different time delays are received after the transmission of an impulse from a transmitter, as shown in **Figure 2**. Such a multipath channel can be viewed as a time varying linear filter having an impulse response $h(\tau, t)$ observed at time $t$, which can be expressed as

$$h(\tau, t) = \sum_{l=0}^{L-1} h_l(t)\delta(\tau - \tau_l), \tag{1}$$

where $L$ is the number of resolvable paths, $h_l(t)$ and $\tau_l$ are respectively the path gain and time delay of the $l$-th path and $\delta(t)$ is the delta function. In urban areas,

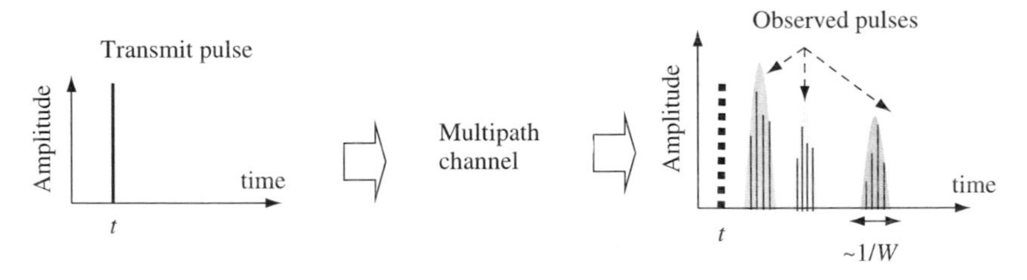


*Figure 2.* Channel impulse response

since $h_l(t)$ is the contribution from many irresolvable paths, it is known to be characterized as the complex Gaussian process (resulting in the Rayleigh fading). $\Omega(\tau) = E[|h_l(t)|^2]\delta(\tau - \tau_l)$ is the so-called power delay profile. Often used model is an exponentially decaying power delay profile. Power delay profile with decay

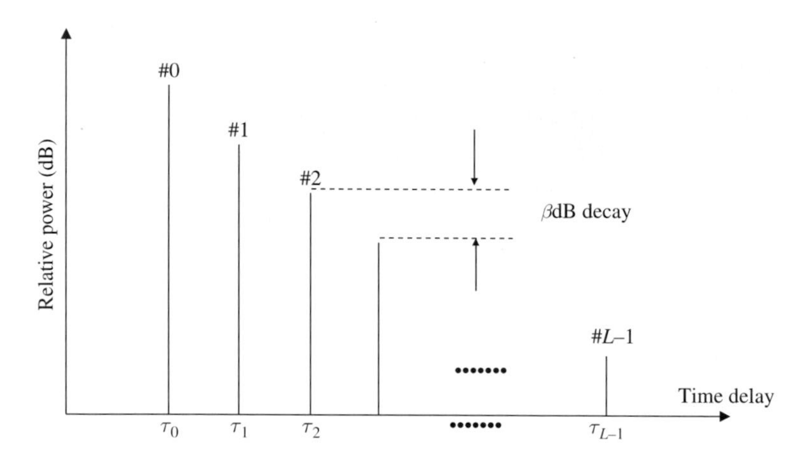


*Figure 3.* Power delay profile of multipath channel

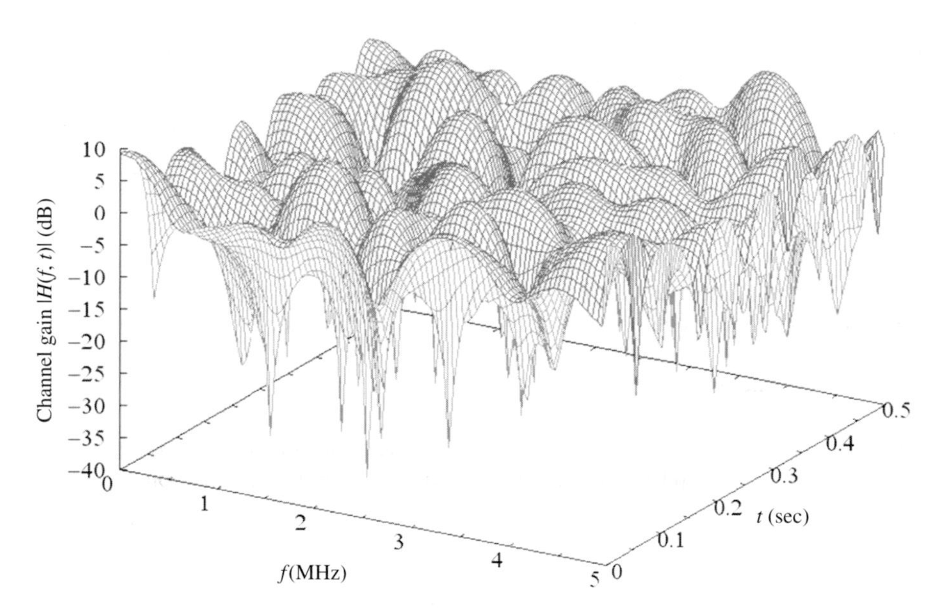


*Figure 4.* Transfer function of multipath channel

factor $\beta$ is illustrated in **Figure 3**. The transfer function $H(f, t)$ of such a multipath channel at time $t$ is expressed as

$$H(f,t) = \sum_{l=0}^{L-1} h_l(t)\exp(-j2\pi f\tau_l), \tag{2}$$

which is not anymore constant over the signal bandwidth and results in the so-called frequency-selective channel. Furthermore, the channel transfer function varies rapidly in a random manner over a short distance. When a transmitter or/and receiver travels, doubly (frequency-time) selective fading channel is produced. **Figure 4** shows how the channel transfer function varies in the frequency- and time-domain for an $L = 16$-path channel with uniform power delay profile ($\beta = 0$dB) and a time delay separation of 100ns between adjacent paths. A carrier frequency of 2GHz and a terminal speed of 4km/h are assumed.

## 3. MULTI-CARRIER CDMA AND SINGLE-CARRIER CDMA

Promising multiple access techniques under a severe frequency- and time-selective fading environment are SC-CDMA and MC-CDMA. The former uses the time-domain spreading technique, while the latter uses the frequency-domain spreading technique. First, we start with the overview of SC-CDMA with rake combining and then, MC-CDMA.

**Figure 5** illustrates the transmitter/receiver structure for SC-CDMA. At the transmitter, after the binary information data is channel-encoded and interleaved, the encoded information data sequence is transformed into data-modulated symbol sequence. The resulting symbol sequence is spread (time-domain spreading) by a spreading chip sequence, $c(t)$, with $SF$ times higher rate $1/T_c$ than symbol rate $1/T$. The spreading factor $SF$ is defined as $SF = T/T_c$. The bandwidth of spread signal is $(1+\alpha)/T_c$, where $\alpha$ is the roll-off factor (typically $\alpha = 0.5$) of the chip shaping filter. At the receiver, the SC-CDMA signal is received via an $L$-path channel. The receiver matched filter (MF) consists of $L$ correlators; each correlator multiplies the received signal with the locally generated chip sequence, which is time-synchronized to the time delay of each propagation path, and integrates over one symbol period. The MF outputs are coherently summed up based on maximal ratio combining (MRC), which is called as coherent rake combining, followed by despreading. Finally, the coherent rake combiner output sequence is demodulated, de-interleaved and passed to the channel decoder to obtain the decoded binary information sequence.

In MC-CDMA, a number of narrowband orthogonal subcarriers are used for parallel transmission and simple one-tap frequency-domain equalization (FDE) is used. **Figure 6** shows the transmitter/receiver structure for MC-CDMA with $N_c$ subcarriers. A difference from SC-CDMA transmitter is the introduction of $N_c$-point inverse fast Fourier transform (IFFT) after time-domain spreading and the guard interval (GI) insertion. The use of serial-to-parallel (S/P) conversion followed by the IFFT transforms the time-domain spread signal into a frequency-domain

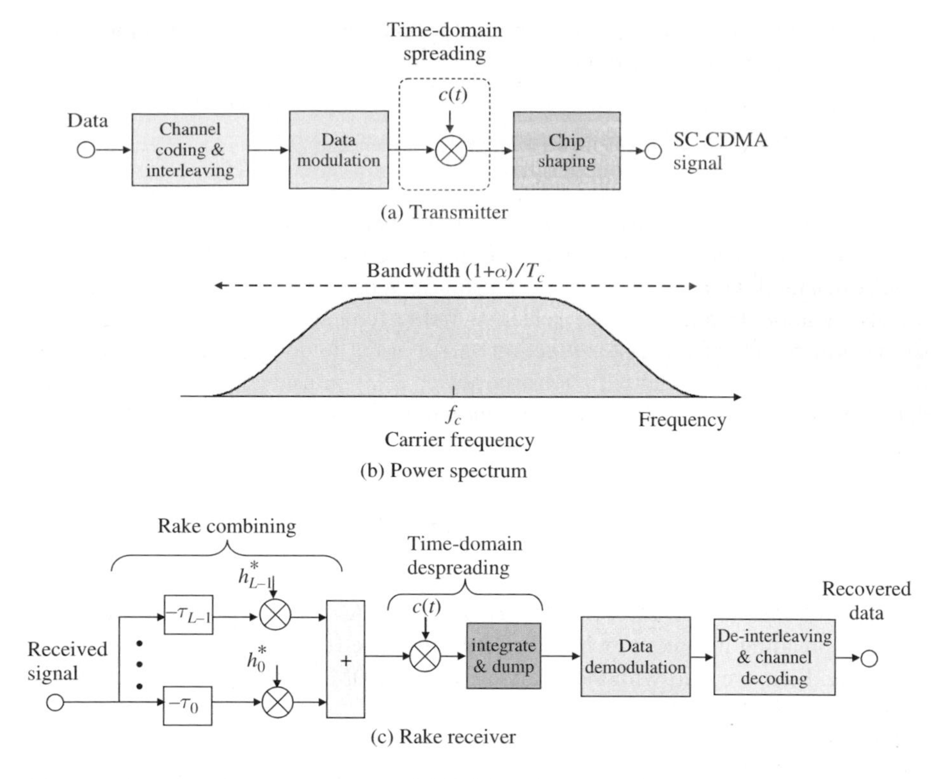


*Figure 5.* Transmitter/receiver structure for SC-CDMA with rake combining

spread signal, resulting in the MC-CDMA signal. MC-CDMA with $SF = 1$ is OFDM. The GI insertion is necessary to avoid the orthogonality destruction among $N_c$ subcarriers due to the presence of multipaths with different time delays. The GI length needs to be larger than the maximum time delay difference among multipaths. At the receiver, after removing the GI, the received signal is decomposed by $N_c$-point FFT into $N_c$ subcarrier components. The distortion of the signal spectrum due to frequency-selective fading is compensated by using one-tap FDE. The equalized subcarrier components are parallel-to-serial (P/S) converted into the time-domain spread signal, followed by despreading as in SC-CDMA receiver. FDE can be jointly used with antenna diversity reception for further performance improvement in MC-CDMA. Among various FDE weights, it was shown that the use of minimum mean square error (MMSE) weight provides the best bit error rate (BER) performance. This is because the MMSE weight can provide the best compromise between the noise enhancement and suppression of frequency-selectivity. MC-CDMA with MMSE-FDE provides much better BER performance than SC-CDMA with coherent rake combining. Because of this, until recently, research attention was shifted from SC techniques to MC techniques such as MC-CDMA and OFDM ($SF = 1$). But, as will be shown in this chapter, FDE can

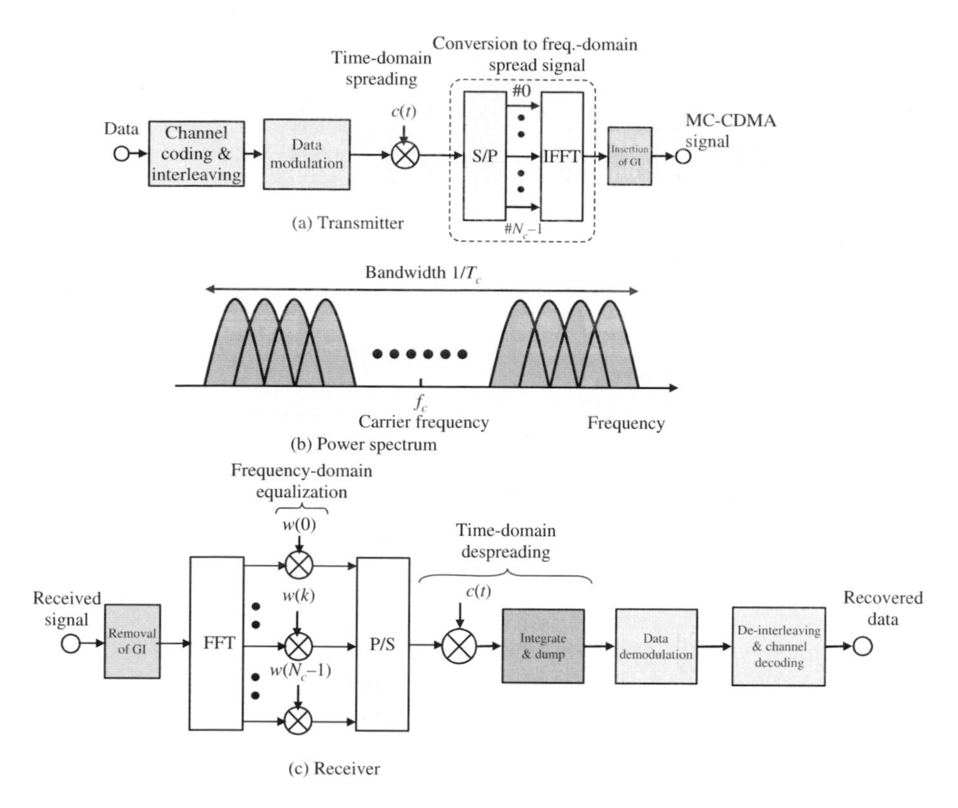


*Figure 6.* Transmitter/receiver structure for MC-CDMA

also be applied to SC-CDMA with much improved performance compared to rake combining. SC-CDMA is considered again as a promising access technique similar to MC-CDMA.

## 4. FREQUENCY-DOMAIN EQUALIZATION

The application of MMSE-FDE to SC-CDMA can replace the coherent rake combining with much improved BER performance. First, FDE for SC-CDMA is shown. However, the residual inter-chip interference (ICI) is present after MMSE-FDE and this will limit the BER performance improvement. The ICI cancellation can be used to reduce the residual ICI and hence improve the BER performance. These are presented here.

### 4.1 MMSE Equalization

Transmitter/receiver structure of multicode SC-CDMA with FDE is illustrated in **Figure 7**. We assume that $C$ data streams are simultaneously transmitted. At the

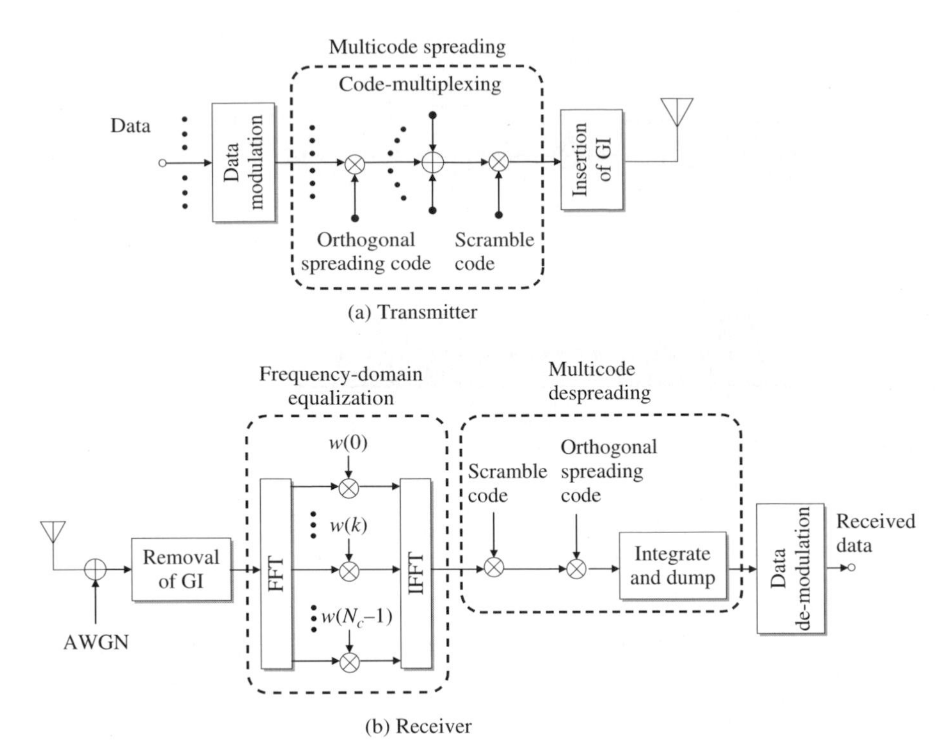


*Figure 7.* Multicode SC-CDMA transmitter/receiver structure

transmitter, the $u$th binary data sequence is transformed into a data modulated symbol sequence $\{d_u(n);\ n = 0 \sim N_c/SF - 1\}$, $u = 0 \sim C - 1$, and then spread by multiplying an orthogonal spreading sequence $c_u(t)$ with spreading factor $SF$. The resulting $C$ chip sequences are added and further multiplied by a common scramble sequence $c_{scr}(t)$ to make the resulting multicode SC-CDMA chip sequence white-noise like. $C$ is called code-multiplexing order. This is called multicode spreading. Then, the orthogonal multicode SC-CDMA chip sequence is divided into a sequence of blocks of $N_c$ chips each and then the last $N_g$ chips of each block are copied as a cyclic prefix and inserted into the GI placed at the beginning of each block as shown in **Figure 8**. The GI-inserted multicode SC-CDMA chip sequence

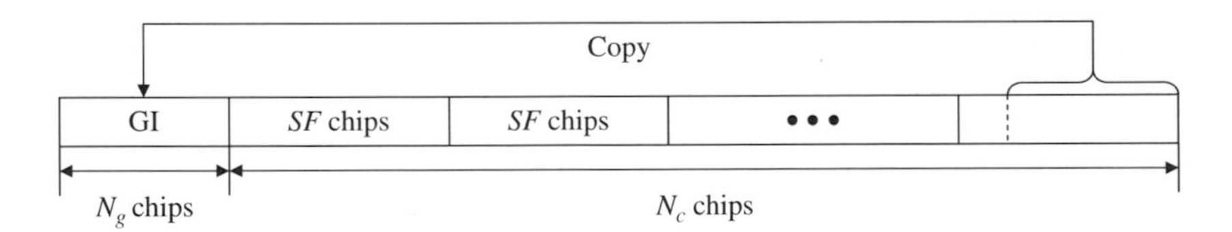


*Figure 8.* Block structure

$\{\hat{s}(t);\, t=-N_g \sim N_c-1\}$ in a block can be expressed, using the equivalent lowpass representation, as

$$\hat{s}(t)=\sqrt{\frac{2E_c}{T_c}}s(t \bmod N_c), \tag{3}$$

where $E_c$ and $T_c$ denote the chip energy and the chip duration, respectively, and $s(t)$ is given by

$$s(t)=\left[\sum_{u=0}^{C-1} d_u(\lfloor t/SF \rfloor)\, c_u(t \bmod SF)\right] c_{scr}(t) \tag{4}$$

for $t=0\sim N_c-1$, where $|c_u(t)|=|c_{scr}(t)|=1$ and $\lfloor x \rfloor$ represents the largest integer smaller than or equal to $x$.

The chip block $\{\hat{s}(t);\; t=-N_g\sim N_c-1\}$ is transmitted over a frequency-selective fading channel and received by a receiver. After the removal of the GI, the received chip sequence $\{r(t); t=0\sim N_c-1\}$ in a block is decomposed by $N_c$-point FFT into $N_c$ subcarrier components $\{R(k);\, k=0\sim N_c-1\}$ (the terminology "subcarrier" is used for explanation purpose although subcarrier modulation is not used). The $k$th subcarrier component $R(k)$ can be written as

$$\begin{aligned} R(k) &= \sum_{t=0}^{N_c-1} r(t)\exp\left(-j2\pi k\frac{t}{N_c}\right) \\ &= \sqrt{\frac{2E_c}{T_c}}H(k)S(k)+\Pi(k) \end{aligned}, \tag{5}$$

where $S(k)$, $H(k)$ and $\Pi(k)$ are the $k$th subcarrier components of $s(t)$, the channel gain and the noise component due to the additive white Gaussian noise (AWGN), respectively. $H(k)$ corresponds to $H(f,t)$ defined by Eq. (2), but with $f=k/(N_cT_c)$; time dependency of the channel gain is dropped since we are assuming very slow fading channel for simplicity.

FDE is carried out similar to MC-CDMA. $R(k)$ is multiplied by the FDE weight $w(k)$ as

$$\begin{aligned} \hat{R}(k) &= w(k)R(k) \\ &= \sqrt{\frac{2E_c}{T_c}}S(k)\hat{H}(k)+\hat{\Pi}(k) \end{aligned}, \tag{6}$$

where $\hat{H}(k)=w(k)H(k)$ and $\hat{\Pi}(k)=w(k)\Pi(k)$ are the equivalent channel gain and the noise component after performing FDE, respectively. As the FDE weight, maximal ratio combining (MRC), zero forcing (ZF), equal gain combining (EGC) and minimum mean square error (MMSE) weights are considered. They are given by

$$w(k) = \begin{cases} H^*(k) & \text{for MRC} \\ H^*(k)\big/|H(k)|^2 & \text{for ZF} \\ H^*(k)/|H(k)| & \text{for EGC,} \\ H^*(k)\Big/\left(|H(k)|^2 + \left(\frac{C}{SF}\frac{E_s}{N_0}\right)^{-1}\right) & \text{for MMSE} \end{cases} \tag{7}$$

where $E_s/N_0$ ($=E_cSF/N_0$) is the average received signal energy per data symbol-to-AWGN power spectrum density ratio and * denotes the complex conjugate operation.

One-shot observation of the equivalent channel gain $\hat{H}(k)$ and the noise $\hat{\Pi}(k)$ for MMSE, ZF and MRC weights are illustrated in **Figure 9**. An $L = 16$-path fading channel is assumed. Also plotted in the figure is the original channel gain $H(k)$. The MRC weight enhances the frequency-selectivity of the channel after equalization. Using the ZF weight, the frequency-nonselective channel can be perfectly restored after equalization (of course, only if the channel estimation is ideal), but the noise enhancement is produced at the subcarrier where the channel gain drops. However, the MMSE weight can avoid the noise enhancement by giving up the perfect restoration of the frequency-nonselective channel (the MMSE weight minimizes the mean square error between $S(k)$ and $\hat{R}(k)$). Among these FDE weights, the MMSE weight can provide the best compromise between the noise enhancement and suppression of frequency-selectivity and therefore, gives the best BER performance.

After MMSE-FDE, $N_c$-point IFFT is applied to obtain the time-domain multicode SC-CDMA chip sequence as

$$\begin{aligned} \hat{r}(t) &= \frac{1}{N_c}\sum_{k=0}^{N_c-1}\hat{R}(k)\exp\left(j2\pi t\frac{k}{N_c}\right) \\ &= \sqrt{\frac{2E_c}{T_c}}\left(\frac{1}{N_c}\sum_{k=0}^{N_c-1}\hat{H}(k)\right)s(t) + \hat{\mu}(t) + \hat{\eta}(t) \end{aligned}, \tag{8}$$

where $s(t)$ in the first term represents the transmitted chip sequence, $\hat{\mu}(t)$ is the residual inter-chip interference (ICI) component and $\hat{\eta}(t)$ is the noise component. $\hat{\mu}(t)$ can be expressed as

$$\hat{\mu}(t) = \sqrt{\frac{2E_c}{T_c}}\frac{1}{N_c}\sum_{k=0}^{N_c-1}\hat{H}(k)\left[\sum_{\substack{\tau=0 \\ \neq t}}^{N_c-1} s(\tau)\exp\left(j2\pi k\frac{t-\tau}{N_c}\right)\right]. \tag{9}$$

Note that if $\hat{H}(k) = \text{constant}$, $\hat{\mu}(t) = 0$ (i.e., this is the case of ZF-FDE and no ICI is produced). The residual ICI degrades the achievable BER performance (this is

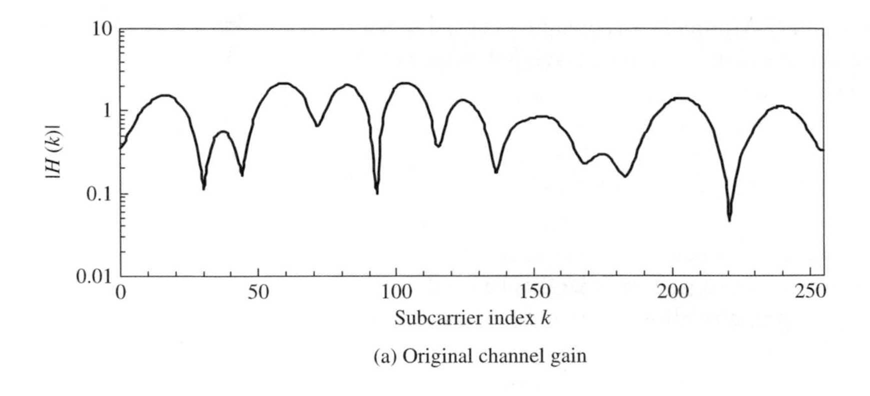


(a) Original channel gain

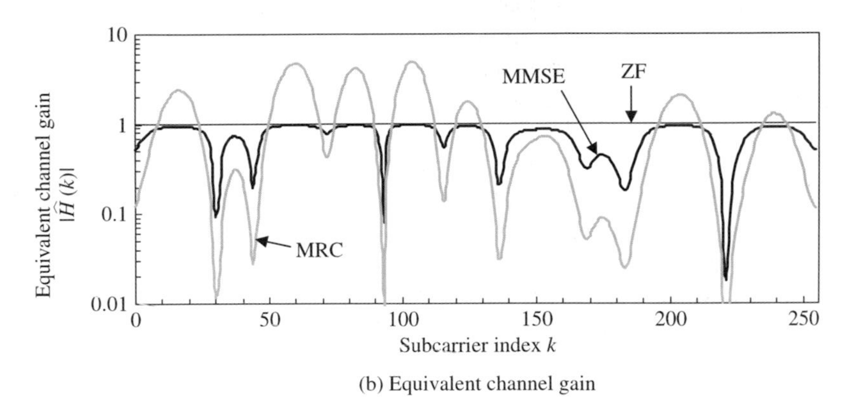


(b) Equivalent channel gain

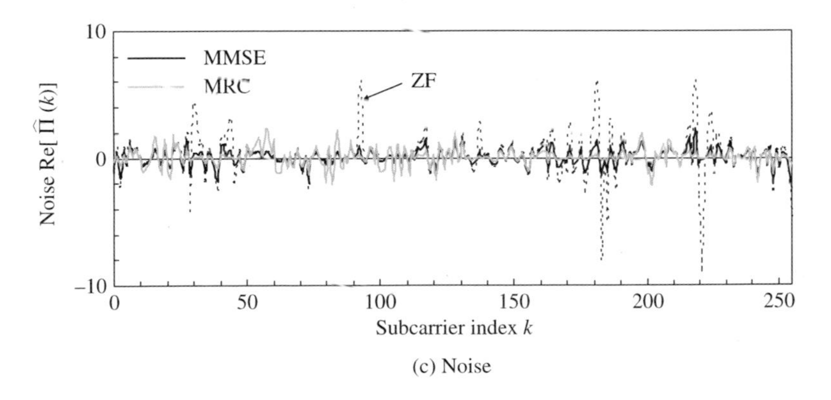


(c) Noise

*Figure 9.* One-shot observation of equivalent channel gain and noice after FDE

explained later). Multicode despreading is carried out on $\hat{r}(t)$ to obtain the decision variable for the data modulated symbol sequence $\{d_u(n);\ n = 0 \sim N_c/SF - 1\}$, $u = 0 \sim C - 1$, as

$$\hat{d}_u(n) = \frac{1}{SF} \sum_{t=nSF}^{(n+1)SF-1} \hat{r}(t) c_u^*(t \bmod SF) c_{scr}^*(t), \tag{10}$$

based on which data demodulation is done.

An arbitrary spreading factor $SF$ can be used for the given value of FFT window size $N_c$. This property allows variable rate transmission even when FDE is used in SC-CDMA systems.

**Figure 10** plots the BER performance of multicode SC-CDMA using MMSE-FDE for $SF = 16$, obtained by computer simulation, as a function of the average received bit energy-to-AWGN noise power spectrum density ratio $E_b/N_0$. QPSK data modulation and an $L = 16$-path frequency-selective Rayleigh fading channel having a uniform power delay profile ($E[|h_l|^2] = 1/L$) are assumed. For comparison,

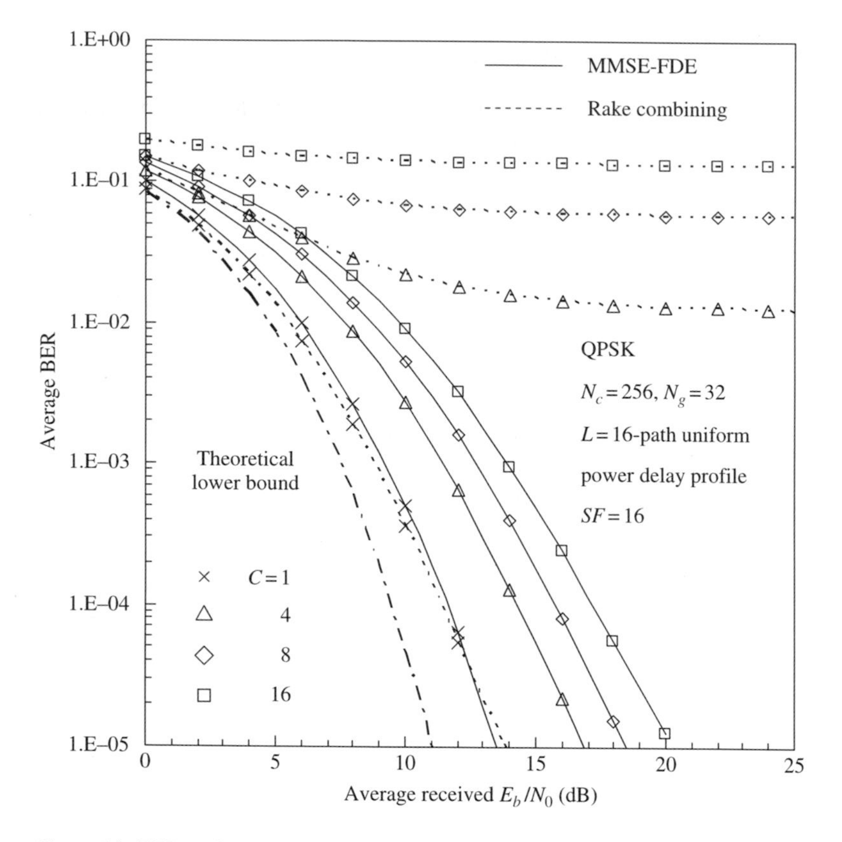


*Figure 10.* BER performance of multicode SC-CDMA with MMSE-FDE

the BER performance of coherent rake combining and theoretical lower-bound are also plotted. When $C = 1$, MMSE-FDE and rake combining can achieve almost the same BER performance. However, when $C \geq 4$, the BER performance using rake combining significantly degrades due to strong ICI and exhibits large BER floors. MMSE-FDE can always achieve better BER performance than rake combining and no BER floors are seen. However, although MMSE-FDE provides much better BER performance, the BER performance degrades as the code-multiplexing order $C$ increases since the orthogonality distortion among codes is produced due to the residual ICI $\hat{\mu}(t)$. As the frequency-selectivity becomes stronger (or $L$ increases), the complexity of the rake receiver increases since more correlators are required for collecting enough signal power for data demodulation. However, unlike rake receiver, the complexity of MMSE-FDE receiver is independent of the channel frequency-selectivity. The use of FDE can alleviate the complexity problem of the rake receiver arising from too many paths in a severe frequency-selective channel. These suggest that SC-CDMA with MMSE-FDE is a promising broadband access as MC-CDMA for 4G wireless networks.

### 4.2 Inter-chip Interference (ICI) Cancellation

Although MMSE-FDE can significantly improve the BER performance of orthogonal multicode SC-CDMA, there is still a big performance gap to the theoretical lower-bound as shown in **Figure 10**. This is due to the residual ICI after MMSE-FDE, given by Eq. (9). An ICI cancellation technique can be introduced into MMSE-FDE to improve the BER performance. The ICI in SC-CDMA with $SF=1$ is equivalent to the inter-symbol interference (ISI) in the non-spread (i.e., $SF=1$) SC transmissions; the ISI cancellation techniques can be found in the literature. Similar to ISI cancellation for MC-CDMA, ICI cancellation for SC-CDMA can be carried out either in the time-domain or in the frequency-domain after performing MMSE-FDE.

For the frequency-domain ICI cancellation, the replicas of frequency components $\{M(k);\ k = 0 \sim N_c - 1\}$ of the residual ICI $\hat{\mu}(t)$ in Eq. (9) are subtracted from $\{\hat{R}(k) : k = 0 \sim N_c - 1\}$ after MMSE-FDE. $M(k)$ is given by

$$M(k) = \sum_{t=0}^{N_c-1} \hat{\mu}(t) \exp\left(-j2\pi k \frac{t}{N_c}\right) = \sqrt{\frac{2E_c}{T_c}} \left\{ \hat{H}(k) - \frac{1}{N_c} \sum_{k'=0}^{N_c-1} \hat{H}(k') \right\} S(k) \quad (11)$$

A joint MMSE-FDE and ICI cancellation is repeated in an iterative fashion so as to improve the accuracy of the ICI replica generation. **Figure 11** shows the structure of joint MMSE-FDE and ICI cancellation.

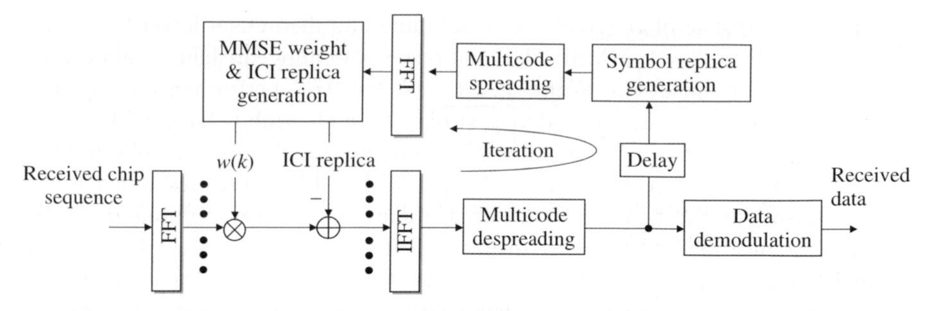


*Figure 11.* Joint MMSE-FDE and ICI cancellation

The *i*th iteration is described below. After performing MMSE-FDE with the MMSE weight $w^{(i)}(k)$ , ICI cancellation is performed in the frequency-domain as

$$\tilde{R}^{(i)}(k) = \hat{H}^{(i)}(k) - \tilde{M}^{(i)}(k), \tag{12}$$

where $\hat{H}^{(i)}(k)\left(= w^{(i)}(k)H(k)\right)$ is the equivalent channel gain and $\tilde{M}^{(i)}(k)$ is the replica of $M(k)$ which is given, from Eq. (11), as

$$\tilde{M}^{(i)}(k) = \begin{cases} 0 & \text{for } i = 0 \\ \sqrt{\dfrac{2E_c}{T_c}}\left\{\hat{H}^{(i)}(k) - A^{(i)}\right\}\tilde{S}^{(i-1)}(k) & \text{for } i > 0, \end{cases} \tag{13}$$

where $\tilde{S}^{(i-1)}(k)$ is the *k*th frequency component of the soft decision transmitted chip block replica $\tilde{s}^{(i-1)}(t)$ (which is generated by feeding back the $(i-1)$th ICI cancellation result) and $A^{(i)}$ is given by

$$A^{(i)} = \frac{1}{N_c}\sum_{k=0}^{N_c-1}\hat{H}^{(i)}(k). \tag{14}$$

$N_c$-point IFFT is performed on $\{\tilde{R}^{(i)}(k); k = 0 \sim N_c - 1\}$to obtain the time-domain chip sequence for multicode despreading.

A series of joint MMSE-FDE and ICI cancellation, $N_c$-point IFFT, multicode despreading, data symbol replica generation, and multicode spreading is repeated a sufficient number of times. Finally, data-demodulation is carried out to obtain the received data.

The MMSE weight $w^{(i)}(k)$ minimizes the mean square error (MSE) $E[|e(k)|^2]$ for the given $H(k)$, i.e., $\partial E[|e(k)|^2]/\partial w^{(i)}(k) = 0$, where $e(k)$ is the equalization error between $\tilde{R}^{(i)}(k)$ after the ICI cancellation and $S(k)$ of the transmitted signal $s(t)$ and is defined as

$$e(k) = \tilde{R}^{(i)}(k) - A^{(i)}S(k). \tag{15}$$

The MMSE weight is given as

$$(16) \qquad w^{(i)}(k) = \frac{H^*(k)}{\rho^{(i-1)}\,|H(k)|^2 + \left(\frac{E_c}{N_0}\right)^{-1}},$$

where $\rho^{(i-1)}$ is an interference factor determined by feeding back the $(i-1)$th iteration result and given by

$$(17) \qquad \rho^{(i-1)} \approx \sum_{t=0}^{N_c-1} \left\{ \left|\bar{s}^{(i-1)}(t)\right|^2 - \left|\tilde{s}^{(i-1)}(t)\right|^2 \right\},$$

where $\bar{s}^{(i-1)}(t)$ is the hard decision replica of transmitted chip block.

The BER performance for the case of $SF = 16$ is plotted in **Figure 12** with the code-multiplexing order $C$ as a parameter. When $C = 1$, the BER performance approaches the theoretical lower-bound by about 0.5 dB. As $C$ increases, the BER

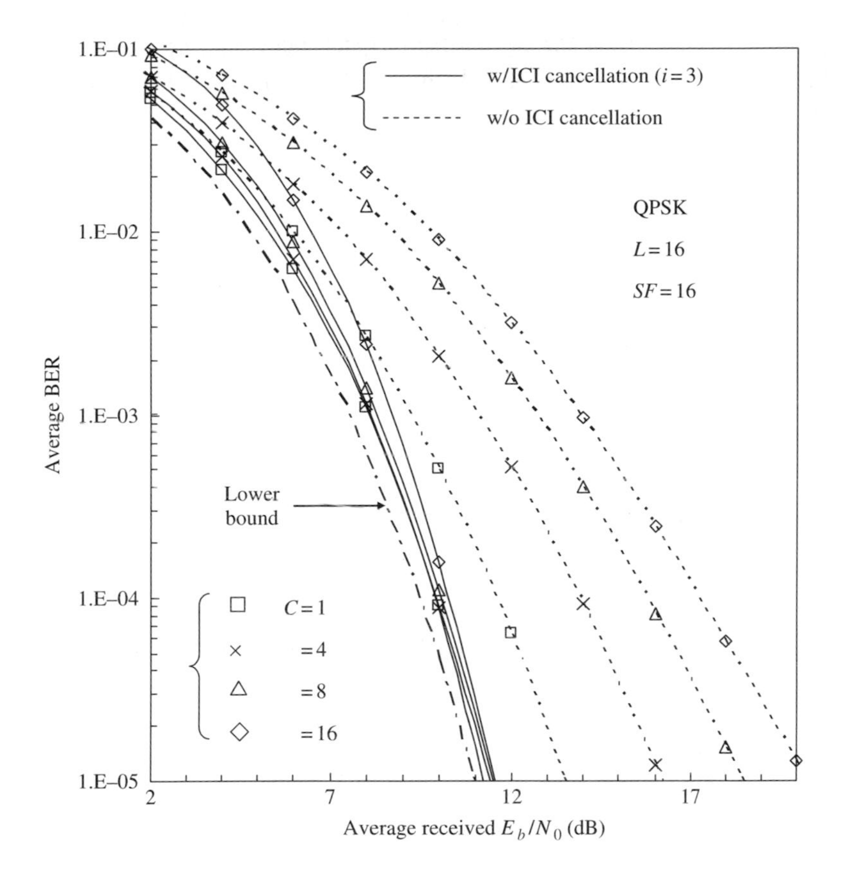


*Figure 12.* Simulated BER performance with joint MMSE-FDE and ICI cancellation

performance without ICI cancellation degrades. This is because a severe orthogonality distortion is produced by the residual ICI. The use of ICI cancellation can improve the BER performance. When $C = 16$, the $E_b/N_0$ reduction from the no ICI cancellation case is as much as 6.9 dB for BER $= 10^{-4}$.

## 5. SPACE-TIME BLOCK CODING

The antenna diversity technique can be used to increase the received signal-to-noise power ratio (SNR) and hence improve the transmission performance. There are two types of antenna diversity: receive diversity and transmit diversity (they can be jointly used). Receive antenna diversity has been successfully used in practical systems. Recently, transmit antenna diversity has been gaining much attention since the use of transmit diversity at a base station can alleviate the complexity problem of mobile receivers.

Space-time block coded transmit diversity (STTD) can achieve the space diversity gain without requiring channel information at the transmitter. In MC-CDMA, each subcarrier component is STTD encoded and then decoded in conjunction with MMSE equalization. This STTD can be applied to SC-CDMA with MMSE-FDE. Here, this is called frequency-domain STTD. In frequency-domain STTD, consecutive chip blocks are encoded in the frequency-domain.

**(a) $N_t = 2$**

STTD encoding for $N_t = 2$ is shown in **Table 1**. Two consecutive chip blocks, $\{s_e(t);\, t = 0 \sim N_c - 1\}$ and $\{s_o(t);\, t = 0 \sim N_c - 1\}$, at even and odd time intervals are decomposed by$N_c$-point FFT into $N_c$ subcarrier components, $\{S_e(k);\, k = 0 \sim N_c - 1\}$ and $\{S_o(k);\, k = 0 \sim N_c - 1\}$, respectively, for STTD encoding. Then, $N_c$-point IFFT is used to obtain the time-domain coded chip blocks. This encoding requires FFT and IFFT operations. An equivalent time-domain STTD encoding that requires no FFT and IFFT operations is shown in. Since

$$\begin{cases} \frac{1}{N_c}\sum\limits_{k=0}^{N_c-1} S_e^*(k)\exp\left(j2\pi t\frac{k}{N_c}\right) = s_e^*\left((N_c - t) \bmod N_c\right) \\ \frac{1}{N_c}\sum\limits_{k=0}^{N_c-1} S_o^*(k)\exp\left(j2\pi t\frac{k}{N_c}\right) = s_o^*\left((N_c - t) \bmod N_c\right) \end{cases}, \tag{18}$$

STTD encoding of **Table 1** can be replaced by equivalent time-domain STTD encoding of **Table 2**. Equivalent time-domain STTD encoding is illustrated in

*Table 1.* Frequency-domain STTD encoding for $N_t = 2$

| Time (in chip block) | Antenna #0 | Antenna #1 |
|---|---|---|
| Even | $\frac{1}{\sqrt{2}}S_e(k)$ | $\frac{1}{\sqrt{2}}S_o(k)$ |
| Odd | $-\frac{1}{\sqrt{2}}S_o^*(k)$ | $\frac{1}{\sqrt{2}}S_e^*(k)$ |

*Table 2.* Equivalent time-domain STTD encoding $N_t = 2$

| Time (in chip block) | Antenna #0 | Antenna #1 |
|---|---|---|
| Even | $\frac{1}{\sqrt{2}} s_e(t)$ | $\frac{1}{\sqrt{2}} s_o(t)$ |
| Odd | $-\frac{1}{\sqrt{2}} s_o^* ((N_c - t) \bmod N_c)$ | $\frac{1}{\sqrt{2}} s_e^* ((N_c - t) \bmod N_c)$ |

**Figure 13** (note that transmit power from each antenna is halved to keep the same total transmit power).

At a receiver, after the removal of the GI, the even and odd chip blocks received by the $n_r$th ($n_r = 0 \sim N_r - 1$) receive antenna are decomposed by $N_c$-point FFT into $N_c$ subcarrier components $\{R_{e,n_r}(k); k = 0 \sim N_c - 1\}$ and $\{R_{o,n_r}(k); k = 0 \sim N_c - 1\}$, respectively. $R_{e,n_r}(k)$ and $R_{o,n_r}(k)$ can be written as

$$(19) \qquad \begin{cases} R_{e,n_r}(k) = \frac{1}{\sqrt{2}} S_e(k) H_{n_r,0}(k) + \frac{1}{\sqrt{2}} S_o(k) H_{n_r,1}(k) + \Pi_{e,n_r}(k) \\ R_{o,n_r}(k) = -\frac{1}{\sqrt{2}} S_o^*(k) H_{n_r,0}(k) + \frac{1}{\sqrt{2}} S_e^*(k) H_{n_r,1}(k) + \Pi_{o,n_r}(k) \end{cases},$$

where $H_{n_r,0}(k)$ (or $H_{n_r,1}(k)$) represents the $N_c$-point Fourier transform of the channel gain between the $n_r$th receive antenna and the 0th (or 1st) transmit antenna and $\Pi_{e,n_r}(k)$ and $\Pi_{o,n_r}(k)$ represent the $k$-th subcarrier components of the noise in the received even and odd chip blocks, respectively. STTD decoding for $N_t = 2$ is carried out jointly with receive antenna diversity combining, in the frequency-domain, jointly with MMSE-FDE as

$$(20) \qquad \begin{cases} \tilde{S}_e(k) = \sum_{n_r=0}^{N_r-1} \{ w_{n_r,0}^*(k) R_{e,n_r}(k) + w_{n_r,1}(k) R_{o,n_r}^*(k) \} \\ \tilde{S}_o(k) = \sum_{n_r=0}^{N_r-1} \{ w_{n_r,1}^*(k) R_{e,n_r}(k) - w_{n_r,0}(k) R_{o,n_r}^*(k) \}. \end{cases}$$

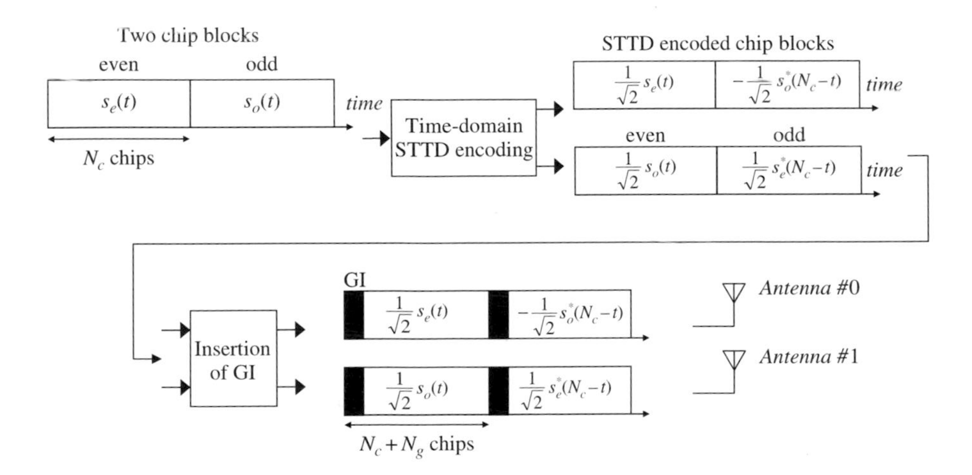


*Figure 13.* Equivalent time-domain STTD encoding for SC-CDMA

In the above, $w_{0,n_r}(k)$ and $w_{1,n_r}(k)$ are the MMSE weights, given by

$$\text{(21)}\qquad \begin{cases} w_{n_r,0}(k) = \dfrac{H_{n_r,0}(k)}{\sum\limits_{n_r=0}^{N_r}\sum\limits_{n_t=0}^{1}|H_{n_r,n_t}(k)|^2+\left(\frac{1}{2}\frac{C}{SF}\frac{E_s}{N_0}\right)^{-1}} \\ w_{n_r,1}(k) = \dfrac{H_{n_r,1}(k)}{\sum\limits_{n_r=0}^{N_r}\sum\limits_{n_t=0}^{1}|H_{n_r,n_t}(k)|^2+\left(\frac{1}{2}\frac{C}{SF}\frac{E_s}{N_0}\right)^{-1}}. \end{cases}$$

where $C$ denotes the code multiplexing order. Finally, $N_c$-point IFFT is applied to $\{\tilde{S}_e(k)\}$ and $\{\tilde{S}_o(k)\}$ to obtain the time-domain chip blocks for despreading and data demodulation.

**(b) $N_t = 3$ and 4**

When $N_t = 3$ and 4, four consecutive chip blocks $\{s_q(t);\, t = 0 \sim N_c - 1\}$, $q = 0 \sim 3$, are encoded. STTD encoding for $N_t = 3$ and 4 can be expressed, using the matrix representation, as

$$\text{(22)}\qquad \begin{pmatrix} s_{0,0}(t) & s_{1,0}(t) & s_{2,0}(t) & s_{3,0}(t) \\ s_{0,1}(t) & s_{1,1}(t) & s_{2,1}(t) & s_{3,1}(t) \\ s_{0,2}(t) & s_{1,2}(t) & s_{2,2}(t) & s_{3,2}(t) \end{pmatrix} = \frac{1}{\sqrt{3}} \begin{pmatrix} s_0(t) & -s_1^*((N_c-t) \bmod N_c) & -s_2^*((N_c-t) \bmod N_c) & 0 \\ s_1(t) & s_0^*((N_c-t) \bmod N_c) & 0 & -s_2^*((N_c-t) \bmod N_c) \\ s_2(t) & 0 & s_0^*((N_c-t) \bmod N_c) & s_1^*((N_c-t) \bmod N_c) \end{pmatrix} \text{ for } N_t = 3$$

and

$$\text{(23)}\qquad \begin{pmatrix} s_{0,0}(t) & s_{1,0}(t) & s_{2,0}(t) & s_{3,0}(t) \\ s_{0,1}(t) & s_{1,1}(t) & s_{2,1}(t) & s_{3,1}(t) \\ s_{0,2}(t) & s_{1,2}(t) & s_{2,2}(t) & s_{3,2}(t) \\ s_{0,3}(t) & s_{1,3}(t) & s_{2,3}(t) & s_{3,3}(t) \end{pmatrix} = \frac{1}{2} \begin{pmatrix} s_0(t) & -s_1^*((N_c-t) \bmod N_c) & -s_2^*((N_c-t) \bmod N_c) & 0 \\ s_1(t) & s_0^*((N_c-t) \bmod N_c) & 0 & -s_2^*((N_c-t) \bmod N_c) \\ s_2(t) & 0 & s_0^*((N_c-t) \bmod N_c) & s_1^*((N_c-t) \bmod N_c) \\ 0 & s_2(t) & -s_1(t) & s_0(t) \end{pmatrix} \text{ for } N_t = 4,$$

where $\{s_{q,n_t}(t);\, t = 0 \sim N_c - 1\}$ is the coded chip block to be transmitted from the $n_t$th transmit antenna in the $q$th time interval.

STTD decoding are carried out, in the frequency-domain, jointly with MMSE-FDE as

$$\text{(24)}\qquad \begin{pmatrix} \tilde{S}_0(k) \\ \tilde{S}_1(k) \\ \tilde{S}_2(k) \end{pmatrix} = \begin{pmatrix} \sum\limits_{n_r=0}^{N_r-1}\{R_{0,n_r}(k)w_{n_r,0}^*(k) + R_{1,n_r}^*(k)w_{n_r,1}(k) + R_{2,n_r}^*(k)w_{n_r,2}(k)\} \\ \sum\limits_{n_r=0}^{N_r-1}\{R_{0,n_r}(k)w_{n_r,1}^*(k) - R_{1,n_r}^*(k)w_{n_r,0}(k) + R_{3,n_r}^*(k)w_{n_r,2}(k)\} \\ \sum\limits_{n_r=0}^{N_r-1}\{R_{0,n_r}(k)w_{n_r,2}^*(k) - R_{2,n_r}^*(k)w_{n_r,0}(k) - R_{3,n_r}^*(k)w_{n_r,1}(k)\} \end{pmatrix} \text{ for } N_t = 3$$

and

$$(25)\quad \begin{pmatrix} \tilde{S}_0(k) \\ \tilde{S}_1(k) \\ \tilde{S}_2(k) \end{pmatrix} = \begin{pmatrix} \sum_{n_r=0}^{N_r-1} \begin{Bmatrix} R_{0,n_r}(k)w^*_{n_r,0}(k) + R^*_{1,n_r}(k)w_{n_r,1}(k) \\ +R^*_{2,n_r}(k)w_{n_r,2}(k) + R_{3,n_r}(k)w^*_{n_r,3}(k) \end{Bmatrix} \\ \sum_{n_r=0}^{N_r-1} \begin{Bmatrix} R_{0,n_r}(k)w^*_{n_r,1}(k) - R^*_{1,n_r}(k)w_{n_r,0}(k) \\ -R_{2,n_r}(k)w^*_{n_r,3}(k) + R^*_{3,n_r}(k)w_{n_r,2}(k) \end{Bmatrix} \\ \sum_{n_r=0}^{N_r-1} \begin{Bmatrix} R_{0,n_r}(k)w^*_{n_r,2}(k) + R_{1,n_r}(k)w^*_{n_r,3}(k) \\ -R^*_{2,n_r}(k)w_{n_r,0}(k) - R^*_{3,n_r}(k)w_{n_r,1}(k) \end{Bmatrix} \end{pmatrix} \quad \text{for } N_t = 4,$$

where $R_{q,n_r}(k)$ is the $k$th frequency component of the chip block received by the $n_r$th receive antenna in the $q$th time interval, and $w_{n_r,n_t}(k)$ is the MMSE weight given as

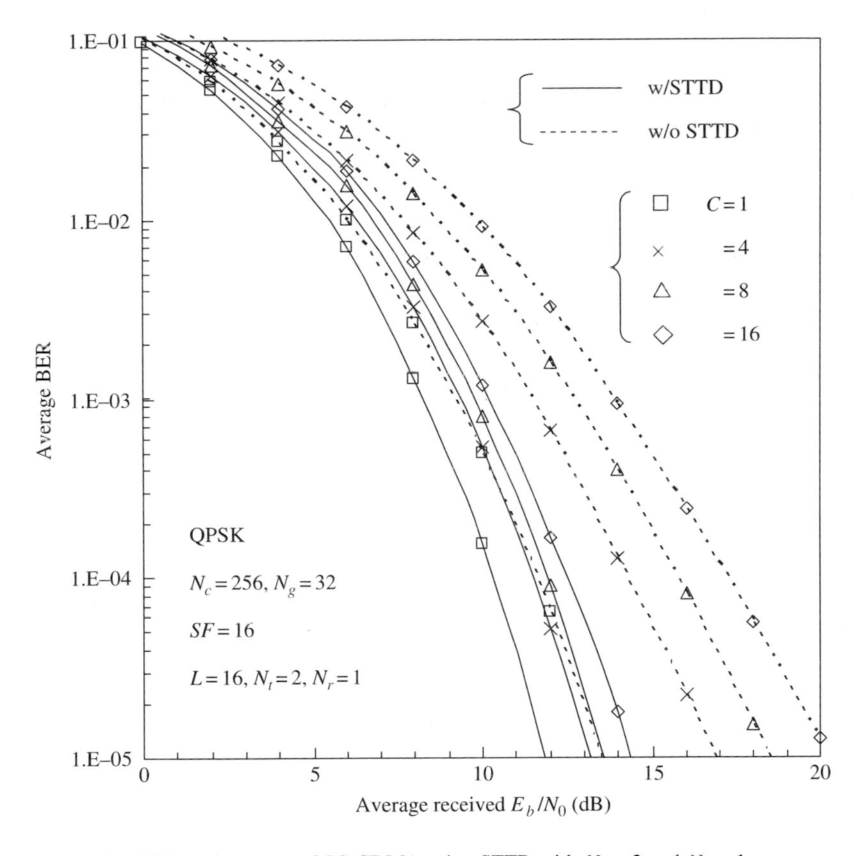


*Figure 14.* BER performance of SC-CDMA using STTD with $N_t = 2$ and $N_r = 1$

$$(26) \qquad w_{n_r,n_t}(k) = \frac{H_{n_r,n_t}(k)}{\frac{1}{N_t}\sum_{n_t=0}^{N_t-1}\sum_{n_r=0}^{N_r-1}|H_{n_r,n_t}(k)|^2 + \left(\frac{C}{SF}\frac{E_s}{N_0}\right)^{-1}}.$$

Frequency-domain STTD with $N_t = 2$ and $N_r = 1$ is evaluated by the computer simulation. We assume $N_c = 256$, $N_g = 32$, coherent QPSK data-modulation, and a chip-spaced $L = 16$-path frequency-selective block Rayleigh fading channel with uniform power delay profile ($\beta = 0$ dB), and ideal channel estimation. The BER performance using frequency-domain STTD is plotted in **Figure 14** for $SF=16$. For comparison, the single transmit antenna case ($N_t = 1$) is also plotted. The transmit diversity gain similar to that of two-antenna receive diversity ($N_r = 2$) using MRC is obtained, but with a 3dB power penalty (this is because the transmit power from each antenna is halved to keep the same total transmit power).

## 6. MIMO SPACE DIVISION MULTIPLEXING

High-speed data services of 100M~1Gbps are demanded in the next generation wireless systems. However, the available bandwidth is limited. Space division multiplexing (SDM) is a promising technique to achieve highly spectrum-efficient transmission. In SDM, different data sequences are transmitted in parallel from different transmit antennas using the same carrier frequency. At a receiver, a superposition of different data sequences transmitted from different antennas is received. A lot of research attention has been paid to the signal separation/detection schemes, e.g., maximum likelihood detection (MLD), ZF detection, MMSE detection and vertical-Bell Laboratories layered space-time architecture (V-BLAST). For high-speed data transmissions, the channels become severely frequency-selective and the BER performance of SC-CDMA using SDM degrades due to the inter-chip interference (ICI) and interference from other antennas. Therefore, the receiver must have two tasks: signal separation/detection and channel equalization.

### 6.1 Transmit/Receive Signal Representation

Orthogonal multicode SC-CDMA is considered. **Figure 15** illustrates the transmitter/receiver structure of ($N_t$, $N_r$)SDM, where $N_t$ and $N_r$ denote the number of

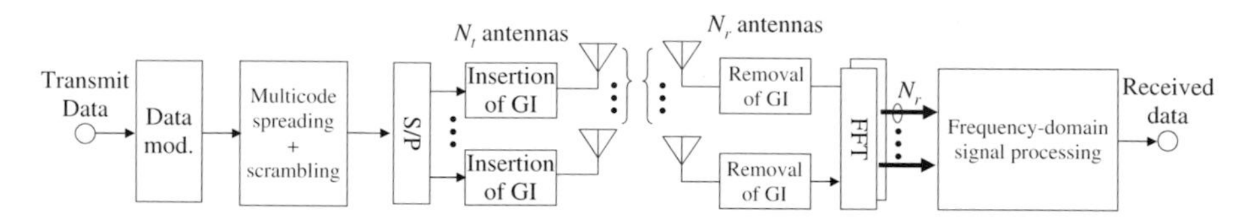


*Figure 15.* Transmitter/receiver structure for ($N_t$, $N_r$) SDM

transmit antennas and that of receive antennas, respectively. At the transmitter, a binary information sequence is data-modulated and converted to $C$ parallel streams by serial/parallel (S/P) conversion. Then, each stream is spread by using a different orthogonal spreading code with the spreading factor $SF$. The resulting $C$ parallel chip streams are added and multiplied by a scramble sequence for making the resulting orthogonal multicode SC-CDMA signal noise-like. The full code-multiplexed SC-CDMA (i.e., $C=SF$) has the same data rate as the non-spread SC transmission. The code-multiplexed SC-CDMA signal is converted by S/P converter to $N_t$ parallel streams $\{s_{n_t}(t)\}$, $n_t=0\sim N_t-1$. Each stream is divided into a sequence of blocks of $N_c$ chips each. After inserting the GI, $N_t$ chip blocks are transmitted simultaneously from $N_t$ transmit antennas.

A superposition of $N_t$ transmitted chip blocks is received by $N_r$ receive antennas. After the removal of GI, the chip block $\{r_{n_r}(t);\ t=0\sim N_c-1\}$ received on the $n_r$th receive antenna is decomposed by $N_c$-point FFT into $N_c$ frequency components $\{R_{n_r}(k);\ k=0\sim N_c-1\}$ as

$$R_{n_r}(k)=\sum_{t=0}^{N_c-1} r_{n_r}(t)\exp\left(-j2\pi k\frac{t}{N_c}\right),$$
$$=\sqrt{\frac{2E_c}{T_c}}\sum_{n_t=0}^{N_t-1} H_{n_r,n_t}(k)S_{n_t}(k)+\Pi_{n_r}(k) \tag{27}$$

where $E_c$ is the chip energy, $T_c$ is the chip length, $H_{n_r,n_t}(k)$ is the complex channel gain between the $n_t$th transmit antenna and the $n_r$th receive antenna, $S_{n_t}(k)$ is the $k$th frequency component of $s_{n_t}(t)$, and $\Pi_{n_r}(k)$ is the noise.

## 6.2 Signal Separation/Detection

Since the channel is frequency-selective, signal separation/detection and frequency-domain equalization need to be jointly performed. Below, frequency-domain MLD (FD-MLD), two dimensional (2D)-ZF FDE detection, 2D-MMSE FDE detection, FD V-BLAST and iterative joint MMSE-FDE/FD-parallel interference cancellation (PIC) are introduced.

**(a) FD-MLD**

MLD computes the log-likelihood metric as

$$\lambda=\sum_{k=0}^{N_c-1}\sum_{n_r=0}^{N_r-1}\left|R_{n_r}(k)-\sqrt{\frac{2E_c}{T_c}}\sum_{n_t=0}^{N_t-1} H_{n_r,n_t}(k)\tilde{S}_{n_t}(k)\right|^2, \tag{28}$$

where $\tilde{S}_{n_t}(k)$ is the $k$th frequency component of the candidate chip block $\{\tilde{s}_{n_t}(t);\ t=0\sim N_c-1\}$. MLD finds the best combination of $N_t$ transmitted chip blocks which provides the smallest log-likelihood metric, i.e., $\{\hat{s}_0(t), ..., \hat{s}_{n_t}(t), ..., \hat{s}_{N_t-1}(t)\} = \min_{\{\tilde{s}_{n_t}(t)\}} \lambda$. After de-spreading and de-scrambling, the most reliable symbol sequence is obtained. MLD provides the best transmission performance; however, it has a drawback of quite large computational complexity since the number of metric computations is as much as $2^{N_t \cdot N_c \cdot B}$, where $B$ is the number of bits per symbol.

**(b) 2D-ZF FDE detection and 2D-MMSE FDE detection**

In 2D-ZF FDE detection and 2D-MMSE FDE detection, the $k$th frequency component $\hat{R}_{n_t}(k)$ of the $n_t$th transmitted chip block is obtained as

$$\hat{R}_{n_t}(k) = \mathbf{w}_{n_t}(k)\mathbf{R}(k), \tag{29}$$

where $\mathbf{R}(k) = [R_0(k), \cdots, R_{N_r-1}(k)]^T$ is the $N_r$-by-1 received signal vector and $\mathbf{w}_{n_t}(k) = [w_{0,n_t}(k), \cdots, w_{N_r-1,n_t}(k)]$ is the 1-by-$N_r$ weight vector. 2D-ZF FDE weight can be derived by Moore-Penrose generalized inversed matrix. The MMSE weight minimizes the MSE $E[|e(k)|^2]$ between the signal transmitted from the $n_t$th antenna and the received signal after performing FDE, where $e(k)$ is defined as

$$e(k) = \sqrt{\frac{2E_c}{T_c}} S_{n_t}(k) - \mathbf{w}_{n_t}(k)\mathbf{R}(k). \tag{30}$$

$\mathbf{w}_{n_t}(k)$ can be derived from as

$$\mathbf{w}_{n_t}(k) = \begin{cases} \left[\mathbf{H}^H(k)\mathbf{H}(k)\right]_{n_t}^{-1} \mathbf{H}^H(k) & \text{for 2D-ZF} \\ \mathbf{H}_{n_t}^H(k)\left[\mathbf{H}(k)\mathbf{H}^H(k) + \left(\dfrac{C \cdot E_c}{N_0}\right)^{-1} \mathbf{I}_{N_r}\right]^{-1} & \text{for 2D-MMSE} \end{cases}, \tag{31}$$

where $\mathbf{H}(k) = [\mathbf{H}_0(k), \cdots, \mathbf{H}_{N_t-1}(k)]$ is the $N_r$-by-$N_t$ complex channel gain matrix whose element of the $n_r$th row and the $n_t$th column is $H_{n_r,n_t}(k)$, $[\mathbf{H}^H(k)\mathbf{H}(k)]_{n_t}^{-1}$ is the $n_t$th row vector of the inverse matrix of $\mathbf{H}^H(k)\mathbf{H}(k)$ and $\mathbf{I}_{N_r}$ is the $N_r$-by-$N_r$ identity matrix. $N_c$-point IFFT is performed on $\{\hat{R}_{n_t}(k);\ k = 0 \sim N_c - 1\}$ to obtain the time-domain chip block. After performing despreading and de-scrambling, the received symbol sequence is recovered.

2D-ZF FDE weight gives the perfect separation of transmitted chip blocks since a frequency-nonselective channel is restored (if the channel estimation is ideal). However, the BER performance using 2D-ZF FDE detection degrades due to the noise enhancement. The 2D-MMSE FDE weight can reduce simultaneously the ICI and the interference from other antennas while suppressing the noise enhancement. Therefore, 2D-MMSE FFDE gives much better transmission performance.

**(c) FD V-BLAST**

In FD V-BLAST, the signal detection and interference cancellation are repeated until all the transmitted signals are detected according to the descending order of the received signal reliability. Without loss of generality, the transmit antenna having the highest reliability is assumed to be the 0th transmit antenna, followed by the 1st, 2nd, ..., and $(N_t - 1)$th antennas. The $n_t$th signal component $\hat{R}_{n_t}(k)$ is obtained, by performing 2D-MMSE FDE detection (or 2D-ZF FDE detection), as

$$\hat{R}_{n_t}(k) = \mathbf{w}'_{n_t}(k)\mathbf{R}'(k), \tag{32}$$

where $\mathbf{R}'(k) = [R'_0(k), \cdots, R'_{N_r-1}(k)]^T$ is the received signal vector after interference cancellation of the signals transmitted from the $0 \sim (n_t - 1)$th antennas and $\mathbf{w}'_{n_t}(k) = [w'_{0,n_t}(k), \cdots, w'_{N_r-1,n_t}(k)]$ is the 1-by-$N_r$ MMSE FDE or ZF FDE weight vector given by

$$\mathbf{w}'_{n_t}(k) = \begin{cases} \left[\mathbf{H}'^H(k)\mathbf{H}'(k)\right]^{-1}_{n_t} \mathbf{H}'^H(k) \text{ for 2D-ZF} \\ \mathbf{H}'^H_{n_t}(k)\left[\mathbf{H}'(k)\mathbf{H}'^H(k) + \left(\frac{C \cdot E_c}{N_0}\right)^{-1}\mathbf{I}_{N_r}\right]^{-1} \text{ for 2D-MMSE} \end{cases}, \tag{33}$$

where $\mathbf{H}'(k) = [\mathbf{H}'_{n_t}(k), \cdots, \mathbf{H}'_{N_t-1}(k)]$ is the $N_r$-by-$(N_t - n_t)$ channel gain matrix obtained by deleting the $0 \sim (n_t - 1)$th channel gain column vector from the $N_r$-by-$N_t$ channel gain matrix $\mathbf{H}(k)$. The $n_t$th time-domain received chip block is obtained by carrying out $N_c$-point IFFT on $\{\hat{R}_{n_t}(k);\ k = 0 \sim N_c - 1\}$. After performing despreading and descrambling, the received symbol sequence is recovered.

For the detection of the signal transmitted from the $(n_t+1)$th antenna, the interference replicas $\tilde{S}_{n'_t}(k)$, $n'_t = 0 \sim n_t$, are generated and subtracted from $R_{n_r}(k)$ as

$$R'_{n_r}(k) = R_{n_r}(k) - \sqrt{\frac{2E_c}{T_c}} \sum_{n'_t=0}^{n_t} H_{n_r,n'_t}(k)\tilde{S}_{n'_t}(k) \tag{34}$$

for the detection of the chip block transmitted from the $(n_t + 1)$th antenna. The above operation is repeated, until all the transmitted signals are detected.

**(d) Iterative joint MMSE-FDE/FD-PIC**

The interference from other transmit antennas are suppressed by joint MMSE-FDE and FD-PIC. However, since the interference suppression is not sufficient, joint MMSE-FDE and FD-PIC is repeated. At the initial stage ($i = 0$), 2D-MMSE FDE is used. The interfering signal replicas $\{\tilde{S}^{(i-1)}_{n'_t}(k);\ k = 0 \sim N_c - 1\}$, $n'_t = 0 \sim N_t - 1, \neq n_t$, at the $(i - 1)$th iteration are generated and are subtracted from the received signal $R_{n_r}(k)$ as

$$R'^{(i)}_{n_r,n_t}(k) = R_{n_r}(k) - \sqrt{\frac{2E_c}{T_c}} \sum_{n'_t=0 \neq n_t}^{N_t-1} H_{n_r,n'_t}(k)\tilde{S}^{(i-1)}_{n'_t}(k). \tag{35}$$

Since the resulting signal is close to that for the single antenna transmission case, 1D-MMSE FDE is used instead of 2D-MMSE FDE for $i \geq 1$. Joint 1D-MMSE FDE is performed as

$$\hat{R}^{(i)}_{n_t}(k) = \mathbf{w}'^{(i)}_{n_t}(k)\mathbf{R}'^{(i)}_{n_t}(k), \tag{36}$$

where $\mathbf{R'}^{(i)}_{n_t}(k) = [R'^{(i)}_{0,n_t}(k), \cdots, R'^{(i)}_{N_r-1,n_t}(k)]^T$ and $\mathbf{w'}^{(i)}_{n_t}(k) = [w'^{(i)}_{0,n_t}(k), \cdots, w'^{(i)}_{N_r-1,n_t}(k)]$ is the 1D-MMSE FDE weight vector, given by

$$\mathbf{w'}^{(i)}_{n_t}(k) = \mathbf{H}^H_{n_t}(k)\left[\mathbf{H}^H_{n_t}(k)\mathbf{H}_{n_t}(k) + \left(\frac{C \cdot E_c}{N_0}\right)^{-1}\right]^{-1}. \tag{37}$$

## 6.3 BER Performance

The BER performance of full code-multiplexed SC-CDMA using (4,4)SDM is plotted in **Figure 16** as a function of the average received $E_b/N_0$ per receive antenna. $N_t \times N_r$ channels are independent Rayleigh fading channels having an $L = 16$-path uniform power delay profile ($\beta = 0$dB). Iterative joint MMSE-FDE/FD-PIC is superior to 2D-ZF FDE detection, 2D-MMSE FDE detection, and FD V-BLAST. For comparison, the BER performance of the perfect PIC (i.e., the interference from

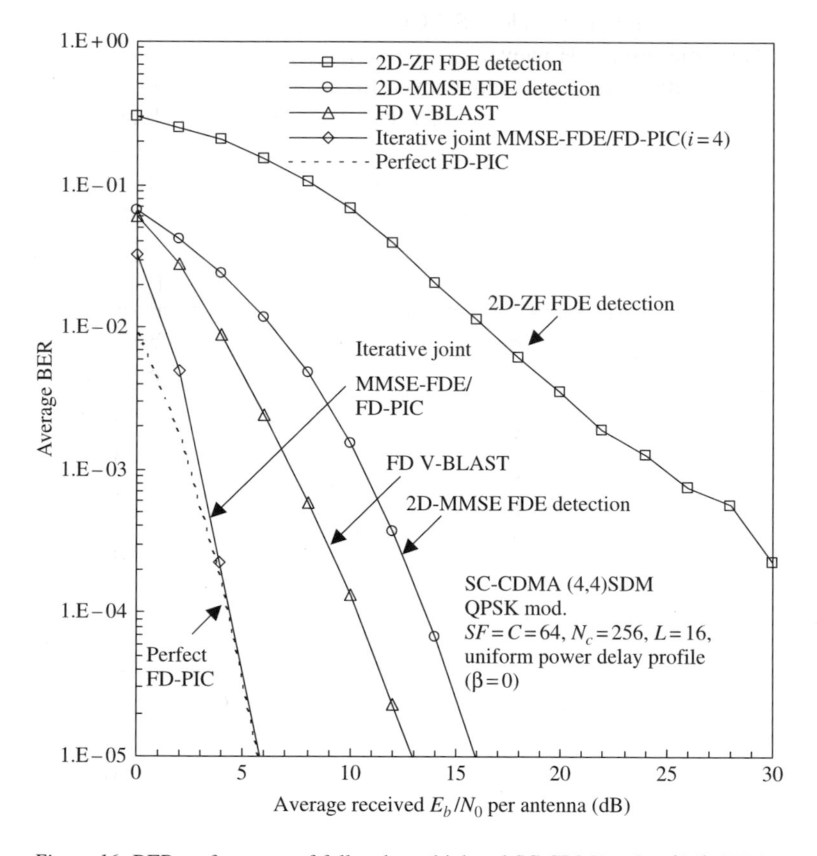


*Figure 16.* BER performance of full code-multiplexed SC-CDMA using (4,4) SDM

other antennas is perfectly cancelled) is also plotted. The $E_b/N_0$ degradation, for BER = $10^{-4}$, of iterative joint MMSE-FDE/FD-PIC from the perfect PIC is only about 0.1dB. A reason for this superiority of iterative FD-PIC is discussed below.

Only an $(N_r - N_t + 1)$-order antenna diversity gain can be obtained by 2D-ZF FDE detection and 2D-MMSE FDE detection. The BER performance with 2D-ZF FDE detection degrades due to the noise enhancement. In FD V-BLAST, the transmitted signals are detected according to the descending order of signals' reliability. After performing signal detection, the detected signal is subtracted from the received signal by using its replica and then the corresponding channel gain vector is deleted from the $N_r$-by-$N_t$ channel gain matrix. For the detection of the signal transmitted from the $n'_t$th antenna $(n'_t = 0 \sim N_t - 1)$, the $N_r$-by-$(N_t - n'_t)$ channel gain matrix is used. As a consequence, a diversity order of between $(N_r - N_t + 1)$ and $N_r$ is obtained by FD V-BLAST. On the other hand, iterative joint MMSE-FDE/FD-PIC can always obtain the $N_r$th-order diversity gain.

## 7. BLOCK SPREAD CDMA

Relying on orthogonal spreading codes, SC-CDMA allows simultaneous transmissions from multiple users. However, as the chip rate increases, multipath channels become time-dispersive and frequency-selective fading is produced. The frequency-selective fading causes ICI. In the downlink (base-to-mobile) transmission, different users' data sequences are spread by orthogonal spread codes and are code-multiplexed. The ICI destroys the code orthogonality at an MS receiver and gives rise to the downlink multi-access interference (MAI). The downlink MAI severely limits the performance of single-user rake receivers. The use of MMSE-FDE can exploit the channel frequency-selectivity as well as suppressing the downlink MAI and therefore improve the downlink BER performance as discussed in **Section 4**. However, in the uplink (mobile-to-base) transmission, different users' signals go through different channels and are asynchronously received, thereby producing the uplink MAI. Unfortunately, the uplink MAI cannot be sufficiently suppressed by single-user MMSE-FDE. Multiuser detection (MUD), can suppress the uplink MAI. However, its computational complexity grows exponentially with the number of users. Block spreading can be used to convert the MUD problem into a set of equivalent single-user equalization problems.

### 7.1 One-dimensional Block Spreading

One-dimensional (1D) block spreading technique proposed in is shown in **Figure 17**. The $u$th user's data block consisting of $N_c$ symbols, $\mathbf{d}_u = [d_u(0),\ldots,d_u(N_c - 1)]^T$, is block spread by using a spreading code $\mathbf{c}_u = [c_u(0),\ldots,c_u(SF_u - 1)]^T$. The result can be expressed as an $SF_u \times N_c$ matrix $\mathbf{S}_u$ as

$$\mathbf{S}_u = \mathbf{c}_u \mathbf{d}_u^T. \tag{38}$$

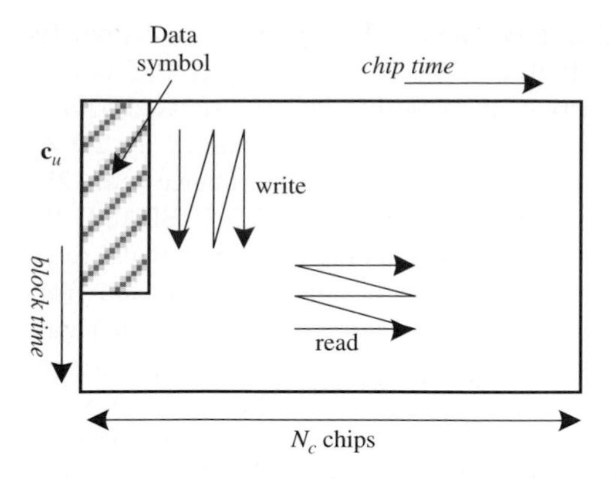


*Figure 17.* 1D block spreading

Chips from matrix $\mathbf{S}_u$ are transmitted row-by-row over $SF_u$ block periods, which means that $N_c$ chips are transmitted in each block. The $m$th chip block is represented by the $N_c \times 1$ vector as

$$\text{(39)} \qquad \mathbf{x}_u(m) = c_m(m)\,[d_u(0), \cdots, d_u(N_c - 1)]^T.$$

The signal to be transmitted can be expressed, using equivalent lowpass representation, as

$$\text{(40)} \qquad \mathbf{s}_u = \sqrt{\frac{2E_c}{T_c}}\left[\mathbf{x}_u^T(0), \cdots, \mathbf{x}_u^T(SF_u - 1)\right]^T,$$

where $E_c$ is the chip energy, $T_c$ is the chip duration. After the insertion of $N_g$-chip GI, the $u$th user's signal is transmitted.

A superposition of $U$ users' faded signals is received, via $L$-path fading channels, by the BS's receiver. The sum of the maximum time delay of the channel and the timing offsets among different users is assumed to be within the GI. After the GI removal, the received signal vector $\mathbf{r} = [r(0), \ldots, r(SF_u N_c - 1)]^T$ with length $SF_u N_c$ chips is given by

$$\text{(41)} \qquad \mathbf{r} = \sqrt{\frac{2E_c}{T_c}} \sum_{u=0}^{U-1} \tilde{\mathbf{H}}_u \mathbf{s}_u + \boldsymbol{\eta},$$

where $\tilde{\mathbf{H}}_u = diag\,\{\mathbf{H}_u(0), \ldots, \mathbf{H}_u(SF_u - 1)\}$ is the $u$th user's channel matrix with $\mathbf{H}_u(m)$ being the $N_c \times N_c$ channel matrix at the $m$th block time and $\boldsymbol{\eta} = [\eta(0), \ldots, \eta(SF_u N_c - 1)]^T$ is the noise vector with zero-mean and variance $2N_0/T_c$ ($N_0$ is the AWGN one-sided power spectrum density). Because of the GI insertion, $\mathbf{H}_u(m)$ is a circulant Toeplitz matrix with the first column given as $[h_{u,0}(m),\ldots,h_{u,L-1}(m),0,\ldots,0]^T$, where $h_{u,l}(m)$ is the $l$th path gain of the $u$th user's channel at the $m$th block time. $\mathbf{r}$ is written into an $SF_u \times N_c$ matrix row-by-row as $\mathbf{R}_u$ and then despread by using $\mathbf{c}_u$ as

$$\begin{aligned}\hat{\mathbf{r}}_u &= \frac{1}{SF_u}\left(\mathbf{c}_u^H \mathbf{R}_u\right)^T \\ &= \sqrt{2E_c/T_c}\overline{\mathbf{H}}_{\mathbf{u}}\mathbf{d}_u + \sum_{\substack{u'=0 \\ u'\neq u}}^{U-1} \sqrt{2E_c/T_c}\overline{\mathbf{H}}_{\mathbf{u}'}\mathbf{d}_{u'} + \overline{\boldsymbol{\eta}},\end{aligned} \tag{42}$$

where the 1st and 2nd terms are respectively the desired signal and the MAI with

$$\begin{cases}\overline{\mathbf{H}}_{u'} = \dfrac{1}{SF_u}\sum\limits_{m=0}^{SF_u-1} c_u^*(m)c_{u'}(m)\mathbf{H}_{u'}(m) \\ \overline{\boldsymbol{\eta}} = \dfrac{1}{SF_u}\left(\mathbf{c}_u^H \Pi\right)^T\end{cases}. \tag{43}$$

Here, $\Pi$ is the $SF_u \times N_c$ noise matrix, whose $x$th row and $y$th column element is $\eta(x \cdot SF_u + y)$, and $\overline{\boldsymbol{\eta}}$ is an $N_c \times 1$ vector whose elements are independent zero-mean complex Gaussian variables with variance $2N_0/T_c/SF_u$. If the channel is time-invariant (i.e., $\mathbf{H}_u(m) = \mathbf{H}_u(0)$ for $m = 0 \sim SF_u - 1$), the MAI is removed since $\mathbf{c}_u^H \mathbf{c}_{u'} = SF_u \delta(u - u')$, where $\delta(\cdot)$ is the delta function.

## 7.2 Two-dimensional Block Spreading

1D block spread SC-CDMA is a single-rate transmission. In the next generation mobile communications, a flexible support of low-to-high multi-rate services is required. The suppression of MAI to increase the link capacity in a mutli-rate environment is a challenging task. Two-dimensional (2D) block spreading using orthogonal variable spreading factor (OVSF) codes can remove the MAI while allowing multi-rate transmissions. 2D block spreading also achieves the time- and frequency-domain diversity gains.

A transmitter/receiver structure of 2D block spread SC-CDMA is illustrated in **Figure 18**. The data symbol to be transmitted is spread in two dimensions, as shown in **Figure 19**. The overall spreading factor is $SF_u = SF_u^t \times SF_u^f$, where $SF_u^f$ is the spreading factor of chip-time spreading and $SF_u^t$ is that of block-time spreading. 2D block spread SC-CDMA includes the conventional SC-CDMA and 1D block spread SC-CDMA as its special cases; 2D block spread SC-CDMA becomes the conventional SC-CDMA when $SF_u^t = 1$, while it becomes 1D block spread SC-CDMA when $SF_u^f = 1$.

Let's consider a block transmission of $N_c/SF_u^f$ data symbols. The data symbol vector $\mathbf{d}_u = [d_u(0), ..., d_u(N_c/SF_u^f - 1)]^T$ is spread by a 2D block spreading code $\mathbf{C}_u$ with the overall spreading factor $SF_u$ as

$$\mathbf{S}_u = \mathbf{C}_u \otimes \mathbf{d}_u^T, \tag{44}$$

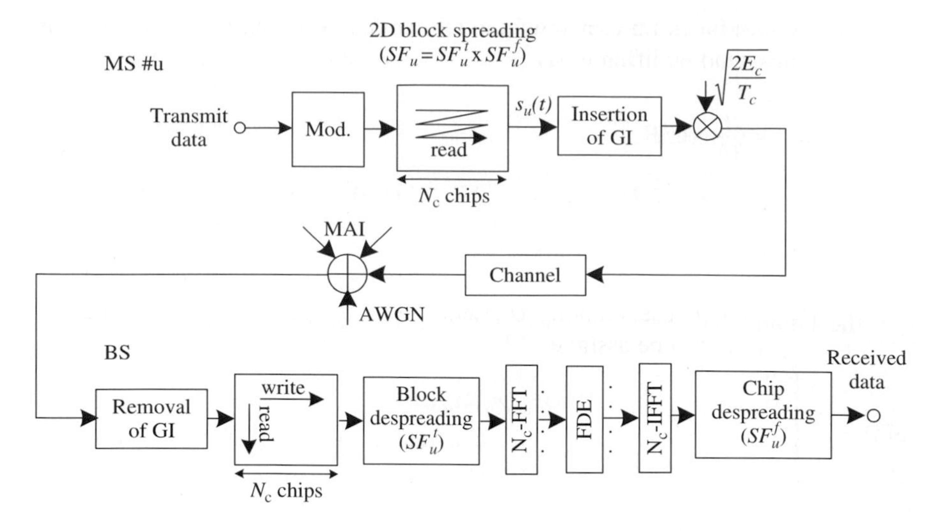


*Figure 18.* 2D block spread SC-CDMA transmitter/receiver

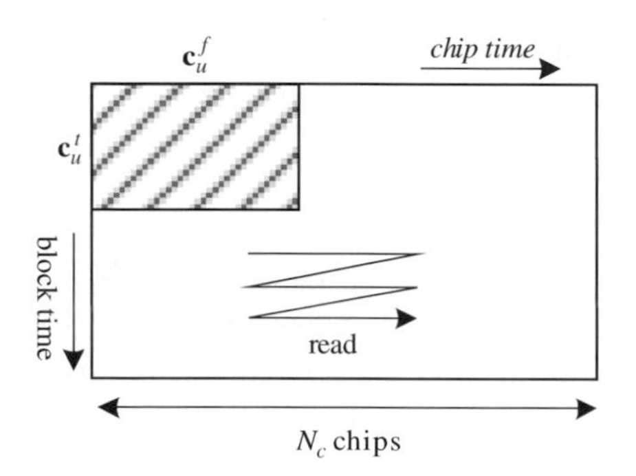


*Figure 19.* 2D block spreading

where $\otimes$ denotes the Kronecker product and $\mathbf{C}_u$ is an $SF_u^t \times SF_u^f$ matrix and can be written as

$$(45) \qquad \mathbf{C}_u = \mathbf{c}_u^t \left(\mathbf{c}_u^f\right)^T$$

with $\mathbf{c}_u^t$ and $\mathbf{c}_u^f$ being respectively the column and row spreading codes, given by

$$(46) \qquad \begin{cases} \mathbf{c}_u^t = [c_u^t(0), \ldots, c_u^t(SF_u^t - 1)]^T \\ \mathbf{c}_u^f = \left[c_u^f(0), \ldots, c_u^f(SF_u^f - 1)\right]^T \end{cases}.$$

$\mathbf{c}_u^f$ is used for multi-rate transmissions and $SF_u^f$ can be arbitrarily set according to the requested data rate independently of the FFT block size $N_c$, but $SF_u^f \leq N_c$. $\mathbf{c}_u^t$

is used for orthogonal multi-user multiplexing without MAI. In order to maintain the orthogonality among different users, $SF_u^t$ should be as small as possible.

## 7.3 Code Assignment

$\mathbf{c}_u^t$ and $\mathbf{c}_u^f$ can be selected from OVSF codes. OVSF codes are generated by the code tree, as shown in **Figure 20**. The data rates may be different for different users. $\mathbf{c}_u^t$ should be orthogonal for different users while $\mathbf{c}_u^f$ is not necessary orthogonal. If the number $U$ of users is $U=2^k$ ($k=0, 1, \ldots$), all users can be assigned $SF_u^t = 2^k$. If $2^{k-1} < U < 2^k$, $(2^k - U)$ users among $U$ users can be assigned $SF_u^t = 2^{k-1}$ and the other $(2U-2^k)$ users can be assigned $SF_u^t = 2^k$. After setting the value of $SF_u^t$, $SF_u^f$ can be set equal to $SF_u^f = SF_u/SF_u^t$ for the given overall spreading factor $SF_u$. By doing so, all $U$ users' signals can be orthogonal.

For simplicity, we assume that the data rate is the same for all users and the overall spreading factor is $SF_u = SF$. Hence, we use $(SF_u^t, SF_u^f) = (U, SF/U)$. Consider $SF = 16$. If $U=8$, $\mathbf{c}_u^t$ is selected from $\{\mathbf{c}_m^8;\, m = 0 \sim 7\}$, e.g., $\mathbf{c}_u^t = \mathbf{c}_5^8 = [1, -1, 1, -1, -1, 1, -1, 1]^T$ in **Figure 20**, and $\mathbf{c}_u^f$ can be selected from $\{\mathbf{c}_m^2;\, m = 0 \sim 1\}$. If $U=4$, $\mathbf{c}_u^t$ is selected from $\{\mathbf{c}_m^4;\, m = 0 \sim 3\}$, e.g., $\mathbf{c}_u^t = \mathbf{c}_2^4 = [1, -1, 1, -1]^T$, and $\mathbf{c}_u^f$ can also be selected from $\{\mathbf{c}_m^4;\, m = 0 \sim 3\}$. If $U=2$, then $\mathbf{c}_u^t$ is selected from $\{\mathbf{c}_m^2;\, m = 0 \sim 1\}$, e.g., $\mathbf{c}_u^t = \mathbf{c}_1^2 = [1, -1]^T$, and $\mathbf{c}_u^f$ can be selected from $\{\mathbf{c}_m^8;\, m = 0 \sim 7\}$. When $U = 1$, 2D block spreading reduces to the conventional 1D spreading.

## 7.4 BER Performance

The BER performance of 2D block spread SC-CDMA using single-user MMSE-FDE is compared, in **Figure 21**, with conventional SC-CDMA using MMSE-MUD

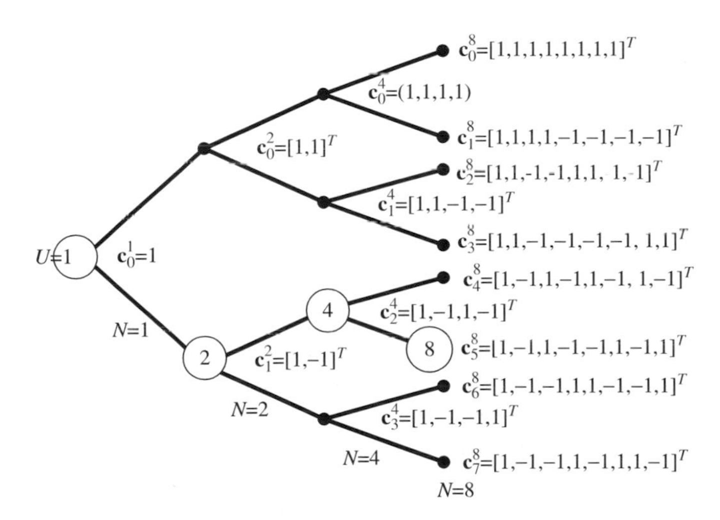


*Figure 20.* OVSF code tree

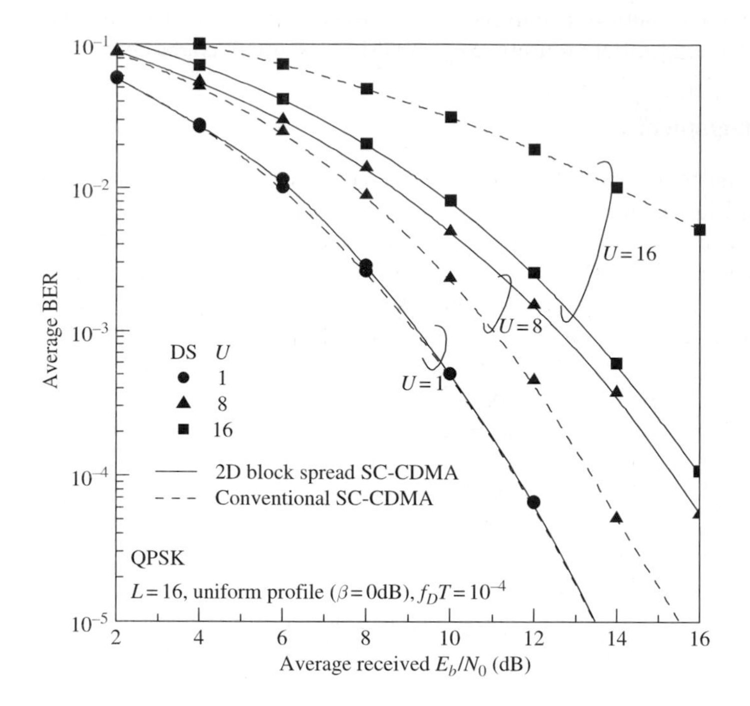


*Figure 21.* BER performance comparison of 2D block spread and conventional SC-CDMA when $SF = 16$

when $SF = 16$. For block spread SC-CDMA, $(SF_u^t, SF_u^f) = (U, 16/U)$ is assumed for all users. When the system is lightly loaded (i.e., $U = 8$), conventional SC-CDMA using MUD exhibits better BER performance since the MAI is less severe. However, when the system is heavily loaded (i.e., $U \approx SF$), the BER performance of conventional SC-CDMA with MUD severely degrades. Even when $U = 16$, 2D block spread SC-CDMA provides better BER performance than conventional SC-CDMA.

The block spreading technique can also be applied to MC-CDMA. The chip-time spreading in 2D block spread SC-CDMA corresponds to the frequency-domain spreading in the case of 2D block spread MC-CDMA. **Figure 22** compares the BER performances of 2D block spread SC- and MC-CDMA when $SF = 16$. In 2D block spread SC-CDMA, the data symbol is always spread over all subcarriers, yielding large frequency-diversity gain irrespective of $SF_u^f$. However, SC-CDMA suffers from ICI. On the other hand, in the case of MC-CDMA, the data symbol is spread over only $SF_u^f$ subcarriers. When $U=1$, 2D block spread MC-CDMA performs slightly better than SC-CDMA since no ICI is present in MC-CDMA. As $U$ increases, the value of $SF_u^f$ must be decreased and therefore, frequency-diversity gain becomes smaller in MC-CDMA and the ICI suppression become weaker in SC-CDMA. When $U = 16$

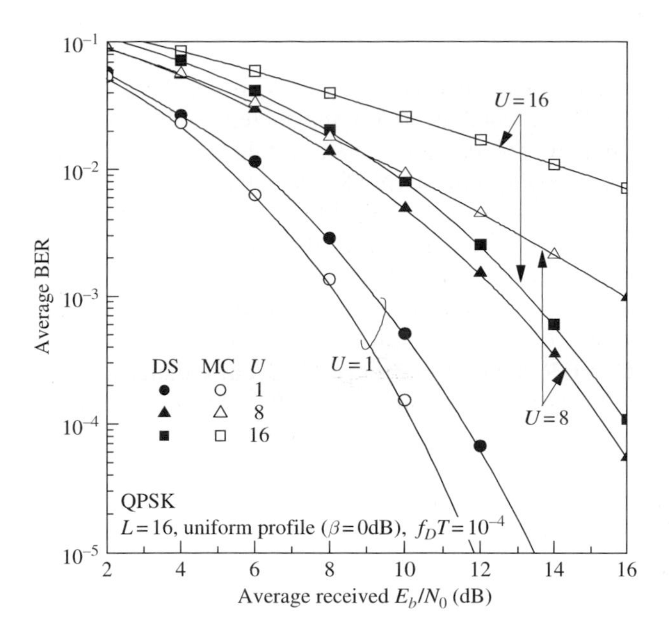


*Figure 22.* BER performance comparison of 2D block spread SC- and MC-CDMA when $SF = 16$

(i.e., $SF_u^f = 1$), the frequency-diversity gain cannot be obtained in MC-CDMA, but larger frequency-diversity gain can still be obtained in SC-CDMA. Therefore, 2D block spread SC-CDMA provides much better BER performance than MC-CDMA.

## 8. HYBRID AUTOMATIC REPEAT REQUEST

For packet transmission, some form of error control mechanism is necessary to satisfy the quality requirement. There are two techniques for controlling transmission errors: the forward error correction (FEC) scheme that adds redundancy to the information to be transmitted and errors are corrected at the receiver, and the backward error correction or the automatic repeat request (ARQ) scheme in which error control is accomplished using error detection and retransmission. ARQ is simple and provides high system reliability. However, ARQ systems have a severe drawback - their throughput falls rapidly with increasing channel error rate. Systems using FEC maintain constant throughput regardless of the channel error rate. However, FEC systems cannot achieve high system reliability without using long powerful codes. This makes the decoding hard to implement and the throughput is considerably low in case of low error channels. Drawbacks in both ARQ and FEC could be overcome if two error control schemes are properly combined. Such a combination of two error control schemes is referred to as hybrid ARQ (HARQ).

## 8.1 Hybrid ARQ

An HARQ system consists of an FEC subsystem contained in an ARQ system. Popular HARQ strategies are Chase combing (CC) and incremental redundancy (IR). In HARQ using CC strategy, the previously transmitted packet is retransmitted following a negative acknowledgement (NAK); the retransmitted packets are combined to increase the received signal power. In CC, a fixed number of parity bits for error correction are transmitted even if all of them are not needed under good channel conditions. However, in IR, the parity bits are transmitted only when requested. In IR the coding rate decreases and the error correction power becomes stronger as the redundancy increases with each retransmission. A conceptual diagram for CC and IR is shown in **Figure 23**. Cyclic redundancy code (CRC) is the most commonly used error detection code in ARQ. At the receiver, the processing is shown for the case when a retransmitted packet is received following an NAK.

HARQ with turbo coding has gained importance as a promising error control scheme. Since the invention of turbo codes, they have been widely accepted and incorporated in many communications systems. Turbo encoder consists of recursive systematic convolutional (RSC) encoders connected in parallel with interleavers between them. The simplest and the most widely studied and used turbo encoder/decoder consist of two RSC component encoders and decoders resulting in a rate 1/3 code. The turbo encoder with a constraint length 4 and (13, 15) RSC component encoders followed by a puncturer (to adjust the coding rate) is illustrated in **Figure 24 (a)**. The turbo decoder is an iterative decoder with component decoders associated with the constituent RSC encoders in the turbo encoder. At the receiver, each component decoder computes the reliability (soft value) for each information bit from the coherently detected received signal and the extrinsic information generated by the other component decoder, except for the first decoder in the first iteration which uses only the coherently detected received signal. The iterative turbo decoder is illustrated in **Figure 24 (b)**. Several algorithms are available to generate the extrinsic estimation: soft-output Viterbi algorithm (SOVA), maximum-a-posteriori (MAP) algorithm, log-MAP algorithm, and max-log-MAP algorithm. The performance improvement is known to be minimal after about 8 decoding iterations. Turbo codes with coding rates higher than 1/3 are generated by puncturing the parity sequences at the transmitter and replacing the unsent bits at the receiver by a zero channel value.

## 8.2 Combination of HARQ and Frequency-domain Equalization

The data rate can be increased in a SC-CDMA system by assigning $C$ orthogonal spreading codes to the same user, resulting in what is commonly referred to as multicode SC-CDMA. The data-modulated symbol sequence is converted into $C$ parallel streams and spread using $C$ different orthogonal codes. The spread signals are multiplexed and divided into a sequence of chip blocks of $N_c$ chips each. After

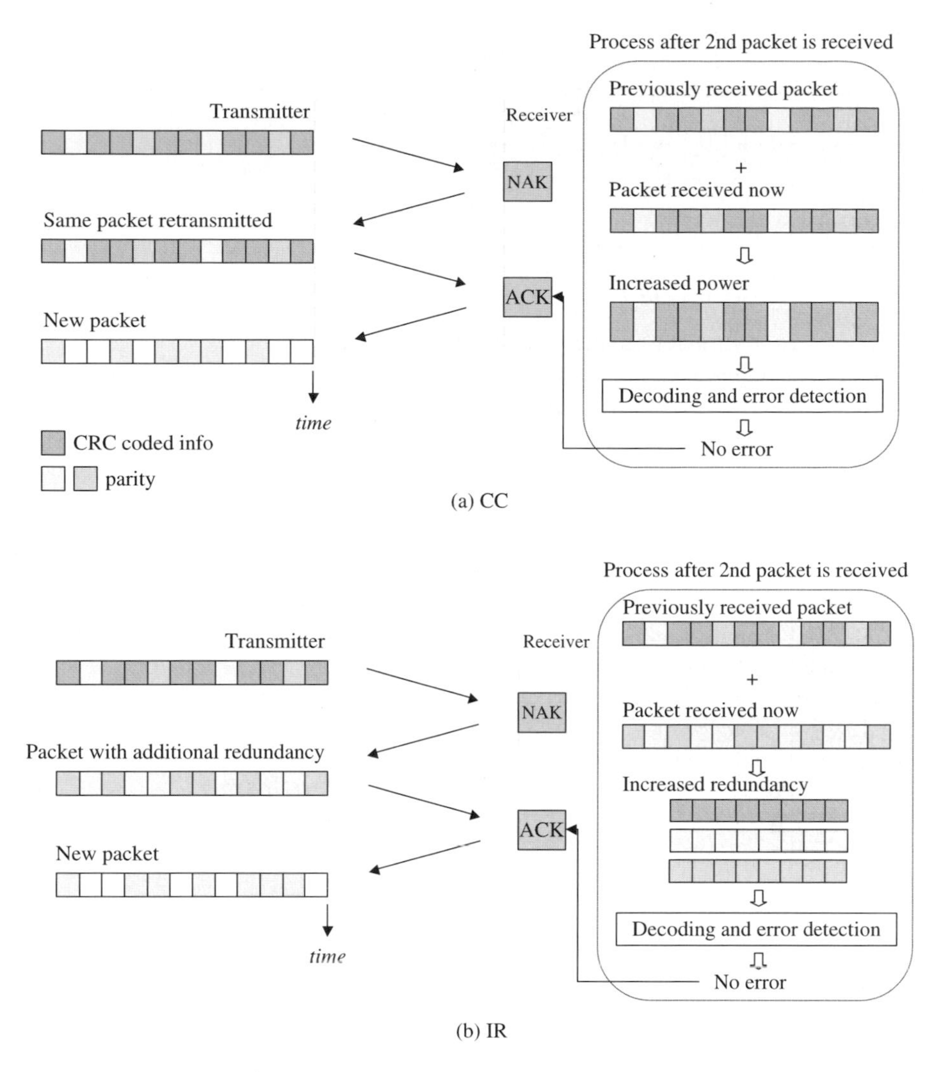


*Figure 23.* Conceptual diagram for Chase combining (CC) and incremental redundancy (IR)

the insertion of $N_g$-sample GI in each block, a sequence of GI-inserted chip blocks is transmitted. At the receiver, after GI removal, each block is first decomposed by applying $N_c$-point FFT into $N_c$ frequency components for MMSE-FDE. The time-domain signal is obtained by applying $N_c$-point IFFT. Then, the time-domain signal is despread using the $C$ different orthogonal codes. When the number $C$ of codes multiplexed is taken to be the same as the spreading factor $SF$, the transmission rate is the same as that of a non-spread system.

The component decoders of the turbo decoder requires soft values as input. The soft value for turbo decoding can be generated using the log-likelihood ratio (LLR)

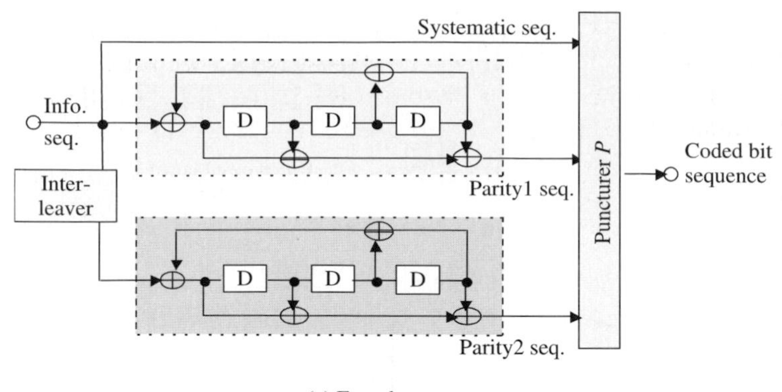


(a) Encoder

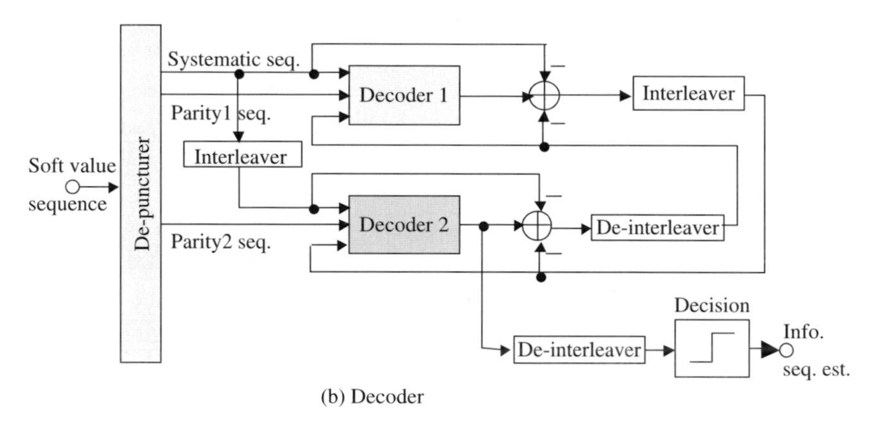


(b) Decoder

*Figure 24.* Turbo encoder and decoder

for each bit. It is necessary to know the statistics of the received symbol to properly calculate the LLR. Hence it is important to know the statistics of the despread signal that is input to the turbo decoder when turbo coding is used in the SC-CDMA system using FDE.

The LLR values should be computed for each bit taking into account its equivalent channel gain and the residual ICI after FDE. The residual ICI can be approximated as a complex Gaussian noise and therefore the sum of residual ICI and noise can be treated as a new complex Gaussian noise process with variance $2\sigma^2$. The LLR is given by

$$L(b) = \ln \left[ \frac{\sum_{\{\hat{s}:b=1\}} \frac{1}{\sqrt{2\pi\sigma^2}} \exp\left( -\frac{1}{2\sigma^2} \left| \hat{d}_c(n) - \sqrt{\frac{2E_c}{T_c}} \tilde{H}\hat{s} \right|^2 \right)}{\sum_{\{\hat{s}:b=0\}} \frac{1}{\sqrt{2\pi\sigma^2}} \exp\left( -\frac{1}{2\sigma^2} \left| \hat{d}_c(n) - \sqrt{\frac{2E_c}{T_c}} \tilde{H}\hat{s} \right|^2 \right)} \right], \tag{47}$$

where $\hat{d}_c(n)$ is the despreader output corresponding to the $n$th symbol for the $c$th parallel stream ($c = 0 \sim C-1$), $\hat{s}$ is the candidate symbol with the $b$th bit as 0 or 1 ($b = 0 \sim B-1$ with $B = 2$, 4 or 6 for QPSK, 16QAM and 64QAM, respectively), $E_c$ is the chip energy per parallel stream, $T_c$ is the chip period, and $\tilde{H} = \frac{1}{N_c}\sum_{k=0}^{N_c-1}\hat{H}(k)$, where $\hat{H}(k) = w(k)H(k)$ is the equivalent channel gain introduced in Eq. (6). The variance $2\sigma^2$ of the Gaussian approximated residual ICI plus noise is given by

$$2\sigma^2 = 2\frac{N_0}{T_c}\left[\begin{array}{l}\frac{1}{N_c}\sum_{k=0}^{N_c-1}|w(k)|^2 \\ +\left(\frac{C}{SF}\frac{E_s}{N_0}\right)\left\{\frac{1}{N_c}\sum_{k=0}^{N_c-1}|\hat{H}(k)|^2 - \left|\frac{1}{N_c}\sum_{k=0}^{N_c-1}\hat{H}(k)\right|^2\right\}\end{array}\right], \tag{48}$$

where $N_0$ is the one-sided noise power spectrum density, $SF$ is the spreading factor, and $E_s = SF \cdot E_c$ is the symbol energy. The denominator and numerator inside the brackets of Eq. (47) can be approximated as

$$\begin{array}{l}\sum_{\{\hat{s}:b=1 \text{ or } 0\}}\frac{1}{\sqrt{2\pi\sigma^2}}\exp\left(-\frac{1}{2\sigma^2}\left|\hat{d}_c(n)-\sqrt{\frac{2E_c}{T_c}}\tilde{H}\hat{s}\right|^2\right) \\ \approx \max_{\{\hat{s}:b=1 \text{ or } 0\}}\frac{1}{\sqrt{2\pi\sigma^2}}\exp\left(-\frac{1}{2\sigma^2}\left|\hat{d}_c(n)-\sqrt{\frac{2E_c}{T_c}}\tilde{H}\hat{s}\right|^2\right)\end{array}. \tag{49}$$

Hence, the LLR can be written as

$$L(b) = \frac{1}{2\sigma^2}\left(\left|\hat{d}_c(n)-\sqrt{\frac{2E_c}{T_c}}\tilde{H}\hat{s}_{0,\min}\right|^2 - \left|\hat{d}_c(n)-\sqrt{\frac{2E_c}{T_c}}\tilde{H}\hat{s}_{1,\min}\right|^2\right) \tag{50}$$

for the $b$th bit in the $n$th symbol, where $\hat{s}_{0,\min}$ (or $\hat{s}_{1,\min}$) is the symbol that has the minimum Euclidean distance from $\hat{d}_c(n)$ among all candidate symbols with the $b$th bit as 0 (or 1). The LLR values are computed for $c = 0 \sim C-1$ and for all the bits in the symbol. Turbo decoding is performed using these LLR values as soft input after rate matching. Then, error detection is performed and retransmission is requested if errors are detected.

The throughput performance in bps/Hz of SC-CDMA with MMSE-FDE when turbo coded HARQ is used are shown in **Figure 25** for $f_D T_{blk} = 0.001$ ($f_D$ denotes the maximum Doppler frequency of the fading channel and $T_{blk} = T_c(N_c + N_g)$ denotes the chip block length). $L$-path Rayleigh fading channel with uniform power delay profile is assumed. The equivalent spreading factor $SF_{eq}$(=$SF/C$) is set to $SF_{eq} = 1$ (i.e., the transmission symbol rate is equal to the non-spread SC). Both CC and IR are considered. The data modulation is fixed to 16QAM. The coding rate is fixed at 3/4 for CC. For IR, the coding rate of the initial transmission is

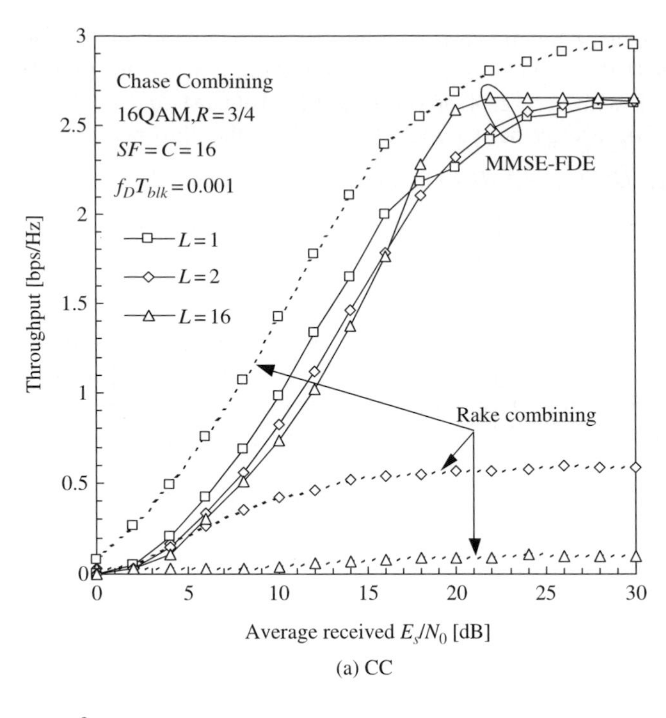


(a) CC

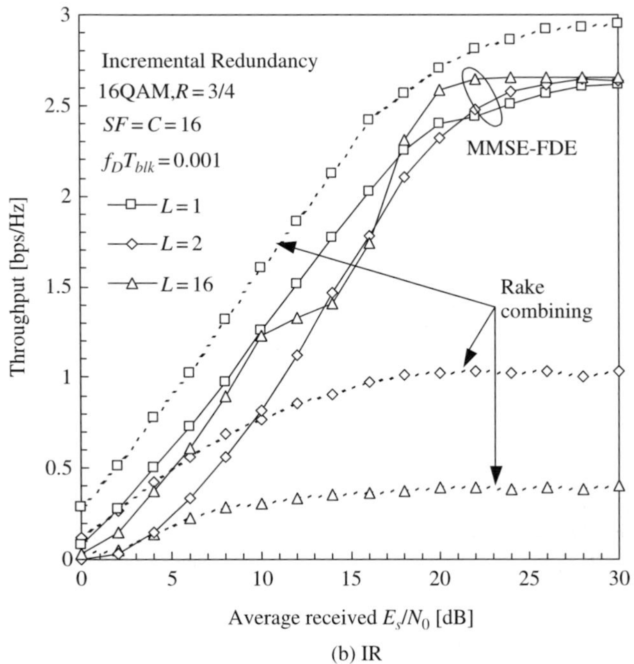


(b) IR

*Figure 25.* Throughput performance

$R = 3/4$ and additional redundancy is transmitted with the second transmission. The following puncturing matrices are used (1 represents that the bit at that position is transmitted and 0 represents that it is not transmitted):

$$P_1 = \begin{bmatrix} 1\,1\,1\,1\,1\,1 \\ 1\,0\,0\,0\,0\,0 \\ 0\,0\,0\,1\,0\,0 \end{bmatrix}, P_2 = \begin{bmatrix} 0\,0\,0\,0\,0\,0 \\ 0\,1\,1\,1\,1\,0 \\ 1\,1\,0\,0\,1\,1 \end{bmatrix} \tag{51}$$

For CC, the same packet with puncturing matrix $P_1$ is transmitted until a positive acknowledge (ACK) is received. For IR, the puncturing matrix $P_1$ is used for the first transmission and $P_2$ for the second transmission and the order repeated for further transmissions. Code combining is employed if the same packet is transmitted more than once. For reference, the throughput obtained with coherent rake combining is also plotted; the throughput degrades drastically when the number $L$ of paths increases. With the increase in $L$, the frequency-selectivity of the channel gets stronger and the orthogonality distortion is severer. Hence, the throughput decreases with the increase in $L$. However, with MMSE-FDE, the throughput is almost insensitive to $L$. This is because MMSE-FDE can partially restore the code orthogonality which is distorted due to the frequency selectivity of the channel and obtain the frequency diversity gain. For $L = 1$, the throughput is lower with MMSE-FDE compared to rake combining, due to the GI insertion loss. However in broadband channels characterized by time- and frequency-selective fading, the MMSE-FDE has a better performance.

## REFERENCES


[1] F. Adachi, M. Sawahashi, and H. Suda, "Wideband DS-CDMA for next generation mobile communications systems," IEEE Commun. Mag., Vol. 36, No. 9, pp. 56–69, Sept. 1998.

[2] Y. Kim, et al., "Beyond 3G: vision, requirements, and enabling technologies," IEEE Commun. Mag., Vol. 41, No. 3, pp.120–124, Mar. 2003.

[3] M. Helard, R. Le Gouable, J-F. Helard and J-Y. Baudais, "Multicarrier CDMA techniques for future wideband wireless networks," Ann. Telecommun., vol. 56, pp. 260–274, 2001.

[4] S. Hara and R. Prasad, "Overview of multicarrier CDMA," IEEE Commun. Mag., Vol. 35, pp. 126–144, Dec. 1997.

[5] B. Sklar, "Rayleigh fading channels in mobile digital communication systems part 1: characterization," IEEE Commun. Mag., pp. 90–100, July 1997.

[6] F. Adachi and T. Sao, "Joint antenna diversity and frequency-domain equalization for multi-rate MC-CDMA," IEICE Trans. Commun., Vol. E86-B, No. 11, pp. 3217–3224, Nov. 2003.

[7] F. Adachi, D. Garg, S. Takaoka, and K. Takeda, "Broadband CDMA techniques," IEEE Wireless Communications Magazine, Vol. 12, No. 2, pp. 8–18, Apr. 2005.

[8] F. Adachi and K. Takeda, "Bit error rate analysis of DS-CDMA with joint frequency-domain equalization and antenna diversity combining," IEICE Trans. Commun., Vol. E87-B, pp. 2991–3002, Oct. 2004.

[9] F. Adachi, T. Sao, and T. Itagaki, "Performance of multicode DS-CDMA using frequency domain equalization in a frequency selective fading channel," Electronics Letters, Vol. 39, pp. 239–241, Jan. 2003.

[10] D. Falconer, S. L. Ariyavistakul, A. Benyamin-Seeyar, and B. Eidson, "Frequency domain equalization for single-carrier broadband wireless systems," IEEE Commun. Mag., Vol. 40, pp. 58–66, Apr. 2002.

[11] N. Benvenuto and S. Tomasin, "On the comparison between OFDM and single carrier modulation with a DFE using a frequency-domain feedforward filter," IEEE Trans. Commun., Vol. 50, No. 6, pp. 947–955, June 2002.

[12] A. M. Chan and G. W. Wornell, "A class of block-iterative equalizers for intersymbol interference channels: fixed channel results," IEEE Trans. Commun., Vol. 49, No. 11, pp. 1966–1976, Nov. 2001.

[13] N. Benvenuto and S. Tomasin, "Block iterative DFE for single carrier modulation," IEE Electronics Letters, Vol. 38, No. 19, pp. 1144–1145, Sept. 2002.

[14] S. Tomasin and N. Benvenuto, "A reduced complexity block iterative DFE for dispersive wireless applications," Proc. 60th IEEE Veh. Technol. Conf. 2004 Fall, Los Angels, U.S.A., 26–29 Sept. 2004.

[15] K. Takeda, K. Ishihara, and F. Adachi, "Downlink DS-CDMA transmission with joint MMSE equalization and ICI cancellation," Proc. 63rd IEEE Veh. Technol. Conf. 2006-Spring, Melbourne, Australia, 7–10 May 2006.

[16] R. T. Derryberry, S. D. Gray, D. M. Ionescu, G. Mandyam, and B. Raghothaman, "Transmit diversity in 3G CDMA systems," IEEE Commun. Mag., Vol. 40, pp. 68–75, Apr. 2002.

[17] S. Alamouti, "A simple transmit diversity technique for wireless communications", IEEE Journal on Selected Areas in Commun., Vol. 16, No. 8, pp. 1451–1458, Oct. 1998.

[18] E. G. Larsson and P. Stoica, *Space–time block coding for wireless communications*, Cambridge Univ. Press, Cambridge, UK, 2003.

[19] D. Garg and F. Adachi, "Performance improvement with space-time transmit diversity using minimum mean square error combining equalization in MC-CDMA," IEICE Trans. Commun., pp. 849–857, Mar. 2004.

[20] N. Al-Dhahir, "Single-carrier frequency-domain equalization for space-time block-coded transmissions over frequency-selective fading channels," IEEE Commun., Lett., Vol. 5, No. 7, pp. 304–306, July 2001.

[21] F. W. Vook, T. A. Thomas, and K. L. Baum, "Cyclic-prefix CDMA with antenna diversity," Proc. 55th IEEE Veh. Technol. Conf. 2002-Spring, pp. 1002–1006, May 2002.

[22] K. Takeda, T. Itagaki, and F. Adachi, "Application of space-time transmit diversity to single-carrier transmission with frequency-domain equalization and receive antenna diversity in a frequency-selective fading channel," IEE Proceedings Communications, Vol. 151, No. 6, pp. 627–632, Dec. 2004.

[23] W. Su, X. G. Xia, and K. J. R. Liu, "A systematic design of high-rate complex orthogonal space-time block codes," IEEE Commun. Lett., Vol. 8, No. 6, pp. 380–382, June 2004.

[24] F. Adachi, "Wireless past and future-evolving mobile communications systems," IEICE Trans. Fundamentals, Vol.E84-A, pp.55–60, Jan. 2001.

[25] G. J. Foschini and M. J. Gans, "On limits of wireless communications in a fading environment when using multiple antennas", Wireless Personal Communications, Kluwer, Vol. 6, No. 3, pp. 311–335, 1998.

[26] G. J. Foschini, "Layered space-time architecture for wireless communication in a fading environment when using multi-element antennas," Bell Lab. Tech. Journal, Vol. 1, No. 2, pp. 41–59, 1996.

[27] T. Matsumoto, J. Ylitalo, and M. Juntti, "Overview and recent challenges of MIMO system," IEEE Vehicular Technology Society News, pp. 4–9, May 2003.

[28] J. G. Proakis, *Digital Communications*, 4th edition, McGraw-Hill, 2001.

[29] P. W. Wolniansky, G. J. Foschini, G. D. Golden, and R. A. Valenzuela, "V-BLAST: an architecture for realizing very high data rates over the rich-scattering wireless channel," Proc. ISSSE, pp. 295–300, 1998.

[30] W. C., Jakes Jr., Ed., Microwave mobile communications, Wiley, New York, 1974.

[31] A. Nakajima, D. Garg, and F. Adachi, "Frequency-domain Iterative Parallel Interference Cancellation for Multicode DS-CDMA-MIMO Multiplexing," Proc. IEEE 62nd Veh. Technol. Conf., Dallas, U.S.A., 26–28 Sept. 2005.

[32] S. Haykin, *Adaptive filter theory*, 4th edition, Prentice Hall, 2001.

[33] Z. Wang and G. B. Giannakis, "Block precoding for MUI/ISI-resilient generalized multicarrier CDMA with multirate capabilities," IEEE Trans. Commun., Vol. 49, No. 11, pp. 2016–2027, Nov. 2001.

[34] S. Tsumura, S. Hara, and Y. Hara, "Performance comparison of MC-CDMA and cyclically prefixed DS-CDMA in an uplink channel," Proc. IEEE VTC'04 Fall, Los Angeles, USA, pp. 414–418, Sept. 2004.

[35] X. D. Wang and H. V. Poor, "Iterative (turbo) soft interference cancellation and decoding for coded CDMA," IEEE Trans. Commun., Vol. 47, No. 7, pp. 1046–1061, July 1999.

[36] S. Zhou, G. B. Giannakis, and C. L. Martret, "Chip-interleaved block-spread code division multiple access," IEEE Trans. Commun., Vol. 50, No. 2, Feb. 2002.

[37] X. Peng, F. Chin, T. T. Tjhung, and A. S. Madhukumar, "A simplified transceiver structure for cyclic extended CDMA system with frequency domain equalization," Proc. IEEE VTC'05 Spring, Sweden, pp. 1565–1569, May 2005.

[38] T. Ottosson and A. Svensson, "On schemes for multirate support in DS/CDMA," J. Wireless Personal Commun., Vol.6, No. 3, pp. 265–287, Mar. 1998.

[39] F. Adachi, M. Sawahashi, and K. Okawa, "Tree-structured generation of orthogonal spreading code with different lengths for foward link of DS-CDMA mobile radio," IEE Electron. Lett., Vol. 33, No. 1, pp. 27–28, Jan. 1997.

[40] L. Liu and F. Adachi, "2-dimensional OVSF spreading for chip-interleaved DS-CDMA uplink transmission," Proc. WPMC'05, Aalborg, Denmark, 19–22 Sept. 2005.

[41] L. Liu and F. Adachi, "2-dimensional OVSF Spread/Chip-interleaved CDMA," IEICE Trans. Commun., conditioned accepted.

[42] R. H. Morelos-Zaragoza, *The art of error correcting codes*, Wiley, 2002.

[43] S. Lin and D. J. Costello, *Error Control Coding: Fundamentals and Applications*, Prentice Hall, Inc., 1983.

[44] D. Chase, "Code combining- A maximum likelihood decoding approach for combing and arbitrary number of noisy packets," IEEE Trans. Commun., Vol. COM-33, No. 5, pp. 385–393, May 1985.

[45] J. Hagenauer, "Rate-compatible punctured convolutional codes (RCPC codes) and their application," IEEE Trans. Commun., Vol. 36, No. 4, pp. 389–400, April 1988.

[46] D. N. Rowitch and L. B. Milstein, "Rate compatible punctured turbo (RCPT) codes in hybrid FEC/ARQ system," Proc. Comm. Theory Mini-conference, IEEE GLOBECOM'97, pp. 55–59, Nov. 1997.

[47] T. Ji and W. E. Stark, "Turbo-coded ARQ schemes for DS-CDMA data networks over fading and shadowing channels: throughput, delay and energy efficiency," IEEE Journal on Selected Areas in Commun., Vol. 18, No. 8, pp. 1355–1364, Aug. 2000.

[48] D. Garg and F. Adachi, "DS-CDMA with frequency-domain equalization for high speed downlink packet access," Journal on Selected Areas in Communications, Vol. 24, No. 1, pp. 161–170, Jan. 2006.

[49] D. Garg and F. Adachi, "Throughput comparison of turbo-coded HARQ in OFDM, MC-CDMA and DS-CDMA with frequency-domain equalization," IEICE Trans. on Commun., Vol. E88-B, No.2, pp. 664–677, Feb. 2005.

[50] C. Berrou, A. Glavieux, and P. Thitimajshima, "Near Shannon limit wrror-correcting coding and ecoding:Turbo codes," Proc. IEEE ICC, pp. 1064–1070, Geneva, May 1993.

[51] C. Berrou, "The ten-year-old turbo codes are entering into service," IEEE Commun. Mag., Vol. 41, No. 8, pp. 110–116, Aug. 2003.

[52] J. P. Woodard and L. Hanzo, "Comparative study of turbo decoding techniques: an overview," IEEE Trans. Veh. Technol., Vol. 49, No. 6, pp. 2208–2233, Nov. 2000.

[53] D. Divsalar and F. Pollara, "Turbo codes for PCS applications," Proc. IEEE ICC'95, pp. 54–59, Seattle, Washington, June 1995.

[54] P. Robertson, E. Villebrum, and P. Hoeher, "A comparison of optimal and sub-optical MAP decoding algorithms operating in the log domain," Proc. IEEE ICC'95, pp. 1009–1013, Seattle WA, June 1995.

[55] J. Hagenauer, E. Offer, and L. Papke, "Iterative decoding of binary block and convolutional codes," IEEE Trans. on Info. Theory, Vol. 42, No. 2, pp. 429–445, Mar. 1999.

[56] B. Sklar, "A primer on turbo code concepts," IEEE Commun. Mag., Vol. 35, No.12, pp. 94–101, Dec. 1997.

[57] C. Heegard and S. B. Wicker, Turbo coding, Kluwer Academic Publishers, 1999.

[58] L. R. Bahl, J. Cocke, F. Jelinek, and J. Raviv, "Optimal decoding of linear codes for minimizing symbol error rate," IEEE Trans. on Inf. Theory, pp. 284–287, March 1974.

[59] F. Adachi, K. Ohono, A. Higuchi, T. Dohi, and Y. Okumura, "Coherent multicode DS-CDMA mobile radio," IEICE Trans. Commun., Vol. E79-B, No. 9, pp. 1316–1325, Sept. 1996.

[60] A. Stefanov and T. Duman, "Turbo coded modulation for wireless communications with antenna diversity," Proc. IEEE VTC99-Fall, pp. 1565–1569, Netherlands, Sept. 1999.

CHAPTER 4

# FUNDAMENTALS OF MULTI-CARRIER CDMA TECHNOLOGIES

SHINSUKE HARA

*Department of Information Systems, Graduate School of Engineering, Osaka City University, Japan*
*hara@info.eng.osaka-cu.ac.jp*

**Abstract:** This chapter introduces and compares two kinds of techniques based on combination of CDMA and multicarrier transmission, such as Multicarrier CDMA and Multicarrier DS/CDMA. Several detection and combining schemes are derived for both the techniques, including a serial interference cancellation in uplink and a rake combining in downlink for MC-DS/CDMA whereas a decorrelating multiuser detection and a minimum mean square error (MMSE) multiuser detection in uplink and an orthogonality restoring combining (ORC), an MMSE combining, a maximum ratio combining (MRC) and an equal gain combining (EGC) in downlink for MC-CDMA. The bit error rate (BER) lower bounds for the two techniques are theoretically analyzed and furthermore the BERs with the several detection/combining schemes are demonstrated by computer simulations

**Keywords:** Multi-carrier transmission, MC-CDMA, MC-DS/CDMA, and maximu ratio combiner

## 1. INTRODUCTION

CDMA technique is robust to frequency-selective fading and it has been successfully introduced in commercial cellular mobile communications systems such as IS-95 and 3G systems. On the other hand, multicarrier transmission technique is also inherently robust to frequency-selective fading and in the name of orthogonal frequency division multiplexing (OFDM), it has been also successfully introduced in commercial wireless systems such as wireless local area networks (LANs) and terrestrial digital video broadcasting (DVB-T). Therefore, it would be quite natural to think of no synergistic effect in combination of these two techniques.

Whether the combination will be beneficial or not depends on a bandwidth and a data transmission rate we intend to support. In fact, for a 2 Mbits/sec-data transmission rate which 3G systems are now supporting, the combination of CDMA and multicarrier transmission techniques brings no benefit at all. However, if we intend to support much higher data transmission than this, such as in future



*Y. Park and F. Adachi (eds.), Enhanced Radio Access Technologies for Next Generation Mobile Communication*, 121–150.

4G systems, the combination does bring a benefit, in other words, it becomes a promising data transmission technique.

This chapter introduces and compares two kinds of combination of CDMA and multicarrier transmission techniques. One is Multicarrier (MC-) CDMA, which was independently proposed by three different research groups in 1993, and another is MC-DS/CDMA, which was also proposed in 1993 and then its variant was proposed in 1996. The difference between the original and variant of MC-DS/CDMA is that the former allows overlapping of subcarrier spectra whereas the latter does not. The subcarrier non-overlapped MC-DS/CDMA is mathematically tractable, so in this chapter, we will use the (subcarrier non-overlapped) MC-DS/CDMA.

This chapter is organized as follows. **Section 2** shows a fatal problem of DS/CDMA in high-speed data transmission and **Section 3** introduces combination of multicarrier transmission and CDMA as a solution of the problem. **Section 4** explains several assumptions required for introducing and comparing MC-CDMA and MC-DS/CDMA. After **Section 5** outlines single-carrier DS/CDMA (In Chapter 3, single-carrier CDMA is referred to as DS/CDMA. In this chapter, on the other hand, to clearly show the structural difference between multi-carrier signaling and single-carrier signaling, the single-carrier CDMA is called "single-carrier DS/CDMA."), MC-DS/CDMA is first introduced in **Section 6** because MC-DS/CDMA has a similarity to single-carrier DS/CDMA, and then MC-CDMA is introduced in **Section 7**. **Section 8** demonstrates numerical results on the performance of MC-DS/CDMA and MC-CDMA systems, and finally **Section 9** concludes this chapter.

## 2. A FATAL PROBLEM OF DS/CDMA IN HIGH-SPEED DATA TRANSMISSION

Let us assume that a signal is emitted at a DS/CDMA transmitter, it goes through a frequency selective fading channel and then it arrives at a DS/CDMA receiver. **Figure 1** shows a block diagram of the DS/CDMA receiver with four rake finger processors. At the receiver, a received signal is fed to a bandpass filter (BPF), down-converted and then analog-to-digital (A/D) converted with $I$ and $Q$ branches. At each rake finger processor, the A/D-converted baseband samples are despread and integrated by a code generator and a correlator, and the differences in the phases and arrival times among the correlator outputs are compensated for by a phase rotator and a delay equalizer. Finally, a combiner sums up the channel impairment-compensated symbols to recover user data symbols.

The matched filter output, namely, observation of a channel impulse response is very important for DS/CDMA receiver, because it determines the number and positions of the paths captured by the rake combiner to collect the energy of received signal. When a receiver observes a channel, how finely it can analyze the temporal structure of the channel is called "time resolution." Defining the sampling rate as $R_{smp}$ samples/sec, the time resolution $\Delta t$ is given by $1/R_{smp}$ sec, so the number of resolvable paths in an observed impulse response of a channel is in proportion to

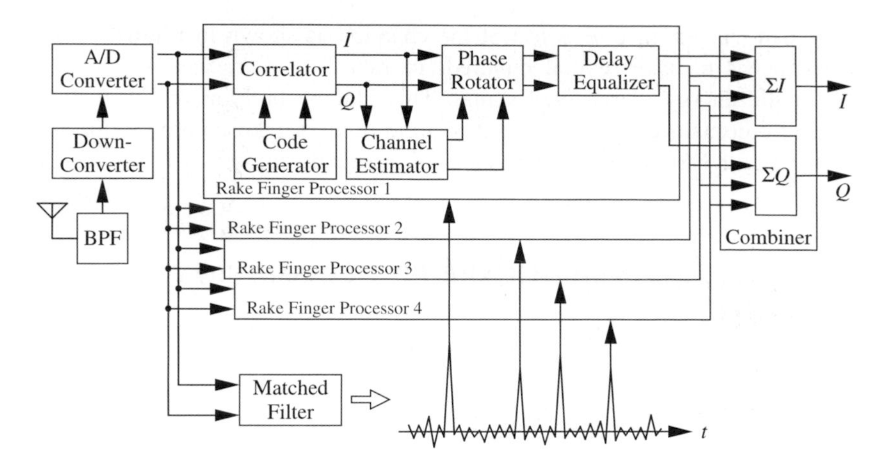


*Figure 1.* A block diagram of a DS/CDMA rake receiver

the sampling rate. For DS/CDMA system, the sampling rate is determined by the chip rate, so consequently, the number of resolvable paths is in proportion to the chip rate.

Let us consider a case where we intend to support a data transmission rate of up to 2 Mbits/sec in a wireless communication channel with carrier frequency of $f_c$. In this case, assuming spreading codes employed in 3G systems, a DS/CDMA receiver always sees less than around ten paths in matched filter outputs of the channel, as shown in **Figure 2 (a)**. Therefore, the receiver can collect almost all part of the received signal energy only with several rake finger processors. As shown in **Figure 1**, roughly speaking, the hardware complexity of DS/CDMA receiver is determined by the number of rake finger processors employed and this *mild* number of rake finger processors was acceptable in terms of cost, size and power consumption of 3G mobile terminals.

Now, let us consider a case where we intend to support a much higher data transmission rate such as 100 Mbits/sec, which is a typical data transmission rate discussed in 4G systems. This means that a DS/CDMA receiver will see several

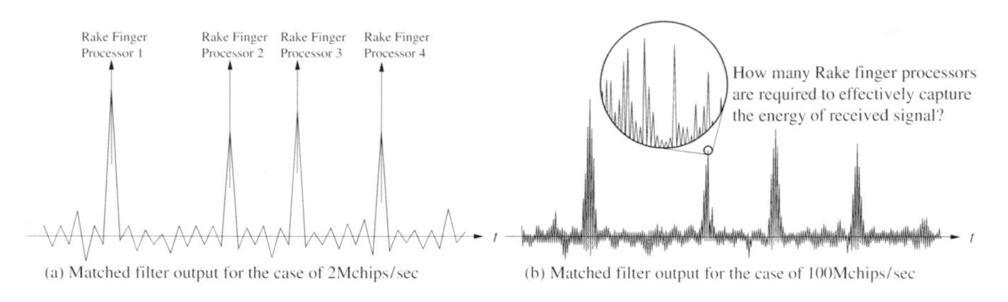


*Figure 2.* Comparison of matched filter output

hundreds of paths in impulse response of the channel, as shown in **Figure 2 (b)**, and hence it needs to have several hundreds of rake finger processors to effectively collect the energy of received signal. This will be prohibitive (Note that, using frequency domain equalizer instead of time domain rake combiner, the bit error rate (BER) of DS/CDMA system can be drastically improved as shown in Chapter 3).

## 3. COMBINATION OF MULTICARRIER TRANSMISSION AND CDMA

Reducing the data transmission rate results in lessening the number of rake finger processors, but it seems contradictory to achieving a high data transmission rate. However, as shown in **Figure 3**, a high data transmission rate is achievable with a number of lower data rate sub-channels with different carrier frequencies. This is the very idea of multicarrier transmission, which is the principle of transmitting data by dividing a data stream into a number of data streams, each of which has a much lower data rate and by using these substreams to modulate subcarriers. In **Figure 3**, the multicarrier system supports $M$ parallel transmissions, reducing the transmission rate over each sub-channel by factor of $M$.

Limiting our discussion within application of CDMA technique to high data rate transmission, there are mainly two ways considered in combination of multicarrier and CDMA techniques. One way is to employ a mild number of sub-channels where there remains a frequency-selective fading in each sub-channel, and another way is to employ a huge number of sub-channels where frequency-selective fading has disappeared in each sub-channel. The former is called "multicarrier (MC)-DS/CDMA," which still requires a DS rake approach to effectively collect the energy of received signal over each sub-channel, whereas the latter is called "multicarrier (MC)-CDMA," which employs a spreading operation across the whole sub-channels to gain frequency diversity effect. **Figure 4** compares the power spectral densities (*PSDs*) among a Single-carrier (SC)-DS/CDMA, MC-DS/CDMA and MC-CDMA waveforms.

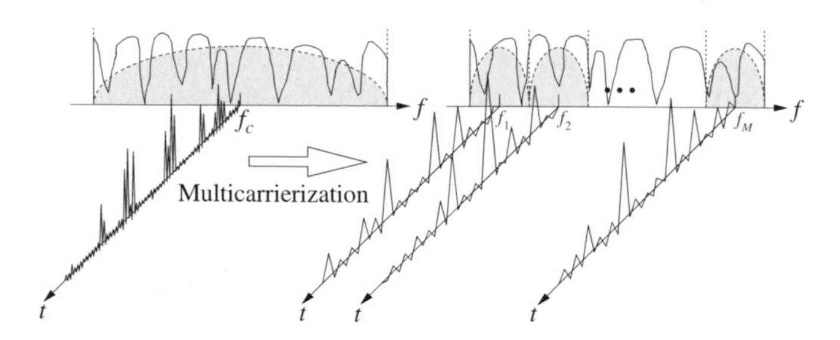


*Figure 3.* Multicarrierization

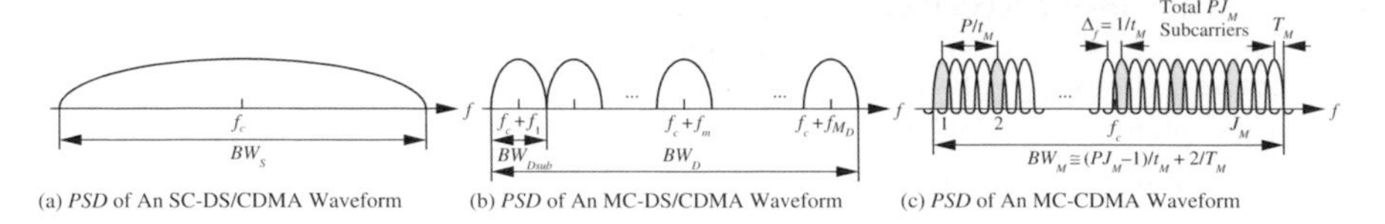


(a) *PSD* of An SC-DS/CDMA Waveform (b) *PSD* of An MC-DS/CDMA Waveform (c) *PSD* of An MC-CDMA Waveform

*Figure 4.* Power spectral densities

# 4. SYSTEM MODEL

## 4.1 Multiplexing/Multiple Access and Spreading Codes

It is assumed that SC-DS/CDMA, MC-DS/CDMA and MC-CDMA systems support $K$ multiplexing/multiple access users employing spreading codes with spreading gain of $J$. In a downlink, a base station multiplexes $K$ signals and then transmits the multiplexed signal to $K$ users. On the other hand, in an uplink, each user transmits its own signal to a base station and the base station receives $K$ signals through different channels. Here, the data symbol duration is defined as $T$ whereas the chip duration as $T_c$. To distinguish the individual systems clearly, the subscripts for showing SC-DS/CDMA, MC-DS/CDMA and MC-CDMA systems are defined as $S$, $D$ and $M$, respectively. In addition, the indices for spreading gain, user and subcarrier are defined as $j$, $k$ and $m$, respectively, and furthermore, the indices for transmitted symbol and path gain in impulse response are defined as $i$ and $l$, respectively.

The $i$-th data symbol for the $k$-th user is defined as $a_{ki}$ for the single-carrier system whereas the $i$-th data symbol transmitted over the $m$-th subcarrier for the $k$-th user is defined as $a_{kim}$ for the multi-carrier systems. Here, defining data symbol vectors $(K \times 1)$ as $\mathbf{a}_{iK} = [a_{1i}, \cdots, a_{Ki}]^T$ and $\mathbf{a}_{imK} = [a_{1im}, \cdots, a_{Kim}]^T$, they are assumed to respectively have the following statistical properties:

$$E\left[\mathbf{a}_{iK}\mathbf{a}_{iK}^H\right] = \mathbf{I}_{K\times K} \tag{1}$$

$$E\left[\mathbf{a}_{imK}\mathbf{a}_{imK}^H\right] = \mathbf{I}_{K\times K} \tag{2}$$

where $E[\cdot]$, $(\cdot)^T$ and $(\cdot)^H$ denote statistical average, transpose and Hermitian transpose of $(\cdot)$, respectively, and $\mathbf{I}_{K\times K}$ denotes the identity matrix with size of $K \times K$.

On the other hand, for spectrum spreading, the random codes are assumed. For the $j$-th chip of the $k$-th user $c_{kj}$, which takes +1 or -1 with the same probability, defining a code vector $(J \times 1)$ and a code matrix $(J \times K)$ as $\mathbf{c}_k = [c_{k1}, \cdots, a_{kJ}]^T$ and $\mathbf{C}_K = [\mathbf{c}_1, \cdots, \mathbf{c}_K]$, respectively, they are assumed to respectively have the following statistical properties:

$$E\left[\mathbf{C}_K\mathbf{C}_K^H\right] = \frac{K}{J}\mathbf{I}_{J\times J} \tag{3}$$

$$E\left[\mathbf{C}_K^H\mathbf{C}_K\right] = \mathbf{I}_{K\times K}. \tag{4}$$

### 4.2 Transmitter/Receiver

The carrier conveying information has a carrier phase $\phi_c$ as well as the frequency $f_c$, but the phase is ignored for the sake of analytical simplicity. In fact, assuming a perfect carrier synchronization, it gives no effect on derivation of the signal to noise power ratio (SNR) and the BER for the CDMA systems. In addition, the received signal is perturbed by different additive Gaussian noise at a base station in an uplink and a user in a downlink, but the same notation $n(t)$ is used in both the uplink and downlink for the sake of analytical simplicity. In fact, it also gives no effect on derivation of the SNR and the BER of the CDMA systems, because they are separately discussed in the uplink and downlink.

### 4.3 Channel and Noise

The channel for the $k$-th user is assumed to be a slowly varying frequency-selective Rayleigh fading one with impulse response of $h_k(t)$. When an SC-DS/CDMA receiver with spreading gain of $J_S$ observes the channel, it sees the impulse response in a vector form with size of $J_S \times 1$. Here, the impulse response is assumed to have only $L$ non-zero components, namely,

$$\text{(5)} \qquad \mathbf{h}_k = [h_{k1}, \cdots, h_{kL}, 0, \cdots, 0]^T$$

where $h_{kl}$ is a zero-mean complex-valued Gaussian-distributed amplitude (called "path"). The auto-correlation matrix of the channel ($J_S \times J_S$) is given by

$$\text{(6)} \qquad E\left[\mathbf{h}_k \mathbf{h}_k^H\right] = \mathbf{H}_k = diag\{\sigma_{sk1}^2, \cdots, \sigma_{skL}^2, 0, \cdots, 0\}$$

where $diag\{\cdots\}$ and $\sigma_{skl}^2$ ($l = 1, \cdots, L$) denote the diagonal matrix with main diagonal elements of $\cdots$ and the $l$-th largest eigenvalues of $\mathbf{H}_k$, namely, the average power of the $l$-th component (path) of $h_k(t)$, respectively.

In addition to the impulse response vector, defining a noise vector ($J_S \times 1$) as $\mathbf{n} = [n_1, \cdots, n_J]^T$, it is assumed to have the following statistical property:

$$\text{(7)} \qquad E\left[\mathbf{n}\mathbf{n}^H\right] = \mathbf{N} = \sigma_n^2 \mathbf{I}_{J\times J}$$

where $\sigma_n^2$ denotes the power of the noise.

## 5. SC-DS/CDMA SYSTEM

**Figure 4 (a)** shows the PSD of a SC-DS/CDMA waveform. If a root Nyquist filter $p_S(t)$ is employed for baseband pulse shaping, the bandwidth $BW_S$ is given by

$$\text{(8)} \qquad BW_S = \frac{1+\alpha_S}{T_{cS}}$$

where $\alpha_S$ denotes a roll-off factor of the root Nyquist filter.

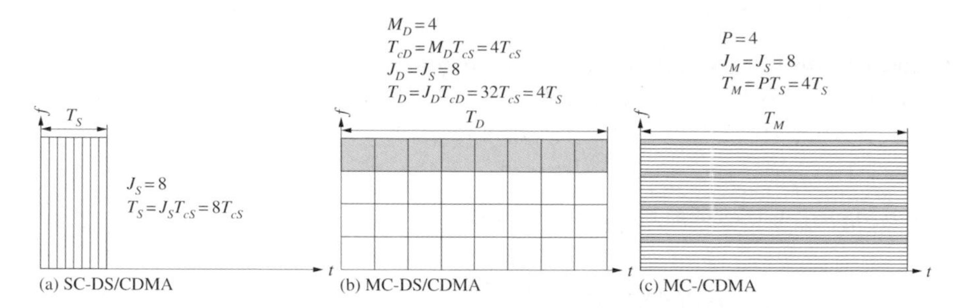


*Figure 5.* Tiling representations on a time-frequency plane

On the other hand, **Figure 5 (a)** shows a tiling representation of a SC-DS/CDMA waveform on a time-frequency plane, where $J_S = 8$ is assumed with $T_S = J_S T_{cS}$.

The structures of SC-DS/CDMA transmitter and receiver are all the same as those of SC-DS/CDMA transmitter and receiver for a certain subcarrier, respectively. Therefore, the BER of SC-DS/CDMA system will be discussed in the next section on MC-DS/CDMA system.

## 6. MC-DS/CDMA SYSTEM

### 6.1 Transmitter

**Figure 4 (b)** shows the PSD of an MC-DS/CDMA waveform, where the entire bandwidth is divided into $M_D$ equi-width frequency sub-channels. Therefore, the entire bandwidth of MC-DS/CDMA waveform is the same as that of SC-DS/CDMA waveform, namely, $BW_D = BW_S$, whereas the bandwidth of each sub-channel is given by

$$BW_{D_{sub}} = \frac{BW_S}{M_D} = \frac{1+\alpha_S}{M_D T_{cS}}. \tag{9}$$

Note that, as compared with the SC-DS/CDMA system, the chip duration over sub-channels is widened into $T_{cD} = M_D T_{cS}$ and hence $T_D = M_D T_S$ if selecting the spreading gain as $J_D = J_S$. **Figure 5 (b)** shows a tiling representation of an MC-DS/CDMA waveform on a time-frequency plane, where $M_D = 4$ and $J_D = J_S = 8$ are assumed.

**Figure 6** shows a block diagram of an MC-DS/CDMA transmitter for the $k$-th user. The data sequence, after spreading and baseband pulse shaping, modulates the $M_D$ subcarrier signals and then is transmitted. The transmitted signal in the uplink is written as

$$s_{Dk}(t) = \sum_{m=1}^{M_D} s_{Dkm}(t) \tag{10}$$

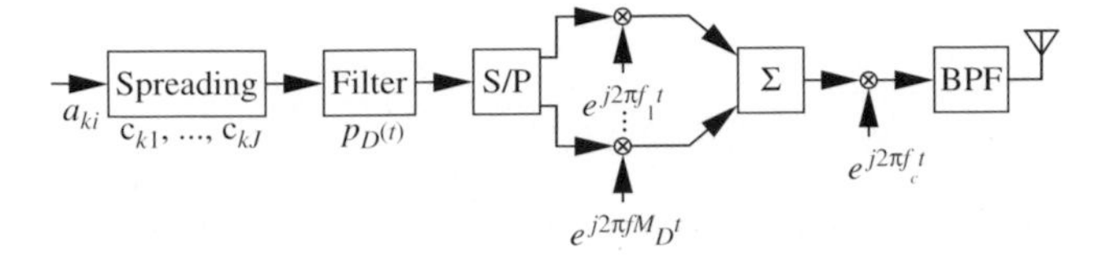


*Figure 6.* A block diagram of an MC-DS/CDMA transmitter for the $k$-th user

$$s_{Dkm}(t) = \sum_{i=-\infty}^{+\infty} \sum_{j=1}^{J_D} a_{kim} c_{kj} p_D(t - iT_D - (j-1)T_{cD}) \cdot e^{j2\pi(f_c+f_m)t} \tag{11}$$

where $s_{Dkm}(t)$ and $f_c + f_m$ denote the signal of the $k$-th user transmitted over the $m$-th subcarrier and the $m$-th subcarrier's center frequency, respectively. On the other hand, the transmitted signal in the downlink is written as

$$s_D(t) = \sum_{m=1}^{M_D} s_{Dkm}(t) \tag{12}$$

$$s_{Dkm}(t) = \sum_{k=1}^{K} \sum_{i=-\infty}^{+\infty} \sum_{j=1}^{J_D} a_{kim} c_{kj} p_D(t - iT_D - (j-1)T_{cD}) \cdot e^{j2\pi(f_c+f_m)t}. \tag{13}$$

## 6.2 Receiver

**Figure 7** shows a block diagram of an MC-DS/CDMA receiver for the $k$-th user. The benefit of multicarrierization is to widen the chip duration by factor of $M_D$, so a quasi-synchronicity among all users can be assumed even in the uplink. In this case, setting the timing offsets among the users to zero, the received signal in the uplink is written as

$$r_D(t) = \sum_{m=1}^{M_D} r_{Dm}(t) \tag{14}$$

$$r_{Dm}(t) = \sum_{k=1}^{K} h_{km}(t) \otimes s_{Dkm}(t) + n_{Dm}(t) e^{j2\pi(f_c+f_m)t} \tag{15}$$

where $r_{Dm}(t)$, $h_{km}(t)$ and $n_{Dm}(t)$ denote the $m$-th subcarrier's received signal, a channel impulse response of the $k$-th user and a baseband Gaussian noise, respectively. On the other hand, the received signal in the downlink is written as

$$r_{Dk}(t) = \sum_{m=1}^{M_D} r_{Dkm}(t) \tag{16}$$

$$r_{Dkm}(t) = h_{km}(t) \otimes s_{Dm}(t) + n_{Dm}(t) e^{j2\pi(f_c+f_m)t}. \tag{17}$$

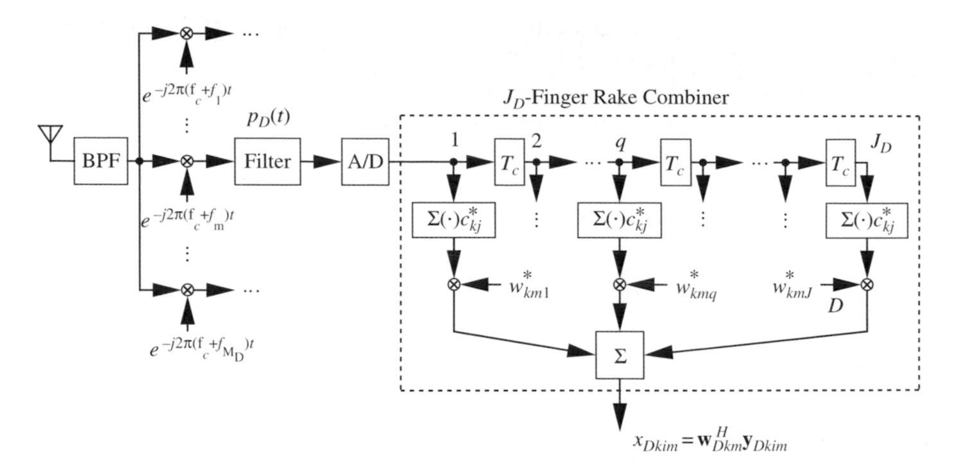


*Figure 7.* A block diagram of an MC-DS/CDMA receiver for the $k$-th user

Assuming that the number of rake finger processors is equal to $J_D$, the $q$-th rake finger output ($q = 1, \cdots, J_D$) for the $i$-th data symbol over the $m$-th subcarrier of the $k'$-th user is given by

$$(18) \qquad y_{Dk'imq} = \sum_{j=1}^{J_D} d_{D\delta imqj} c_{k'j}^*.$$

In (18), $d_{D\delta imqj}$ denotes a received signal after down-conversion, baseband pulse shaping and sampling, which is written as

$$(19) \qquad d_{D\delta imqj} = d_{D\delta m}(t = iT_D + (q-1)T_{cD} + (j-1)T_{cD})$$

where $d_{D\delta m}(t)$ is given by

$$(20) \qquad d_{D\delta m}(t) = p_D(t) \otimes \left(r_{D\delta m}(t) e^{-j2\pi(f_c+f_m)t}\right).$$

Note that, in (18)-(20), the subscript of $\delta$ is dropped for the uplink whereas replaced by $k'$ for the downlink.

It is very important to relate $h_{km}(t)$ with $h_k(t)$ for a fair comparison of the BERs between the SC-DS/CDMA and MC-DS/CDMA systems, but here, we assume that $h_{km}(t)$ ($m = 1, \cdots, M_D$) has $L_m$-path gains when it is observed with chip rate of the MC system, namely, $1/T_{cD} = 1/(M_D T_{cS})$ for a while. The comparison of the BER lower bound between the SC-DS/CDMA and MC-DS/CDMA systems will be shown in the last part of this section, taking into account of the relationship between $h_{km}(t)$ and $h_k(t)$.

In this case, the channel impulse response is defined in a vector form ($J_D \times 1$) as $\mathbf{h}_{km} = [h_{km1}, \cdots, h_{kmL_m}, 0, \cdots, 0]^T$ with the following auto-correlation matrix ($J_D \times J_D$):

$$(21) \qquad E\left[\mathbf{h}_{km}\mathbf{h}_{km}^H\right] = \mathbf{H}_{km} = diag\{\sigma_{skm1}^2, \cdots, \sigma_{skmL_m}^2, 0, \cdots, 0\}$$

where $\sigma^2_{skml}$ $(l = 1, \cdots, L_m)$ denotes the $l$-th largest eigenvalues of $\mathbf{H}_{km}$, namely, the average power of the $l$-th path of $h_{km}(t)$.

Derivation on the rake finger output in the MC-DS/CDMA system is similar to that in the SC-DS/CDMA system. That is, the $q$-th rake finger output for the $k'$-user in the uplink is decomposed as

$$y_{Dk'imq} = g_{Dk'imq} + e_{Dk'imq} \tag{22}$$

$$g_{Dk'imq} = h_{k'mq} a_{k'im} \tag{23}$$

$$\begin{aligned} e_{Dk'imq} = & \sum_{\substack{k=1 \\ k \neq k'}}^{K} h_{kmq} a_{kim} \sum_{j=1}^{J_D} c_{kj} c^*_{k'j} \\ & + \sum_{k=1}^{K} \sum_{l=1}^{q-1} h_{kml} \left( a_{kim} \sum_{j=1}^{J_S - q + l} c_{k(q-l+j)} c^*_{k'j} \right. \\ & \left. + a_{k(i+1)m} \sum_{j=J_S-q+l+1}^{J_D} c_{k(q-l+j-J_S)} c^*_{k'j} \right) \\ & + \sum_{k=1}^{K} \sum_{l=q+1}^{L} h_{kml} \left( a_{k(i-1)m} \sum_{j=1}^{l-q} c_{k(q-l+j+J_S)} c^*_{k'j} \right. \\ & \left. + a_{kim} \sum_{j=l-q+1}^{J_D} c_{k(q-l+j)} c^*_{k'j} \right) \\ & + \sum_{j=1}^{J_D} n'_{Dmqj} c^*_{k'j} \end{aligned} \tag{24}$$

where $n'_{Dmqj}$ is given by

$$n'_{Dmqj} = p_D(t) \otimes n_{Dm}(t)|_{t=iT_D+(q-1)T_{cD}+(j-1)T_{cD}}. \tag{25}$$

Note that the $q$-th rake finger output of the $k$-th user in the downlink is given by replacing $h_{kmq}$ and $h_{kml}$ by $h_{k'mq}$ and $h_{k'ml}$, respectively.

Dealing with $e_{Dk'imq}$ given by (24) jointly as a zero-mean complex-valued Gaussian random variable, no information on the $k$-th user $(k = 1, \cdots, K, k \neq k')$ is required to recover $a_{k'im}$. This is called "a singleuser detection," which is applicable to both the uplink and downlink. Defining the rake finger output vector $(J_D \times 1)$ as $\mathbf{y}_{Dk'im} = [y_{Dk'im1}, \cdots, y_{Dk'imJ_D}]^T$, it is written as

$$\mathbf{y}_{Dk'im} = \mathbf{h}_{k'm} a_{k'im} + \mathbf{e}_{Dk'im} \tag{26}$$

where $\mathbf{e}_{Dk'im}$ denotes an interference/noise vector $(J_D \times 1)$, which is defined as

$$\mathbf{e}_{Dk'im} = [e_{Dk'im1}, \cdots, e_{Dk'imJ_D}]^T. \tag{27}$$

Here, if the auto-correlation matrix $(J_D \times J_D)$ of the interference and noise can be approximated as

$$(28) \qquad E\left[\mathbf{e}_{Dk'im}\mathbf{e}_{Dk'im}^{H}\right] = \mathbf{E} \approx \sigma_e^2 \mathbf{I}_{J_D \times J_D}$$

where $\sigma_e^2$ is the power of the interference/noise, as shown in the Appendix A, a maximum ratio combiner is optimal in the sense of maximizing the SNR of the combiner output. That is, when the following weighted sum is considered:

$$(29) \qquad x_{Dk'im} = \mathbf{w}_{Dk'm}^{H}\mathbf{y}_{Dk'im}$$

where $\mathbf{w}_{Dk'm}$ is a weight vector $(J_D \times 1)$ defined as $\mathbf{w}_{Dk'm} = [w_{Dk'm1}, \cdots, w_{Dk'm1J_D}]^T$, selection of the weight vector as

$$(30) \qquad \mathbf{w}_{Dk'm} = \mathbf{h}_{k'm}$$

maximizes the SNR of the combiner output at the $m$-th subcarrier.

By the way, in the uplink, the base station can know information on the parameters for all the users such as $a_{kim}$, $c_{kj}$ and $h_{kml}$ $(k = 1, \cdots, K)$, so it can recover $a_{k'im}$ with the information. This is called "a multiuser detection." For instance, after finishing re-ordering $k$ such that the received power of the $k$-th user's signal is greater or equal to that of the $(k+1)$-th user's signal, a serial interference cancellation (SIC) starts with decision on the data symbol for the first user with the highest reliability, namely, $a_{1im}$. When the SIC tries to decide $a_{k'im}$, it has decided $a_{kim}$ $(k = 1, \cdots, k'-1)$, so it can cancel the multiple access interference (MAI) associated with the $k$-th user from the received signal before the decision.

Now, to discuss the BER lower bound for the singleuser and multiuser detections, let us set $K = 1$, drop the subscript of $k'$ in (26) and take into consideration only the contribution from the Gaussian noise in (27). In this case, the interference/noise vector given by (27) leads to

$$(31) \qquad \mathbf{e}_{Dim} = [n'_{Dm1}, \cdots, n'_{DmJ_D}]^T$$

with the statistical property of

$$(32) \qquad E\left[\mathbf{n}'_{Dm}\mathbf{n}'^{H}_{Dm}\right] = \mathbf{N}_D = \sigma_n{}^2 \mathbf{I}_{J_D \times J_D}.$$

As shown in the Appendix B, a maximum ratio combiner maximizes the SNR of the combiner output at the $m$-th subcarrier. Namely, the SNR and the BER at the $m$-th subcarrier are respectively given by

$$(33) \qquad \mathrm{SNR}_m = \sum_{l=1}^{L_m} \gamma_{ml}$$

$$(34) \qquad BER_{Dm} = \binom{2L_m - 1}{L_m} \prod_{l=1}^{L_m} \frac{1}{4\overline{\gamma}_{ml}}$$

where $\gamma_{ml} = |h_m l|^2/\sigma_n^2$ and $\overline{\gamma}_{sml} = \sigma_{sml}^2/\sigma_n{}^2$.

Eq. (34) shows the BER lower bound when independent data symbols are transmitted over different subcarriers in the MC-DS/CDMA system, which also means the BER lower bound of an SC-DS/CDMA system, replacing $L_m$ and $\overline{\gamma}_{ml}$ by $L$ and $\overline{\gamma}_l = \sigma_{sl}^2/\sigma_n^2$, respectively. Therefore, the BER lower bound of the SC-DS/CDMA system is given by the BER lower bound is given by

$$(35) \qquad BER_S = \binom{2L-1}{L} \prod_{l=1}^{L} \frac{1}{4\overline{\gamma}_l}.$$

Comparing the BERs between the MC-DS/CDMA and SC-DS/CDMA systems given by (34) and (35), respectively, it is clear that the BER of the MC-DS/CDMA system is inferior to that of the SC-DS/CDMA system, because the diversity order of the MC-DS/CDMA system is far less than that of the SC-DS/CDMA system.

The lowest BER in the MC-DS/CDMA system is achievable when the same data symbol is transmitted over different subcarriers, namely, $a_{ki1} = \cdots, a_{kiM} = a_{ki}$. Therefore, taking into account of the relationship between the channel impulse responses between the SC-DS/CDMA and MC-DS/CDMA systems, let us investigate the BER lower bound.

Defining rake finger output, composite impulse response and noise vectors $(J_S \times 1)$ as

$$(36) \qquad \mathbf{y}_{Di} = [y_{Di1}, \cdots, y_{DiJ_S}]^T$$

$$(37) \qquad \mathbf{h}_D = [\mathbf{h}_1^T, \cdots, \mathbf{h}_M^T]^T$$

$$(38) \qquad \mathbf{n}_D = [\mathbf{n}_1'^T, \cdots, \mathbf{n}_M'^T]^T$$

the rake finger output vector $(J_S \times 1)$ is written as

$$(39) \qquad \mathbf{y}_{Di} = \mathbf{h}_D a_i + \mathbf{n}_D$$

where $\mathbf{n}_D$ has the statistical property of

$$(40) \qquad E\left[\mathbf{n}_D \mathbf{n}_D^H\right] = \sigma_n^2 \mathbf{I}_{J_S \times J_S}.$$

Furthermore, defining a weight vectors $(J_S \times 1)$ as $\mathbf{w}_D = [w_{D1}, \cdots, w_{DJ_S}]^T$, the rake combiner output is written as

$$(41) \qquad x_{Di} = \mathbf{w}_D^H \mathbf{y}_{Di}.$$

In this case, a maximum ratio combiner selecting the weight vector as

$$(42) \qquad \mathbf{w}_D = \mathbf{h}_D$$

maximizes the SNR hence minimizes the BER, that is

$$(43) \qquad \mathrm{SNR}_D = \sum_{l=1}^{J_S} \gamma_{Dl}$$

where $\gamma_{Dl} = |h_{Dl}|^2/\sigma_n^2$. **Figure 8** shows the relationship between the channel impulse response of the SC-DS/CDMA system and the $m$-th channel impulse response of the MC-DS/CDMA system (note here the subscript $k$ is dropped). Defining the $J_S$-point Discrete Fourier Transform (DFT) matrix ($J_S \times J_S$) as $\mathbf{U}_{J_S}$, namely,

$$\mathbf{U}_{J_S} = \{u_{\zeta\eta}\} \quad (\zeta, \eta = 1, \cdots, J_S) \tag{44}$$

$$u_{\zeta\eta} = \frac{1}{\sqrt{J_S}} e^{j2\pi \frac{(\zeta-1)(\eta-1)}{J_S}} \tag{45}$$

the frequency response vector ($J_S \times 1$) for the SC-DS/CDMA system is written as

$$\mathbf{f} = \mathbf{U}_{J_S}\mathbf{h}. \tag{46}$$

By dividing the frequency response over the whole bandwidth into $M_D$ blocks, $M_D$ partial frequency response vectors ($J_D \times 1$) $\mathbf{f}_1, \cdots, \mathbf{f}_M$ are obtained, and then by applying the $J_D$-point inverse DFT to the $m$-th partial frequency response vector $\mathbf{f}_m$,

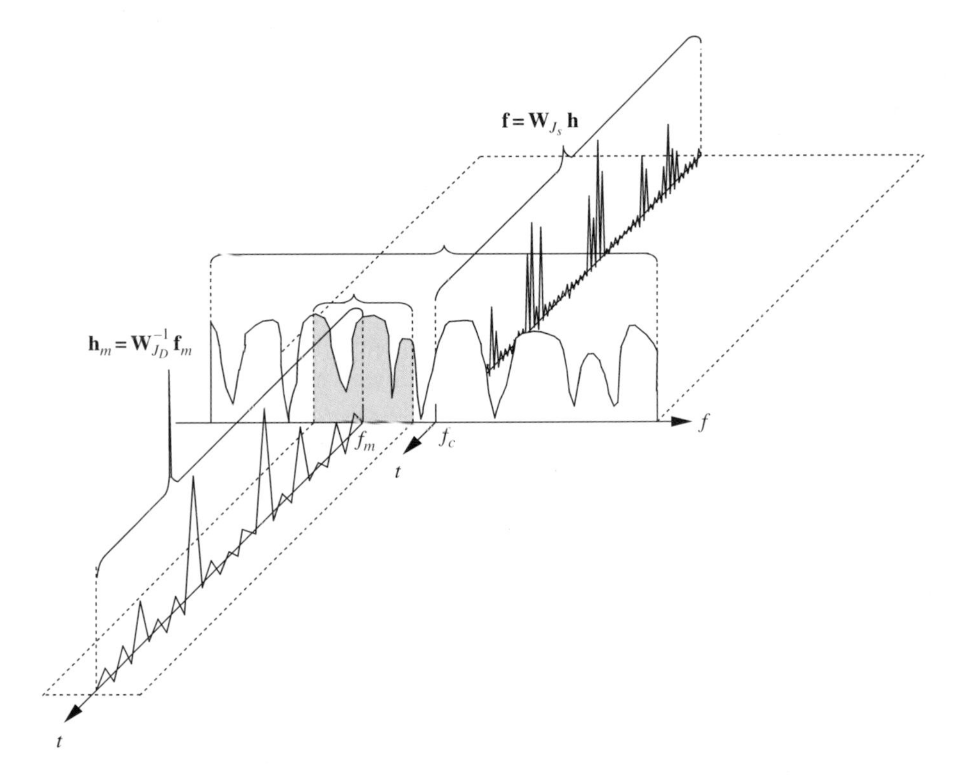


*Figure 8.* Relationship of impulse response between SC- and MC-DS/CDMA systems

the impulse response vector for the $m$-th subcarrier is finally obtained. Therefore, $\mathbf{y}_{Si}$ and $\mathbf{y}_{Di}$ have the following relationship:

$$\mathbf{y}_{Di} = \mathbf{T}_c \mathbf{y}_{Si} \tag{47}$$

$$\mathbf{T}_c = \widetilde{\mathbf{U}}_{J_S}^{-1} \mathbf{U}_{J_S} \tag{48}$$

$$\widetilde{\mathbf{U}}_{J_S}^{-1} = diag\{\underbrace{\mathbf{U}_{J_D}^{-1}, \cdots, \mathbf{U}_{J_D}^{-1}}_{M}\} \tag{49}$$

$$\mathbf{U}_{J_D}^{-1} = \{u'^{-1}_{\zeta'\eta'}\} \quad (\zeta', \eta' = 1, \cdots, J_D) \tag{50}$$

$$u'^{-1}_{\zeta'\eta'} = \frac{1}{\sqrt{J_D}} e^{-j2\pi \frac{(\zeta'-1)(\eta'-1)}{J_D}} \tag{51}$$

where $\widetilde{\mathbf{U}}_{J_S}^{-1}$ $(= \widetilde{\mathbf{U}}_{J_S}^{H})$ and $\mathbf{U}_{J_D}^{-1}$ $(= \mathbf{U}_{J_D}^{H})$ denote the block diagonal matrix composed of $\mathbf{U}_{J_D}^{-1}$ and the $J_D$-point inverse DFT matrix $(J_D \times J_D)$, respectively. It is easy to show that $\mathbf{T}_c$ is a unitary matrix satisfying $\mathbf{T}_c \mathbf{T}_c^H = \mathbf{I}_{J_S \times J_S}$, so as shown in the Appendix C, any unitary transformation cannot change the value of the SNR and thus the BER. Consequently, the achievable BER lower bound of the MC-DS/CDMA system is the same as the BER lower bound of the SC-DS/CDMA system, which is given by

$$BER_D = \binom{2L-1}{L} \prod_{l=1}^{L} \frac{1}{4\overline{\gamma}_l}. \tag{52}$$

## 7. MC-CDMA SYSTEM

### 7.1 Transmitter

MC-CDMA transmitter spreads the input data symbols using a spreading code only in the frequency domain, in other words, a fraction of the data symbol corresponding to a chip of the spreading code is transmitted through a different subcarrier. For the MC-CDMA system, it is essential to have frequency non-selective fading over each subcarrier, so if the original data transmission rate is high enough to become subject to frequency-selective fading, the input data symbol needs to be first serial-to-parallel converted before spreading over the frequency domain.

**Figure 9** shows a block diagram of an MC-CDMA transmitter for the $k$-th user, which is a hybrid of a DS/CDMA spreader and an OFDM modulator. The data sequence is first converted into $P$ parallel sequences, each serial/parallel converter (S/P) output is spread with a spreading code, and then $P$ parallel spread sequences are converted back to a serial data sequence through a parallel/serial converter (P/S). The obtained serial data sequence is fed into the OFDM modulator, namely, it is mapped onto subcarriers through an IDFT, and then a cyclic prefix is inserted in each OFDM symbol to avoid inter-symbol-interference (ISI) caused by multipath fading. Finally, the OFDM symbols are

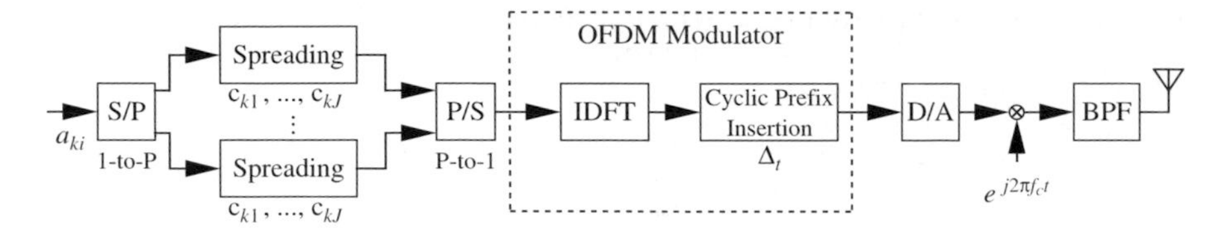


*Figure 9.* A block diagram of an MC-CDMA transmitter for the $k$-th user

transmitted after D/A and up-conversions. The transmitted signal in the uplink is written as

$$s_{Mk}(t) = \sum_{i=-\infty}^{+\infty} \sum_{p=1}^{P} \sum_{j=1}^{J_M} a_{kip} c_{kj} p_M(t - iT_M) \cdot e^{j2\pi(f_c + f_{(j-1)P+p})t} \tag{53}$$

where

$$p_M(t) = \begin{cases} 1 \ (-\Delta_t \le t \le t_M) \\ 0 \ (otherwise) \end{cases} \tag{54}$$

$$f_{(j-1)P+p} = ((j-1)P + p - PJ_M/2)\Delta_f \tag{55}$$

$$t_M = T_M - \Delta_t \tag{56}$$

$$\Delta_f = \frac{1}{t_M} \tag{57}$$

$$T_M = PT_S = PJ_S T_{cS}. \tag{58}$$

In (53)-(58), $p_M(t)$ is a rectangular pulse waveform, $\Delta_t$ and $t_M$ are a cyclic prefix length and a useful symbol length corresponding to the IDFT window width, respectively, $f_{(j-1)P+p}$ and $\Delta_f$ are the center frequency of the $((j-1)P+p)$-th subcarrier and the subcarrier separation, respectively. On the other hand, the transmitted signal in the downlink is written as

$$s_M(t) = \sum_{k=1}^{K} \sum_{i=-\infty}^{+\infty} \sum_{p=1}^{P} \sum_{j=1}^{J_M} a_{kip} c_{kj} p_M(t - iT_M) \cdot e^{j2\pi(f_c + f_{(j-1)P+p})t}. \tag{59}$$

**Figure 4 (c)** shows the PSD of an MC-CDMA waveform. The bandwidth is given by

$$BW_M = \frac{PJ_M - 1}{T_M - \Delta_t} + \frac{2}{T_M} \approx \frac{J_M/J_S}{T_{cS}(1 - \Delta_t/(PJ_S T_{cS}))} \tag{60}$$

$$\approx \frac{(1+\alpha_M)J_M/J_S}{T_{cS}} \tag{61}$$

where $\alpha_M$ is a cyclic prefix factor, which is defined as

$$\alpha_M = \Delta_t/(PJ_S T_{cS}). \tag{62}$$

If we set $J_M = J_S$, the bandwidth results in

$$BW_M = \frac{(1+\alpha_M)}{T_{cS}} \tag{63}$$

so comparison between (8) and (63) reveals that there is no large difference in the bandwidth in terms of the mainlobe among the SC-DS/CDMA, MC-CDMA, and hence MC-DS/CDMA systems. In **Figure 4 (c)**, the hatched subcarriers convey the same data symbol with different chips of spreading code. The frequency separation between neighboring subcarriers is given by $\Delta_f$ whereas that between subcarriers conveying the same data symbol by $P\Delta_f$. This means that the waveform can obtain the maximum frequency diversity effect when it goes through a frequency-selective fading channel.

On the other hand, **Figure 5 (c)** shows a tiling representation of an MC-CDMA waveform in a time-frequency plane, where $J_M = J_S = 8$ and $P = 4$ are assumed. Here, it should be noted that the symbol duration ($T_M$) is widened into $PT_S$ as shown in (58).

## 7.2 Receiver

Assuming that quasi-synchronicity is established in the uplink, the received signal in the uplink is written as

$$r_M(t) = \sum_{k=1}^{K} h_k(t) \otimes s_{Mk}(t) + n_M(t) \tag{64}$$

whereas that in the downlink is written as

$$r_{Mk}(t) = h_k(t) \otimes s_M(t) + n_M(t). \tag{65}$$

First of all, let us discuss the effect of frequency selective fading on the multicarrier transmission with cyclic prefix. The length of the inserted cyclic prefix is sufficiently larger than that of the channel impulse response, so it is written as

$$h_k(t) = \begin{cases} h_k(t) & (0 \le t \le \Delta_t) \\ 0 & (otherwise) \end{cases}. \tag{66}$$

Therefore, with (53), (54), (64) and (66), the output at the $((j'-1)P+p')$-th subcarrier is written as

$$\begin{aligned} y_{Mij'p'} &= \frac{1}{t_M}\int_{iT_M}^{iT_M+t_M} r_M(t) e^{-2\pi f_{(j'-1)P+p'}(t-iT_M)} dt \\ &= \sum_{k=1}^{K} z_{kj'p'} c_{kj'} a_{kip'} + n'_{Mj'p'} \end{aligned} \tag{67}$$

(68) $$z_{kj'p'} = \int_0^{\Delta_t} h_k(t) e^{-j2\pi f_{(j'-1)P+p'}\tau} d\tau$$

(69) $$n'_{Mj'p'} = \int_0^{t_M} n_M(t) e^{-j2\pi f_{(j'-1)P+p'}\tau} d\tau.$$

Eq. (67) clearly shows that frequency-selective fading can be dealt with as a multiplicative noise at a subcarrier level. This is the very benefit of multicarrier transmission, that is, to compensate for frequency-selective fading at receiver, just one tap equalization effectively works at subcarrier level for multicarrier transmission whereas complicated convolutional operation is required for single-carrier transmission.

Next, let us derive several combining methods for the MC-CDMA system. To this end, it is more convenient to express the received signal and channel impulse response in vector forms, replacing the Fourier Transform in (67)-(69) by the DFT. **Figure 10** shows a block diagram of an MC-CDMA receiver for the $k$-th user. Defining the subcarrier output vector ($J_M \times 1$) for a fixed $p = p'$ as

(70) $$\mathbf{y}_{Mip'} = [y_{Mi1p'}, \cdots, y_{MiJ_Mp'}]^T$$

it is written as

(71) $$\mathbf{y}_{Mip'} = \sqrt{J_M}\widetilde{\mathbf{C}}_{Kp'}\mathbf{a}_{ip'} + \mathbf{n}'_{Mp'}$$

where

(72) $$\widetilde{\mathbf{C}}_{Kp'} = [\widetilde{\mathbf{c}}_{p'1}, \cdots, \widetilde{\mathbf{c}}_{p'K}]$$

(73) $$\widetilde{\mathbf{c}}_{p'k} = [z_{k1p'}c_{k1}, \cdots, z_{kJ_Mp'}c_{kJ_M}]^T$$

(74) $$\mathbf{z}_{kp'} = [z_{k1p'}, \cdots, z_{kJ_Mp'}]^T$$

(75) $$\mathbf{a}_{ip'} = [a_{1ip'}, \cdots, a_{Kip'}]^T$$

(76) $$\mathbf{n}'_{Mp'} = [n'_{M1p'}, \cdots, n'_{MJ_Mp'}]^T.$$

In (72)-(76), $\widetilde{\mathbf{c}}_{p'k}$, $\mathbf{z}_{kp'}$, $\mathbf{a}_{ip'}$ and $\mathbf{n}'_{Mp'}$ are a distorted code vector ($J_M \times 1$), a frequency response vector ($J_M \times 1$), an information vector ($K \times 1$) and a noise vector ($J_M \times 1$),

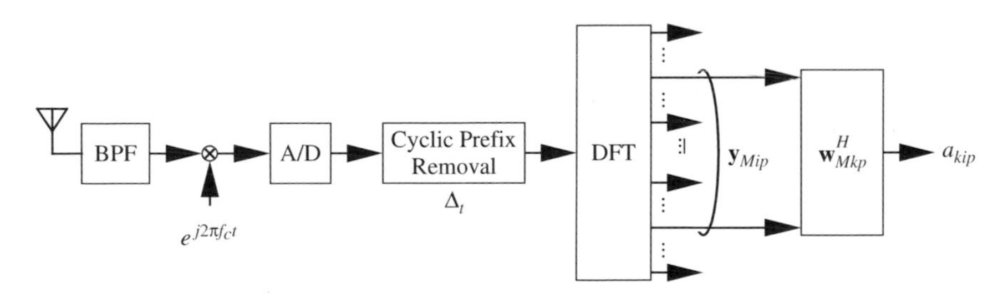


*Figure 10.* A block diagram of an MC-CDMA receiver for the $k$-th user

respectively, and $\widetilde{\mathbf{C}}_{Kp'}$ is a distorted code matrix $(J_M \times K)$. Furthermore, defining an impulse response vector $(PJ_M \times 1)$, a frequency response vector $(PJ_M \times 1)$ and noise vectors $(PJ_M \times 1)$ before/after the DFT over all the subcarriers as

$$\mathbf{h}'_k = [h_k(t=0), \cdots, h_k(t=(PJ_M-1)t_M)]^T \tag{77}$$

$$\mathbf{z}_k = [z_{k1}, \cdots, z_{k(PJ_M)}]^T \tag{78}$$

$$\begin{aligned}\mathbf{n}_M = [&n_M(t=iT_M), \cdots, \\ &n_M(t=iT_M+(PJ_M-1)t_M)]^T\end{aligned} \tag{79}$$

$$\mathbf{n}'_M = [n'_{M1}, \cdots, n'_{M(PJ_M)}]^T \tag{80}$$

they satisfy the following properties:

$$\mathbf{z}_k = \mathbf{U}_{PJ_M}\mathbf{h}'_k \tag{81}$$

$$\mathbf{n}'_M = \mathbf{U}_{PJ_M}\mathbf{n}_M \tag{82}$$

$$E\left[\mathbf{n}'_M\mathbf{n}'^H_M\right] = \sigma_n^2\mathbf{I}_{PJ_M\times PJ_M} \tag{83}$$

where $\mathbf{U}_{PJ_M}$ denotes the $PJ_M$-point DFT matrix $(PJ_M \times PJ_M)$. Paying attention to the fact that $\mathbf{z}_{kp'}$ and $\mathbf{n}'_{Mp'}$ are composed of picking up the $((j-1)P+p')$-th elements $(j = 1, \cdots, J_M)$ from $\mathbf{z}_k$ and $\mathbf{n}'_M$, respectively, it is clear that $\mathbf{n}'_{Mp'}$ satisfies the following properties

$$E\left[\mathbf{n}'_{Mp'}\mathbf{n}'^H_{Mp'}\right] = \sigma_n^2\mathbf{I}_{J_M\times J_M}. \tag{84}$$

Now, with (70), define the following weighted sum:

$$x_{Mik'p'} = \mathbf{w}^H_{Mk'p'}\mathbf{y}_{Mip'} \tag{85}$$

where $\mathbf{w}_{Mk'p'}$ is a weight vector $(J_M \times 1)$, which is defined as

$$\mathbf{w}_{Mk'p'} = \left[w_{Mk'p'1}, \cdots, w_{Mk'p'J_M}\right]^T. \tag{86}$$

In the uplink, Eq.(73) clearly shows that the code orthogonality among the users is totally distorted by the multiplicative noise, namely, the frequency response, so a multiuser detection is required to recover $a_{k'ip'}$ in (71). One method may be to despread the subcarrier output vector given by (70) as

$$\widetilde{\mathbf{C}}^H_{Kp'}\mathbf{y}_{Mip'} = \sqrt{J_M}\widetilde{\mathbf{C}}^H_{Kp'}\widetilde{\mathbf{C}}_{Kp'}\mathbf{a}_{ip'} + \widetilde{\mathbf{C}}^H_{Kp'}\mathbf{n}'_{Mp'} \tag{87}$$

however, $\widetilde{\mathbf{C}}^H_{Kp'}\widetilde{\mathbf{C}}_{Kp'}$ in (87) cannot be the identity matrix, because the orthogonality among the spreading codes is totally distorted. A decorrelating multiuser detection eliminates the cross-correlation among the distorted spreading codes by multiplying (87) with $(\widetilde{\mathbf{C}}_{Kp'}^H \widetilde{\mathbf{C}}_{Kp'})^{-1}$ as

$$(\widetilde{\mathbf{C}}_{Kp'}^H \widetilde{\mathbf{C}}_{Kp'})^{-1} \widetilde{\mathbf{C}}_{Kp'}^H \mathbf{y}_{Mip'} = \sqrt{J_M} \mathbf{a}_{ip'} + (\widetilde{\mathbf{C}}_{Kp'}^H \widetilde{\mathbf{C}}_{Kp'})^{-1} \widetilde{\mathbf{C}}_{Kp'}^H \mathbf{n}'_{Mp'}. \tag{88}$$

This means that the decorrelating weight vector is written as

$$\mathbf{w}_{Mk'p'}^{U,DEC} = \sum_{j=1}^{J_M} \left[ (\widetilde{\mathbf{C}}_{Kp'}^H \widetilde{\mathbf{C}}_{Kp'})^{-1} \right]_{j,j} \widetilde{\mathbf{c}}_{p'k'} \tag{89}$$

where $[\bullet]_{j,j}$ denotes the $(j, j)$-element of matrix $\bullet$. In (88), the noise vector is multiplied with a matrix $(\widetilde{\mathbf{C}}_{Kp'}^H \widetilde{\mathbf{C}}_{Kp'})^{-1} \widetilde{\mathbf{C}}_{Kp'}^H$. This is called "noise enhancement" resulting in degradation of the BER, because low-level subcarriers tend to be multiplied with high gains and hence the noise components are amplified at weaker subcarriers.

Another method is to solve the following minimization problem of the mean square error (MSE) for the $k'$-th user at the base station:

$$\text{miminize } MSE(\mathbf{w}_{Mk'p'}) = E_{\overline{\mathbf{Z}}_{kp'}} \left[ |a_{k'ip'} - \mathbf{w}_{Mk'p'}^H \mathbf{y}_{Mip'}|^2 \right] \tag{90}$$

where $E_{\overline{\mathbf{Z}}_{kp'}}[\cdot]$ denotes statistical average with $\mathbf{z}_{kp'}$ fixed. As the Wiener solution of (90), taking into consideration of the properties given by (2) and (84), the minimum mean square error (MMSE) weight vector is given by

$$\mathbf{w}_{Mk'p'}^{U,MMSE} = \mathbf{R}_{Mk'p'}^{-1} \mathbf{o}_{Mk'p'} \tag{91}$$

$$\mathbf{R}_{Mk'p'} = E_{\overline{\mathbf{Z}}_{kp'}} \left[ \mathbf{y}_{Mip'} \mathbf{y}_{Mip'}^H \right]$$

$$= J_M \widetilde{\mathbf{C}}_{Kp'} \widetilde{\mathbf{C}}_{Kp'}^H + \sigma_n^2 \mathbf{I}_{J_M \times J_M} \tag{92}$$

$$\mathbf{o}_{Mk'p'} = E_{\overline{\mathbf{Z}}_{kp'}} \left[ \mathbf{y}_{Mip'} a_{k'ip'}^H \right] = \sqrt{J_M} \widetilde{\mathbf{c}}_{p'k'} \tag{93}$$

where $\mathbf{R}_{Mk'p'}$ and $\mathbf{o}_{Mk'p'}$ denote the auto-correlation matrix $(J_M \times J_M)$ of the subcarrier outputs and the desired response vector $(J_M \times 1)$ for the $k'$-th user, respectively. This method is called "the MMSE multiuser detection".

On the other hand, in the downlink, setting $z_{kjp'} = z_{k'jp'}$ $(j = 1, \cdots, J_M)$ in (73), (71) changes to the subcarrier output vector of the $k'$-th user as

$$\mathbf{y}_{Mk'ip'} = \sqrt{J_M} \mathbf{Z}_{k'p'} \mathbf{C}_K \mathbf{a}_{ip'} + \mathbf{n}'_{Mp'} \tag{94}$$

$$\mathbf{Z}_{k'p'} = diag\{z_{k'1p'}, \cdots, z_{k'J_M p'}\}. \tag{95}$$

Now, we introduce four combining methods, all of which are categorized into singleuser detection. The first method is obtained by solving the following minimization problem of the MSE for the $k'$-th user:

$$\text{miminize } MSE(\mathbf{w}_{Mk'p'}) = E_{\overline{\mathbf{Z}}_{k'p'}} \left[ |a_{k'ip'} - \mathbf{w}_{Mk'p'}^H \mathbf{y}_{Mk'ip'}|^2 \right]. \tag{96}$$

Like the MMSE weight vector in the uplink, as the Wiener solution of (96), the MMSE weight vector in the downlink is given by

$$\mathbf{w}_{Mk'p'}^{D,MMSE} = \mathbf{R}_{Mk'p'}^{-1}\mathbf{o}_{Mk'p'} \tag{97}$$

$$\mathbf{R}_{Mk'p'} = E_{\overline{\mathbf{Z}}_{k'p'}}\left[\mathbf{y}_{Mk'ip'}\mathbf{y}_{Mk'ip'}^{H}\right] \tag{98}$$

$$\mathbf{o}_{Mk'p'} = E_{\overline{\mathbf{Z}}_{k'p'}}\left[\mathbf{y}_{Mk'ip'}a_{k'ip'}^{H}\right]. \tag{99}$$

From (2), (3), (84), (94) and (95), (98) and (99) respectively result in

$$\begin{aligned}\mathbf{R}_{Mk'p'} &= E_{\overline{\mathbf{Z}}_{k'p'}}\left[(\mathbf{Z}_{k'p'}\mathbf{C}_K\mathbf{a}_{ip'} + \mathbf{n}'_{Mp'})(\mathbf{Z}_{k'p'}\mathbf{C}_K\mathbf{a}_{ip'} + \mathbf{n}'_{Mp'})^H\right]\\ &= \mathbf{Z}_{k'p'}E_{\overline{\mathbf{Z}}_{k'p'}}\left[\mathbf{C}_K\mathbf{a}_{ip'}\mathbf{a}_{ip'}^H\mathbf{C}_K^H\right]\mathbf{Z}_{k'p'}^H + \sigma_n^2\mathbf{I}_{J_M\times J_M}\\ &= diag\{K|z_{j'1p'}|^2+\sigma_n^2,\cdots,K|z_{j'J_Mp'}|^2+\sigma_n^2\}\end{aligned} \tag{100}$$

$$\mathbf{o}_{Mk'p'} = \sqrt{J_M}\widetilde{\mathbf{c}}_{p'k'} \tag{101}$$

so the MMSE weight vector is finally given by

$$\mathbf{w}_{Mk'p'}^{D,MMSE} = \left[\frac{z_{k'1p'}c_{k'1}}{K|z_{k'1p'}|^2+\sigma_n^2},\cdots,\frac{z_{k'J_Mp'}c_{k'J_M}}{K|z_{k'J_Mp'}|^2+\sigma_n^2}\right]^T. \tag{102}$$

The method to recover the transmitted data symbols with (102) is called “the minimum mean square error combining (MMSEC)”.

The denominator of (102) contains the noise power, so it means that calculation of the MMSE weights at users requires its estimation. To avoid estimation of the noise power, the second method ignores the noise term in the denominator, namely, $|z_{k'jp'}|^2 >> \sigma_n^2$ and furthermore ignores a common coefficient to all the elements of (102), that is

$$\mathbf{w}_{Mk'p'}^{D,ORC} = \left[\frac{z_{k'1p'}c_{k'1}}{|z_{k'1p'}|^2},\cdots,\frac{z_{k'J_Mp'}c_{k'J_M}}{|z_{k'J_Mp'}|^2}\right]^T. \tag{103}$$

The method to recover the transmitted data symbols with (103) is called “the orthogonality restoring combining (ORC)”. Indeed, this method can eliminate multiple access interference perfectly as

$$\begin{aligned}\widehat{a}_{k'ip'} &= \mathbf{w}_{Mk'p'}^{D,ORC\,H}\mathbf{y}_{Mip'}\\ &= \sqrt{J_M}a_{k'ip'} + \mathbf{w}_{Mk'p'}^{D,ORC\,H}\mathbf{n}'_{Mp'}\end{aligned} \tag{104}$$

but the noise enhancement degrades the BER, from the same reason in the decorrelating multiuser detection in the uplink.

On the contrary, the third method picks up only the noise term in the denominator, namely, $|z_{k'jp'}|^2 << \sigma_n^2$ and also ignores a common coefficient to all the elements of (102), that is

$$\mathbf{w}_{Mk'p'}^{D,MRC} = [z_{k'1p'}c_{k'1}, \cdots, z_{k'J_Mp'}c_{k'J_M}]^T. \tag{105}$$

The method to recover the transmitted data symbols with (105) is called "the maximum ratio combining (MRC)". This method corresponds to the maximum ratio combining in normal diversity techniques, so it can minimize the BER only for the case of a single user.

Analogous to normal diversity techniques, the fourth method can be considered, which is located at the middle of the ORC and MRC, that is

$$\mathbf{w}_{Mk'p'}^{D,EGC} = \left[\frac{z_{k'1p'}c_{k'1}}{|z_{k'1p'}|}, \cdots, \frac{z_{k'J_Mp'}c_{k'J_M}}{|z_{k'J_Mp'}|}\right]^T. \tag{106}$$

The method to recover the transmitted data symbols with (106) is called "the equal gain combining (EGC)". Indeed, the magnitudes of the weights are kept to be the same, so this method compensates only for the phases of the chips of spreading code distorted by the frequency response of the channel.

We have introduced the two multiuser detection methods for the uplink whereas the four combining methods for the downlink. So now, to discuss the BER lower bound, let us set $P = 1$ and drop the subscripts of $k$ and $p'$ in (94). In this case, (94) leads to

$$\mathbf{y}_{Mi} = \sqrt{J_M}\mathbf{Z}\mathbf{c}a_i + \mathbf{n}'_M. \tag{107}$$

Defining a code matrix ($J_M \times J_M$) as

$$\mathbf{C} = diag\{c_1, \cdots, c_{J_M}\}. \tag{108}$$

Pre-multiplying (107) with the code matrix results in

$$\begin{aligned}
\mathbf{y}'_{Mi} &= \mathbf{C}\mathbf{y}_{Mi} \\
&= \sqrt{J_M}\mathbf{C}\mathbf{Z}\mathbf{c}a_i + \mathbf{C}\mathbf{n}'_M \\
&= \mathbf{U}_{J_M}((\mathbf{h}'/\sqrt{J_M})a_i + \mathbf{n}''_M) \qquad (109)\\
\mathbf{n}''_M &= \mathbf{U}_{J_M}^{-1}\mathbf{C}\mathbf{n}'_M. \qquad (110)
\end{aligned}$$

Here, taking into consideration of

$$\mathbf{C}\mathbf{C}^H = \frac{1}{J_M}\mathbf{I}_{J_M \times J_M} \tag{111}$$

and (83), it is easy to show that $\mathbf{n}''_M$ has the following property:

$$E\left[\mathbf{n}''_M \mathbf{n}''^H_M\right] = E\left[\mathbf{U}_{J_M}^{-1}\mathbf{C}\mathbf{n}'_M \mathbf{U}_{J_M}^{-1}\mathbf{C}\mathbf{n}'^H_M\right]$$

$$= \frac{\sigma_n^2}{J_M}\mathbf{I}_{J_M \times J_M}. \tag{112}$$

Therefore, as shown in the Appendix C, the unitary matrix $\mathbf{U}_{J_M}$ cannot change the value of the SNR and thus the BER is given by

$$BER_M = \binom{2L-1}{L}\prod_{l=1}^{L}\frac{1}{4\overline{\gamma}_l}. \tag{113}$$

## 8. BER PERFORMANCE COMPARISON

From (35), (52) and (113), the BER lower bounds of the SC-DS/CDMA, MC-DS/CDMA and MC-CDMA systems are given by the same equation. That is, once a channel with a bandwidth and a data transmission rate are given, the BER lower bound is uniquely determined by the degree of freedom of the channel. We have assumed a slowly varying fading channel, so in this case, the BER lower bound is determined by the freedom in the frequency response of the channel, in other words, the frequency-selectivity of the channel.

Now, for the MC-DS/CDMA and MC-CDMA systems, let us investigate their BER performances with the multiuser detection schemes in the uplink and the singleuser detection schemes in the downlink. This investigation needs to resort to computer simulation.

As a modulation/detection format, a binary phase shift keying (BPSK)/coherent detection is assumed, and as spreading codes, Gold codes with length of 31 and Walsh-Hadamard codes with length of 32, which approximately satisfy the properties of (4) and (3), are assumed. In addition, the number of subcarriers ($M_D$) in the MC-DS/CDMA system is set to 2 and that of parallelizations ($P$) in the MC-CDMA system is set to 16. This implies that the number of subcarriers for the MC-CDMA system is 512. On the other hand, in a multipath delay profile employed, there are 18 exponentially decaying paths, where the path-by-path decay factor is 1.2 dB. The MC-CDMA system has a 10 % cyclic prefix length, which is assumed to be much larger than the delay spread of the multipath delay profile. In the following figures, the Average SNR is defined as $(\sum_{l=1}^{18}\sigma_{sl}^2)/\sigma_n^2$ and it is set to 10 dB.

**Figures 11 (a)** and **(b)** show the BERs versus the number of users for the MC-DS/CDMA and MC-CDMA systems in the uplink, respectively. An SIC multiuser detection is assumed for the MC-DS/CDMA system whereas a decorrelating multiuser detection and an MMSE multiuser detection are assumed for the MC-CDMA system. For the MC-DS/CDMA system, even if the number of paths is 18 when the impulse response of the channel is observed in the entire bandwidth, the number of paths becomes larger in the impulse response of each subchannel.

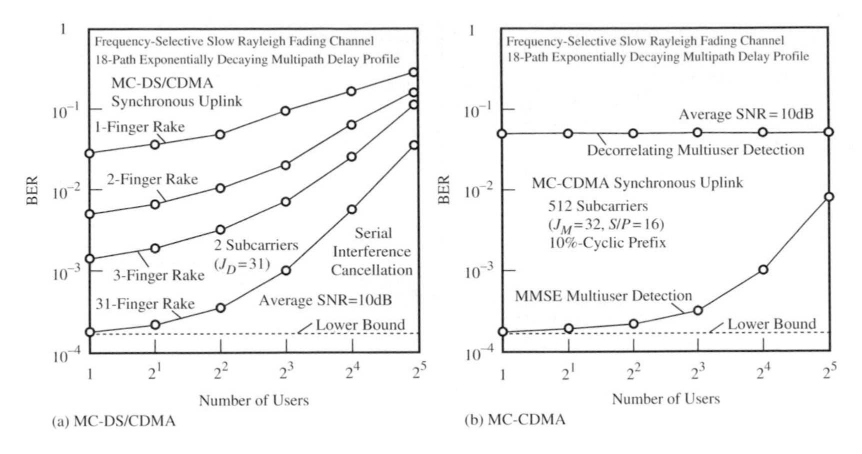


*Figure 11.* BER comparison in uplink: MC-DS/CDMA (a) and MC-CDMA (b)

Therefore, to achieve the lowest BER for the case of a single user, 31-finger rake processors are required in principle. The 1-, 2- and 3-finger rake processors mean picking up to the first, second and third largest paths in instantaneous channel impulse response, respectively. As the number of rake finger processors decreases, the BER performance is degraded because it losses more power of the received signal. For the MC-CDMA systems, on the other hand, the BER of the decorrelating multiuser detection is flat but poor. This is because it can perfectly eliminate the MAI but the noise enhancement introduces a lot of errors in subcarriers. The MMSE multiuser detection performs much better, but it needs to estimate the noise power correctly and also to invert the correlation matrix of the received signal.

**Figures 12 (a)** and **(b)** show the BERs versus the number of users for the MC-DS/CDMA and MC-CDMA systems in the downlink, respectively. For the MC-DS/CDMA system, the number of rake processors decreases, the BER is degraded because of the same reason for the uplink. As compared with the BER curves in the uplink, those in the downlink is more abruptly degraded as the number of users increases. For the MC-CDMA system, on the other hand, the BER of the ORC is flat and poor because of noise enhancement. The MRC is optimal only for the case of a single user, so it performs best only for the case, and as the number of users increases, the BER is abruptly degraded due to MAI. The MMSEC performs best except for the case of a single user but it needs to estimate the noise power. Among the combining methods which requires no estimation of the noise power, the EGC performs best except for the case of a single user.

These four figures may give us an impression that the MC-CDMA systems performs better than the MC-DS/CDMA system almost all region of the number of users. However, it should be noted that these figures are depicted keeping the average *received* SNR to be the same. Taking into consideration that the energy of the received signal in the cyclic prefix is not used for signal detection for the MC-CDMA system, to achieve the same BER performance among the

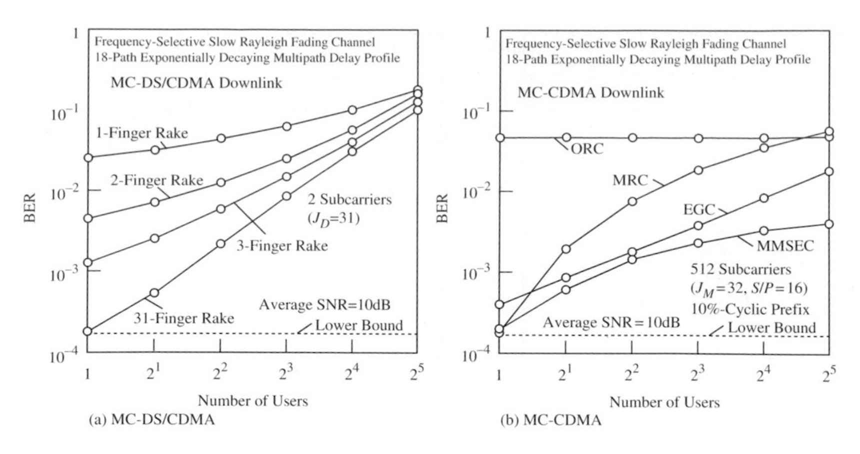


*Figure 12.* BER comparison in downlink: MC-DS/CDMA (a) and MC-CDMA (b)

MC-DS/CDMA and MC-CDMA systems,10 % more energy is required for the MC-CDMA transmitter, in other words, keeping the transmission power to be the same for the MC-DS/CDMA and MC-CDMA transmitters, the MC-CDMA receiver loses 10 % of the received signal power, so all the BER curves of the MC-CDMA system are shifted up a bit.

## 9. CONCLUSIONS

This chapter has presented the principle of Multicarrier based CDMA techniques, such as MC-DS/CDMA and MC-CDMA, and has introduced several detection and combining schemes for the two techniques. Their BER lower bounds have been theoretically analyzed and it has been shown that the BER lower bound is independent of signaling techniques we intended to use and is uniquely determined by the degree of freedom of the channel once a channel is given. Therefore, which signaling method we should use is dependent of hardware complexities of base station and user terminal. When we intend to support a data transmission rate in a channel where we see several tens of multipath components in its impulse response, the combination of multicarrier transmission and CDMA becomes only a solution to achieve a good BER while reducing the hardware complexity.

## APPENDICES

### A: Maximum Ratio Combiner

Assume that a transmitted signal carrying the same information goes through a $J$ diversity channels and arrives at a receiver. Here, the $l$-th channel ($l = 1, \cdots, L$) is assumed to be frequency-nonselective and slowly fading with a zero-mean complex-valued Gaussian-distributed amplitude whereas the signal in the $l$-th channel

$(l = L+1, \cdots, J)$ experiences zero-gain. In addition, the signal in each channel is perturbed by an additive zero-mean complex-valued Gaussian noise, and the fading and noise processes are assumed to be mutually statistically independent, respectively, and also mutually statistically independent each other. The receiver, on the other hand, multiplies the received signal in each channel with an adequate weight and finally combines all the multiplier outputs. The combiner output is written as

$$r = \sum_{l=1}^{J} w_l^* (h_l s + n_l) = \mathbf{w}^H (\mathbf{h}s + \mathbf{n}) \tag{114}$$

where $\mathbf{h}$, $\mathbf{n}$ and $\mathbf{w}$ are the channel gain, noise and weight vectors $(J \times 1)$, respectively, which are given by

$$\mathbf{h} = [h_1, h_2, \cdots, h_L, 0, \cdots, 0]^T \tag{115}$$

$$\mathbf{n} = [n_1, n_2, \cdots, n_J]^T \tag{116}$$

$$\mathbf{w} = [w_1, w_2, \cdots, w_J]^T. \tag{117}$$

The following properties are defined for $\mathbf{h}$, $\mathbf{n}$ and $s$:

$$\mathbf{h}\mathbf{h}^H = \mathbf{H}_{\overline{\mathbf{h}}} \tag{118}$$

$$E\left[\mathbf{h}\mathbf{h}^H\right] = \mathbf{H} = diag\{\sigma_{s1}^2, \sigma_{s2}^2, \cdots, \sigma_{sL}^2, 0, \cdots, 0\} \tag{119}$$

$$E\left[\mathbf{n}\mathbf{n}^H\right] = \mathbf{N} = \sigma_n^2 \mathbf{I}_{J \times J} \tag{120}$$

$$E\left[\mathbf{h}\mathbf{n}^H\right] = \mathbf{0}_{J \times J} \tag{121}$$

$$E\left[|s|^2\right] = 1 \tag{122}$$

where $\mathbf{H}$ and $\mathbf{N}$ are called "the correlation matrix $(J \times J)$ of the diversity channels" with non-zero eigenvalues of $\sigma_{s1}^2, \sigma_{s2}^2, \cdots, \sigma_{sL}^2$" and "the correlation matrix $(J \times J)$ of the noise," respectively, $\mathbf{I}_{J \times J}$ and $\mathbf{0}_{J \times J}$ denote identity and zero matrices $(J \times J)$, respectively, and $\sigma_n^2$ denotes the power of noise. When $\mathbf{h}$ is fixed, the power of the combiner output is written as

$$\begin{aligned} E_{\overline{\mathbf{h}}}\left[|r|^2\right] &= E_{\overline{\mathbf{h}}}\left[(\mathbf{w}^H\mathbf{h}s + \mathbf{w}^H\mathbf{n})(\mathbf{w}^H\mathbf{h}s + \mathbf{w}^H\mathbf{n})^H\right] \\ &= \mathbf{w}^H \mathbf{H}_{\overline{\mathbf{h}}} \mathbf{w} + \sigma_n^2 \mathbf{w}^H \mathbf{w} \end{aligned} \tag{123}$$

where $E_{\overline{\mathbf{h}}}[\cdot]$ denotes statistical average with $\mathbf{h}$ fixed. In (123), the first and second terms mean the powers of signal and noise, respectively, so the signal to noise power ratio (SNR) of the combiner output is written as

$$\gamma = \frac{S}{N} = \frac{\mathbf{w}^H \mathbf{H}_{\overline{\mathbf{h}}} \mathbf{w}}{\sigma_n^2 \mathbf{w}^H \mathbf{w}}. \tag{124}$$

A solution of $\partial\gamma/\partial\mathbf{w}^* = 0$ leads to a weight vector which maximizes the SNR, where $(\cdot)^*$ denotes complex conjugate, but here we take another approach to reach the optimum weight vector.

Let us consider the following eigenvalue problem:

$$\text{(125)} \qquad \mathbf{H}_{\overline{\mathbf{h}}}\mathbf{w} = \lambda\mathbf{w}$$

where $\lambda$ denotes "an eigenvalue of $\mathbf{H}_{\overline{\mathbf{h}}}$." Defining $\lambda_{max}$ and $\mathbf{w}_{max}$ as the maximum eigenvalue and the eigenvector associated with it, respetively, of course, they satisfy

$$\text{(126)} \qquad \mathbf{H}_{\overline{\mathbf{h}}}\mathbf{w}_{max} = \lambda_{max}\mathbf{w}_{max}.$$

Pre-multiplying (126) by $\mathbf{w}_{max}$ leads to

$$\text{(127)} \qquad \lambda_{max} = \frac{\mathbf{w}_{max}^H\mathbf{H}_{\overline{\mathbf{h}}}\mathbf{w}_{max}}{\mathbf{w}_{max}^H\mathbf{w}_{max}}.$$

Consequently, (124) and (127) lead to

$$\text{(128)} \qquad \gamma_{max} = \frac{\lambda_{max}}{\sigma_n^2}.$$

Eq. (128) clearly shows that, *if we select the weight vector as the eigenvector corresponding to the maximum eigenvalue of* $\mathbf{H}_{\overline{\mathbf{h}}}$, *it maximizes the SNR of the combiner output.*

The rank of $\mathbf{H}_{\overline{\mathbf{h}}}$ is 1 because it is defined as $\mathbf{hh}^H$, so the optimum weight vector is $\mathbf{w}_{max} = \mathbf{h}$. In fact, substituting the solution into (125) leads to

$$\text{(129)} \qquad \mathbf{H}_{\overline{\mathbf{h}}}\mathbf{w}_{max} = \mathbf{H}_{\overline{\mathbf{h}}}\mathbf{h} = \mathbf{hh}^H\ \mathbf{h} = |\mathbf{h}|^2\mathbf{w}_{max}$$

therefore, we have

$$\text{(130)} \qquad \lambda_{max} = |\mathbf{h}|^2 = \sum_{l=1}^{L} |h_l|^2.$$

Finally, the maximized SNR of the combiner output is given by

$$\text{(131)} \qquad \gamma_{max} == \frac{\mathbf{h}}{\sigma_n^2} = \sum_{l=1}^{L} \gamma_l$$

$$\text{(132)} \qquad \gamma_l = \frac{|h_l|^2}{\sigma_n^2}.$$

Eqs. (131) and (132) show that, the maximum SNR of the combiner output is the sum of the SNRs of the diversity branches. The combiner with selection of $\mathbf{w}_{max} = \mathbf{h}$ is called "the Maximum Ratio Combiner (MRC)." In this case, the received waves from $L+1$ to $J$ are not used for data demodulation. Therefore, we call this "a diversity system with order of $L$."

**B: Bit Error Rate Expression**

For binary phase shift keying (BPSK)/ coherent detection, it is well known that the bit error rate (BER) is given by

$$BER(\gamma_{max}) = \frac{1}{2} erfc(\sqrt{\gamma_{max}}) \tag{133}$$

where $erfc(x)$ is the complementary error function defined as

$$erfc(x) = \frac{2}{\sqrt{\pi}} \int_{x}^{+\infty} e^{t^2} dt. \tag{134}$$

In the Appendix A, $\gamma_{max}$ in (133) has been derived with $\mathbf{h}$ fixed, so to calculate the average BER for the diversity system with order of $L$ in the fading channel, we average (133) in terms of $\mathbf{h}$, namely, $\gamma_{max}$, because $\gamma_{max}$ is a function of $\mathbf{h}$

$$BER_{fading}^{div\ L} = E_{\mathbf{h}}[BER(\gamma_{max})] = E_{\gamma_{max}}[BER(\gamma_{max})] \tag{135}$$

where $E_{\mathbf{h}}[\cdot]$ and $E_{\gamma_{max}}[\cdot]$ denote statistical averages in terms of $\mathbf{h}$ and $\gamma_{max}$, respectively. If we know the probability density function (*pdf*) of $\gamma_{max}$ as $p(\gamma_{max})$, then we can rewrite (135) as

$$BER_{fading}^{div\ L} = \int_{0}^{+\infty} BER(\gamma_{max}) p(\gamma_{max}) d\gamma_{max}. \tag{136}$$

The amplitude $h_l$ is complex valued-Gaussian-distributed with zero-mean, so $\gamma_l$ given by (132) is exponentially distributed. The *pdf* of $\gamma_l$ and the characteristic function defined as Laplace Transform of the *pdf* are respectively written as

$$p(\gamma_l) = \frac{1}{\overline{\gamma}_l} e^{-\frac{\gamma_l}{\overline{\gamma}_l}} \tag{137}$$

$$\overline{\gamma}_l = \sigma_{sl}^2 / \sigma_n^2 \tag{138}$$

$$\phi(s)_{\gamma_l} = \int_{0}^{+\infty} e^{-s\gamma_l} p(\gamma_l) d\gamma_l = \frac{1/\overline{\gamma}_l}{s + 1/\overline{\gamma}_l} \tag{139}$$

where $\overline{\gamma}_l$ denotes the average SNR. The characteristic function on the sum of independent variables is given by the product of the characteristic function on each variable, so it is written as

$$\phi(s)_{\gamma_{max}} = \prod_{l=1}^{L} \frac{1/\overline{\gamma}_l}{s + 1/\overline{\gamma}_l}. \tag{140}$$

If $\overline{\gamma}_l$ is different each other, by taking the inverse Laplace Transform of (140), the *pdf* of $\gamma_{max}$ becomes

$$p(\gamma_{max}) = \lim_{\beta\to\infty} \frac{1}{2\pi j}\int_{\alpha-j\beta}^{\alpha+j\beta} e^{s\gamma_{max}}\phi(s)_{\gamma_{max}}ds = \sum_{l=1}^{L} Res[\phi(s)_{\gamma_{max}}, \overline{\gamma}_l]$$

$$(141) \qquad = \sum_{l=1}^{L}\prod_{\substack{l'=1\\ l'\neq l}}^{L} \frac{1}{1-\overline{\gamma}_{l'}/\overline{\gamma}_l}\cdot\frac{1}{\overline{\gamma}_l}e^{-\frac{\gamma_{max}}{\overline{\gamma}_l}}$$

where $Res[f(x), y]$ denotes the residue of $f(x)$ at $x = y$. Therefore, substituting (133), (134) and (141) into (136) results in

$$(142) \qquad BER_{fading}^{div\ L} = \sum_{l=1}^{L}\prod_{\substack{l'=1\\ l'\neq l}}^{L} \frac{1}{1-\overline{\gamma}_{l'}/\overline{\gamma}_l}\cdot\frac{1}{2}\left(1-\sqrt{\frac{\overline{\gamma}_l}{1+\overline{\gamma}_l}}\right).$$

Furthermore, taking a Taylor series expansion up to the $L$-th derivative for (142) leads to

$$(143) \qquad BER_{fading}^{div\ L} \approx \binom{2L-1}{L}\prod_{l=1}^{L}\frac{1}{4\overline{\gamma}_l}.$$

Note that (143) is also valid when $\overline{\gamma}_l$ is identical. Taking into consideration of $\overline{\gamma}_l = \sigma_{sl}^2/\sigma_n^2$, it is concluded that the BER is uniquely determined by the number and magnitude of eigenvalues for the channel, namely, the degree of freedom of the channel of interest.

### C: Equivalence Through Linear Transformation

Now, assume that a receiver once linearly transforms the signals through $J$ diversity channels and then combines all the multiplier outputs. In this case, the output of the combiner is written as

$$(144) \qquad r' = \mathbf{v}^H\mathbf{F}(\mathbf{h}s+\mathbf{n}) = \mathbf{v}^H\mathbf{g}s + \mathbf{v}^H\mathbf{F}\mathbf{n}$$

$$(145) \qquad \mathbf{g} = \mathbf{F}\mathbf{h}$$

where $\mathbf{F}$ is any unitary matrix $(J \times J)$ representing the linear transformation as

$$(146) \qquad \mathbf{F}\mathbf{F}^H = \mathbf{I}_{L'\times L'}$$

and $\mathbf{v}$ is a weight vector $(J \times 1)$, which is given by

$$(147) \qquad \mathbf{v} = [v_1, v_2, \cdots, v_J]^T.$$

Similar to the discussion in the previous section, defining $\mathbf{G}_{\bar{\mathbf{h}}}$ as $\mathbf{g}\mathbf{g}^H = (\mathbf{F}\mathbf{h})(\mathbf{F}\mathbf{h})^H$, the SNR of the combiner output is written as

$$\gamma' = \frac{\mathbf{v}^H \mathbf{G}_{\bar{\mathbf{h}}} \mathbf{v}}{\sigma_n^2 \mathbf{v}^H \mathbf{v}} = \frac{(\mathbf{F}^H \mathbf{v})^H (\mathbf{h}\mathbf{h}^H)(\mathbf{F}^H \mathbf{v})}{(\mathbf{F}^H \mathbf{v})^H (\mathbf{F}^H \mathbf{v})} \tag{148}$$

therefore, selecting the weight vector as

$$\mathbf{v}_{max} = \mathbf{F}\mathbf{h} \tag{149}$$

the maximun SNR is obtained, but the value is all the same as the one before the linear transformation, because

$$\begin{aligned} \mathbf{G}_{\bar{\mathbf{h}}} \mathbf{v}_{max} &= (\mathbf{F}\mathbf{h})(\mathbf{F}\mathbf{h})^H \mathbf{F}\mathbf{h} = |\mathbf{h}|^2 \mathbf{v}_{max} \\ &= |\mathbf{h}|^2 \mathbf{v}_{max}. \end{aligned} \tag{150}$$

Consequently, any linear transformation of received signals cannot change the resultant SNR, that is

$$\gamma'_{max} = \sum_{l=1}^{L} \gamma_l. \tag{151}$$

(151) means that the diversity system with $J$ branches has the same BER as that with order of $L$.

## REFERENCES


[1] Y.M.Rhee, *CDMA Cellular Mobile Communications and Network Security*, Upper Saddle River, NJ: Prentice Hall, 1998.

[2] H.Holma and A.Toskala (Editors), *WCDMA for UMTS*, Chichester: John Wiley & Sons, Ltd., 2001.

[3] S.Hara and R.Prasad, *Multicarrier Techniques for 4G Mobile Communications*, Norwood: Artech House, 2003.

[4] N.Yee, J-P.Linnartz and G.Fettweis, "Multi-Carrier CDMA in indoor wireless radio networks," *Proc. of IEEE PIMRC'93*, pp.109–113, Sept. 1993.

[5] K.Fazel and L.Papke, "On the performance of convolutionally-coded CDMA/OFDM for mobile communication system," *Proc. of IEEE PIMRC'93*, pp.468–472, Sept. 1993.

[6] A.Chouly, A.Brajal and S.Jourdan, "Orthogonal multicarrier techniques applied to direct sequence spread spectrum CDMA systems," *Proc. of IEEE GLOBECOM'93*, pp.1723–1728, Nov. 1993.

[7] V.M.DaSilva and E.S.Sousa, "Performance of Orthogonal CDMA Codes for Quasi-Synchronous Communication Systems," *Proc. of IEEE ICUPC'93*, pp.995–999, Oct. 1993.

[8] S.Kondo and L.B.Milstein, "Performance of Multicarrier DS CDMA System," *IEEE Trans. on Commun.*, Vol.44, No.2, pp.238–246, Feb. 1996.

[9] R.Prasad and S.Hara, "An overview of Multi-Carrier CDMA," *Proc. of the 4th IEEE International Symposium on Spread Spectrum Techniques and Applications (ISSSTA'96)*, pp.107–114, Sept. 1996.

[10] S.Hara and R.Prasad, "Overview of Multicarrier CDMA," *IEEE Communications Magazine*, Vol.35, No.12, pp.126–133, Dec. 1997.

[11] S.Hara and R.Prasad, "Design and performanceof Multicarrier CDMA system in frequency-selective Rayleigh fading channels," *IEEE Trans. on Vehi. Technol.*, Vol.48, No.9, pp.1584–1595, Sept. 1999.

[12] S.Haykin, *Adaptive Filter Theory, 4th Ed.*, Upper Saddle River, NJ: Prentice Hall, 2002.

[13] J.G.Proakis, *Digital Communications, 3rd Ed.*, New York: Mc-Graw Hill, 1995.

[14] M. Schwartz, W. R. Bennett and S. Stein, *Communication Sytems and Techniques*, Piscataway: NJ, IEEE PRESS, 1996.

CHAPTER 5

# CDMA2000 1X & 1X EV-DO

SE HYUN OH[1] AND JONG TAE LHM[2]

[1] *Senior Vice President, Strategy Technology Group, SK Telecom, Korea*

[2] *Vice President Mobile Device & Access Network R&D Center, SK Telecom, Korea*

**Abstract:** In this chapter, we will discuss CDMA2000 1x and 1x EV-DO systems. We will talk about channel structure, transmission scheme, call processing, protocol layer and etc. under the topic of radio access technology. And we will talk about coverage & LBA, capacity, scheduling strategy, quality management and etc. under the topic of Engineering & Operation technology. Also, core network structure of CMDA2000 1x system will be mentioned, explanation focused on its main functional elements. Also, characteristics and advancements of next generation technology of 1x EV-DO, EV-DO Rev A and EV-DO Rev-B, will be described in this chapter. Its relation to HSDPA and WiMAX will be looked into

**Keywords:** Mobile communication; CDMA2000 1X, Channel Structure, Transmission Scheme, Call Processing, Protocol Layers, Forward link, Reverse link, Dedicated channel, Common control channel, LAC, MAC, Engineering, Operation, Coverage, LBA, Capacity, Scheduling Strategy, QoS, LBA, Core network, IMS, EV-DO Rev A, EV-DO Rev-B, HSDPA, Mobile WiMAX

## 1. INTRODUCTION

CDMA technology was first proposed by Qualcomm as the standard for the digital cellular services in North America. Then, the CDMA technology was authorized as an IS-95 standard of the Telecommunications Industry Association (TIA) in July 1993. IS-95A revision was published in May 1995 and is the basis for many of the commercial 2G CDMA systems around the world. In addition to voice services, IS-95A provides circuit-switched data connections at 14.4 kbps. The IS-95B revision, also termed TIA/EIA-95, combines IS-95A, ANSI-J-STD-008 and TSB-74 into a single document and offers up to 115kbps packet-switched data

CDMA2000 1X is an ITU-approved as 3G standard. It can double voice capacity of IS-95A networks and delivers peak packet data speeds of 153 kbps (Release 0) or 307 kbps (Release A) in mobile environments in a single 1.25 MHz channel. To provide higher peak data rate to subscribers, 1xEV-DO technology, which part of a family cdma2000 1X digital wireless standards, stands for "Evolution, Data-only"



*Y. Park and F. Adachi (eds.), Enhanced Radio Access Technologies for Next Generation Mobile Communication*, 151–190.

and delivers forward link data rate up to 2.4 Mbps in a single 1.25 MHz channel, addressing data only-not voice. 1xEV-DO is based on a technology initially known as "HDR" (High Data Rate), developed by Qualcomm and the standard is known as IS-856.

CDMA technology shares a block of spectrum through the use of a spreading code (pseudo-random noise or PN code), which is unique to the individual use. It transmits data spread in a full available spectrum reducing the need to guard bands and increasing efficiency use. The CDMA technology accommodates users 10 to 20 times larger than those of the AMPS using FDMA. In addition, CDMA technology is the strong to high frequency selective fading characteristics due to multiple-path signals. So CDMA technology is suitable for areas with high user density or an urban area where high-rise buildings are concentrated.

The Korean government adopted CDMA as the official standard for mobile digital communication through the notice of the Ministry of Communication in November 1993. SK Telecom, a Korean cellular service provider, introduced commercial cellular service based on IS-95A technology for the first time in the world in 1996. And also, SK telecom commercialized CDMA2000 1x Service in October, 2000 and 1xEV-DO Service in February, 2002. 3GPP2, the Third Generation Partnership Project 2, is responsible for establishing specifications related to the synchronization-type CDMA2000, and to keep reflecting next-generation technologies (MIMO and OFDM. etc) regarding specifications to upgrade the data rate.

## 2. RADIO ACCESS NETWORK

### 2.1 CDMA2000 1x

- Channel Structure

– Structure and Characteristics of Forward Link Channel

As shown **Figure 1**, channels for fast data transmission and control channels for efficient signaling control have been added to the forward link in the IS-95 standard.

The forward link channels are divided into the dedicated channels and common channels. The *dedicated* channels are used for specific users, and include a fundamental channel for low-speed rate transmission, a supplemental channel for fast data transmission, and a dedicated control channel for the delivery of mobile-specific control information. And also, a dedicated Auxiliary pilot channel is used with antenna beam-steering techniques to increase the coverage or data rate for a particular user. One common channel includes the pilot channel that measures channel strength and supports coherent detection and hand-off. Handoff is a procedure where a mobile phone with an on-going call changes channel and/or base station under a supervisory system. Other common channels include the sync channel that transmits data necessary for synchronization between terminal and system, and the paging channel that provides system information and paging information. There is also the broadcast control channel added to provide broadcast

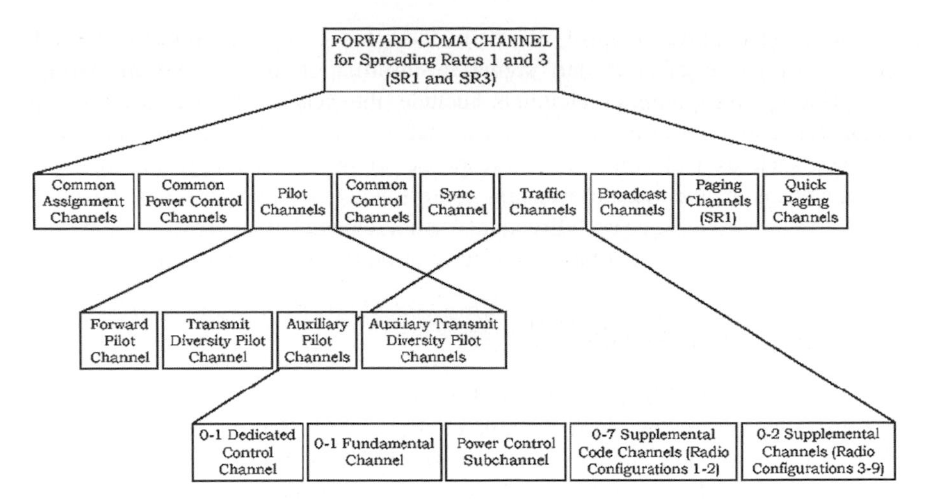


*Figure 1.* 1x Forward link channel structure

system-specific and cell-specific overhead data, and the quick paging channel that improves paging operations in slotted-mode.

– Structure and Characteristics of Reverse Link Channel

The below **Figure 2** shows the pilot channel, the data transmission dedicated channel, and the improved access channel to be used to transmit moderate-sized data packets have been added to the reverse link in the IS-95 standard.

Like the forward link channels, the reverse link channels consist of dedicated channels and common channels, both of which function similarly to dedicate and common channels in the forward link. Reverse dedicated channels include the fundamental channel, the supplemental channel, the dedicated control channel, and the power control sub-channel, which transmits the power control part

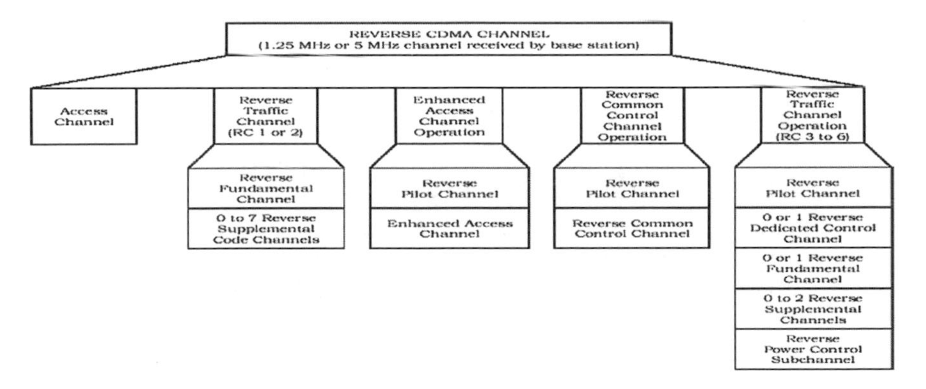


*Figure 2.* 1x Reverse link channel structure

related to reverse power control. Reverse dedicated control channels are used for the transmission of user and signaling information to the system during a call. The reverse common channels include the access channel used by a terminal for communicating to the base station for short signaling message exchanges, such as call originations, response to pages, and registrations, the common control channel used to transmit control data, an enhanced access channel that provides improved accessibility and a pilot channel that provides a phase reference for coherent demodulation and may provide a means for signal strength measurement.

- Transmission Scheme and Characteristics

– Transmission Channel Structure of the Forward

Major improvements to the forward link in the CDMA2000 1X are fast power control that can support up to 800Hz, increased capacity through Orthogonal Transmit Diversity (OTD), enhanced battery life by quick paging channel, and dedicated channel for fast data transmission.

**Figure 3** shows the transmission scheme of the 9.6kbps fundamental channel. The CRC and the tail bits are added to a data bit to create a 9.6kbps bit stream. At this time, the bit stream passes through an encoder and the interleaver for power control puncturing. And then, orthogonal spreading and complex scrambling is made through the Walsh. A long PN code scrambles the channel. The rate of scrambling code depends on the code rate of input. And only PCH(Paging Channel), DCCH(Dedicated Control Channel), FCH(Fundamental Channel) and SCH(Supplemental Channel) are scrambled. A Walsh code running at the chip rate (1.2288Mcps) multiplies the data. The same code is used for both In-Phase and Quadrate components. Each channel is assigned a different Walsh code and might be of different lengths, to adjust to the spreading factor of the data required. The data is then complex PN multiplied, also at the chip rate.

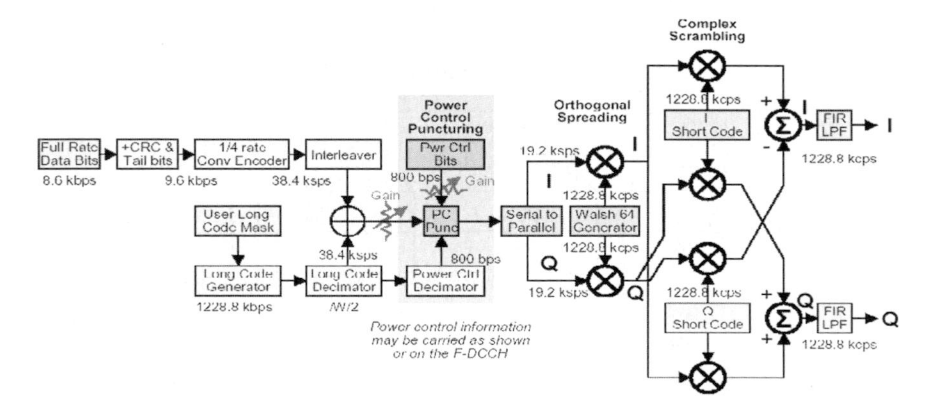


*Figure 3.* 9.6kbps FCH Transmission scheme

The transmission scheme of the SCH (153.6Kbps) is similar to that of the FCH. However, with 19.2Kbps or higher, the Turbo code is used for fast data transmission instead of a convolution code.

– Transmission Channel Structure in Reverse Link

One of the characteristics of the reverse link structure in CDMA2000 1X is that it uses the dedicated signaling channel. In IS-95, one channel is used to logically separate the frame. However, in the CDMA 2000 1X, a dedicated signaling channel is used so that several Walsh codes identify channels. The Walsh code applied to the reverse link has different lengths depending on the channel as shown in **Table 1**.

The following **Figure 4** illustrates the reverse link modulation and spreading process. The channel illustrated here is the Reverse Dedicated Channel. It consists of a Reverse Pilot Channel, which is always present, an R-FCH, an R-SCH, and an R-DCCH. The reverse link uses reverse pilot, hence greatly improving detection performance by only using the preamble. As shown in **Figure 4**, the reverse channels are spread by different-sized Walsh codes, and after gain scaling, the data is transmitted to I and Q channels. The gain scaling is used to apply the relative

*Table 1.* Walsh codes applied to reverse channel

| Channel Type | Walsh Function |
|---|---|
| Reverse Pilot Channel | $W_0^{32}$ |
| Enhanced Access Channel | $W_2^{8}$ |
| Reverse Common Control Channel | $W_2^{8}$ |
| Reverse Dedicated Control Channel | $W_8^{16}$ |
| Reverse Fundamental Channel | $W_4^{16}$ |
| Reverse Supplemental Channel 1 | $W_1^{2}$ Or $W_2^{4}$ |
| Reverse Supplemental Channel 2 | $W_2^{4}$ Or $W_6^{8}$ |

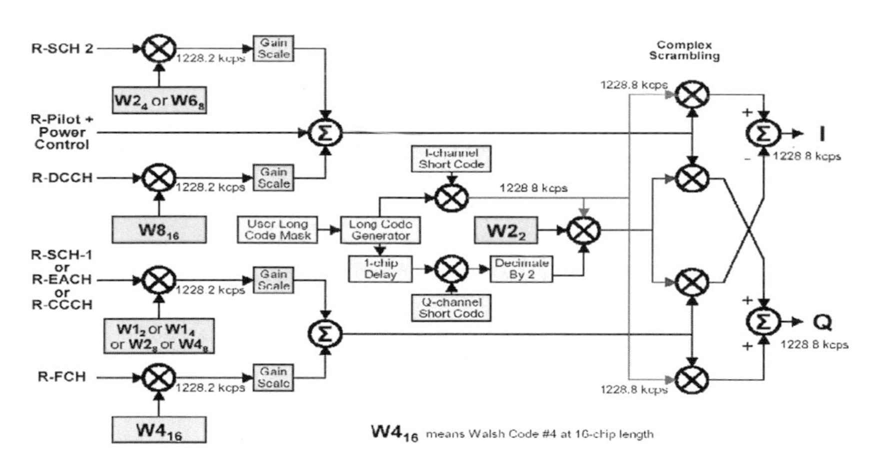


*Figure 4.* Reverse link transmission scheme

offset to each channel based on pilot channel power. The spread Pilot channel and the R-DCCH are mapped to the In-Phase components. The Spread R-FCH and R-SCH are mapped to the Quadrate components.

– RC, P_REV, MO and SO

The CDMA2000 1X system defines RC, P_REV, multiplex options, and service options that are related to the transmission.

Depending on Radio Configuration (RC), the transmission rate and the modulation characteristics are differed at the physical layer. P_REV (Protocol Revision) refers to the protocol version that the system and the terminal use while MIN_P_REV means the minimum protocol version that can be processed. The Multiplex Option specifically defines the traffic channel transmission method; rate set, maximum data block, data block size, the MUX PDU type, etc. The Service Option (SO) is an agreement to negotiate transmission media for communications. The SO identifies various service types, and currently the service types include voice call and data communication.

• Call Processing

The terminal passes through the initialization state, the idle state, and the access state to enter into the traffic state after initial power-up as shown in **Figure 5**.

When the user powers up the terminal, the terminal will enter into the initialization state where the terminal brings necessary information internally stored to decide the system to use and to receive the pilot channel and the sync channel to synchronize with the system. In its idle state, the terminal receives all of the system information and keeps monitoring the paging channel. The terminal in its idle state transits into the system access state through originating or receiving a call or registration and performing a series of operations to access the system. The information necessary for access to the system is received on the paging channel of Forward link. After successfully accessing the system, the terminal transits into the traffic state and

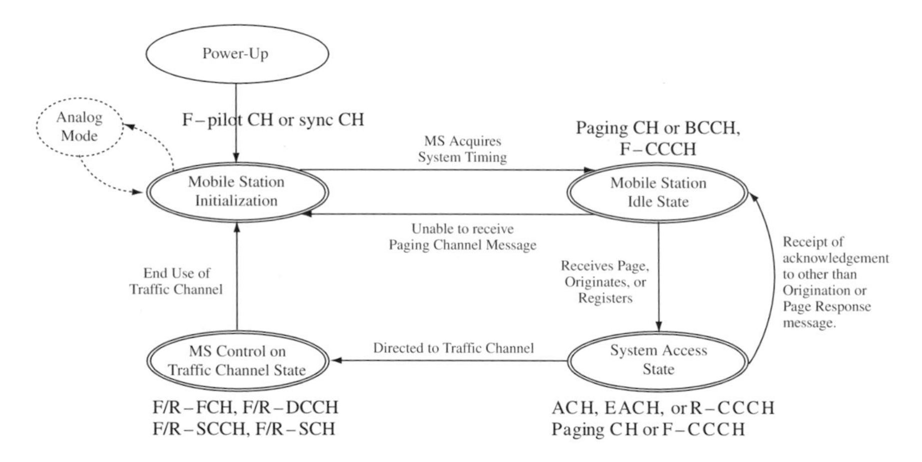


*Figure 5.* MS State transition diagram

establishes a voice call or a data communication. If a call is terminated, the terminal enters into, first the initialization state and then the idle state in order.

When a terminal in a traffic state crosses the cell boundary, the terminal performs a handoff to keep the call. During the handoff process, the user's serving cell may be changed. Major parameters used during the handoff process include Ec/Io of the base station pilot signal strength and T_add and T_drop sent from the system to the terminal, using system parameter messages over the paging channel Pilot Strength Measurement Message (PSMM), the Extended Handoff Direction Message (EHDM) and the Handoff Completion Message (HCM) are the main messages relating with the handoff. With handoffs, CDMA2000 1X system supports two types of handoff; make-before-break, known as soft handoff and break-before-make, known as hard handoff. The below **Figure 6** is an example about handoff procedure; T-add & T-drop Procedure.

- Protocol Layer

**Figure 7** shows the protocol stack of the CDMA2000 1X data. In most cases, the mobile network does not use all OSI 7 layers defined in ISO1 The Protocol stacks that can be managed by CDMA2000 1X are Layers 1~2.

– Physical Layer

The physical layer manages hardware operations that are related to the transmission between terminal and BTS. The physical layer transmits the data coming from the MAC layer through the air through coding, modulation, spreading, and interleaving processes, and sends back the data coming from the air link to the upper MAC layer.

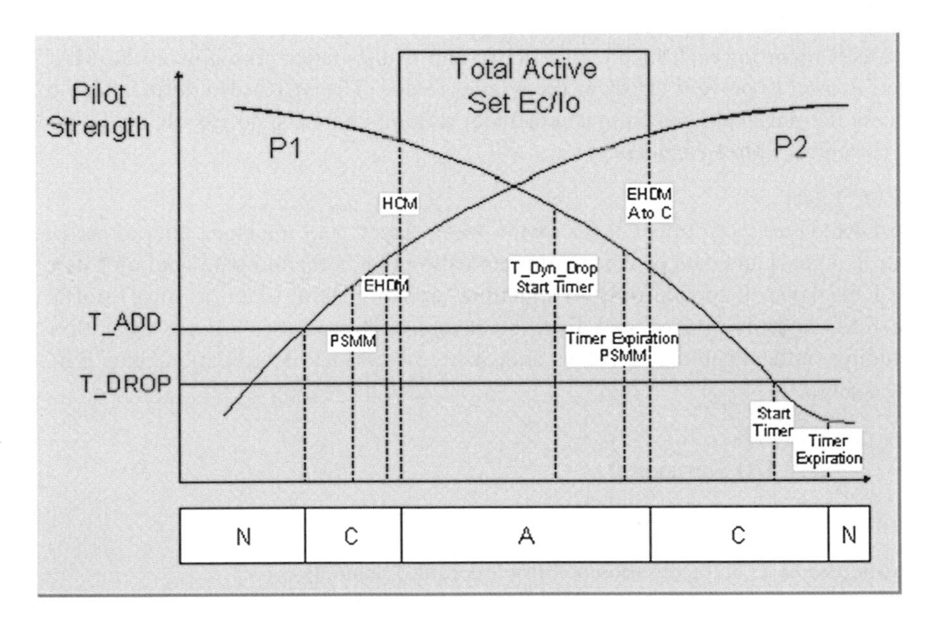


*Figure 6.* CDMA2000 1X Handoff procedure

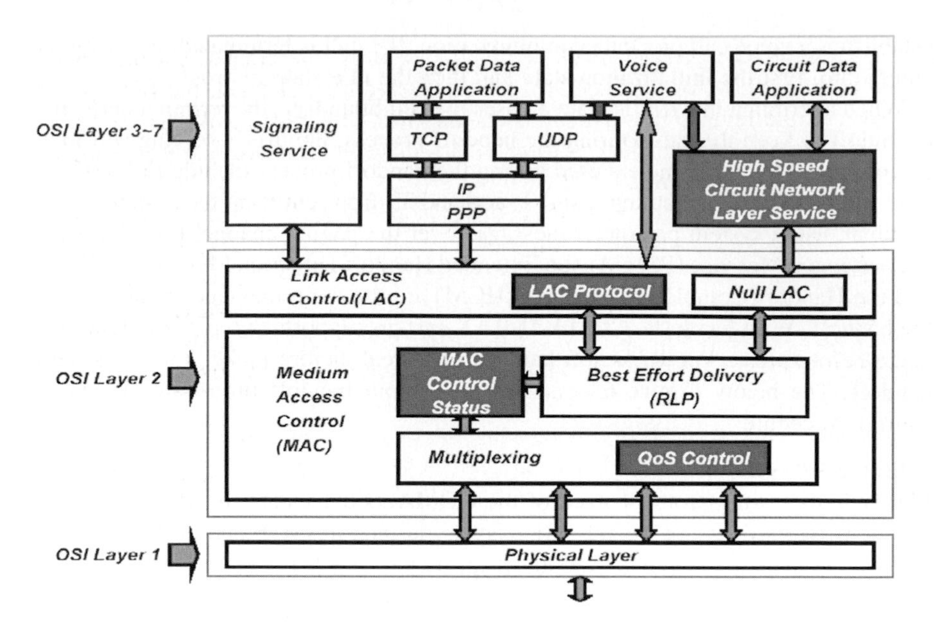


*Figure 7.* Protocol layer

– MAC Layer
The MAC layer classifies and manages (signaling and traffic) data, and performs operations such as multiplexing, Radio Link Protocol (RLP), and Quality of Service (QoS). The MAC layer properly allocates resources that various entities2 need and processes them for each media. The RLP, one of the major protocols at the MAC layer, is used to prevent errors in the wireless node. The MAC also defines how to process the physical layer in sync channels, paging channels, access channels, and the common control channels.

– LAC Layer
The LAC layer is an upper layer of the MAC layer, and manages operations for control, signals and data or voice communication that is the ultimate goal of traffic. The LAC layer is applied only to signaling, not to general (data or voice) traffic. The LAC layer is divided into five sub layers, and processes various operations including authentication, delivery, addressing, message identification, and CRC processing.

## 2.2 1xEV-DO Revision 0

• Channel Structure and Characteristics

– Structure and Characteristics of Forward Link Channels
The forward link channel of 1xEV-DO has been created based on the HDR of Qualcomm that supports 2.4Mbps of high data transmission. The transmission

channels include the pilot channel, the MAC channel, the control channel, and the traffic channel. MAC channels include reverse activity channel, the DRC lock channel, and the reverse power control channel. These forward link channels are shown in **Figure 8** and the EV-DO forward link channel structures are very simple relatively to those of CDMA2000 1x..

The pilot channel is used to aid not only as the coherent demodulation reference for the traffic channel and MAC channel, but also as a sampling reference for the channel state. The reverse activity channel of the MAC channel transmits system load data for a reverse link, and the DRC lock channel checks errors in the DRC channel and informs the terminal of a threshold-crossing state. The reverse power control channel is used to transmit power control information concerning a reverse link.

The control channel which can be transmitted at a data rate of 38.4Kbps or 76.8kbps transmits the overhead message and controls the terminal with a broadcast or directed message.

– Structure and Characteristics of Reverse Link Channel

The 1xEV-DO reverse link channels consist of Access channels and Traffic channels as shown **Figure 9**. The Access channels include pilot channels and the data

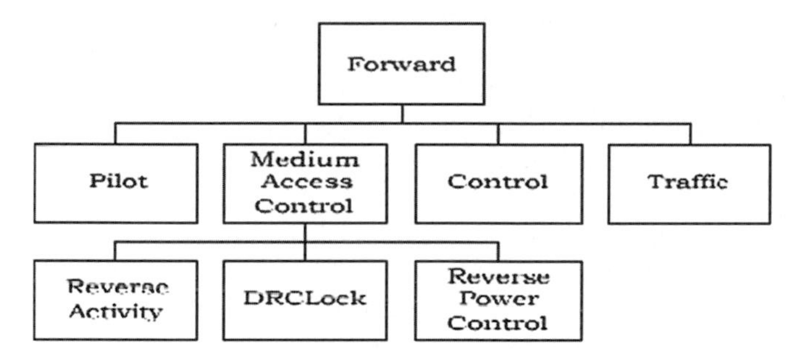


*Figure 8.* 1xEV-DO Rev. 0 Forward link channel structure

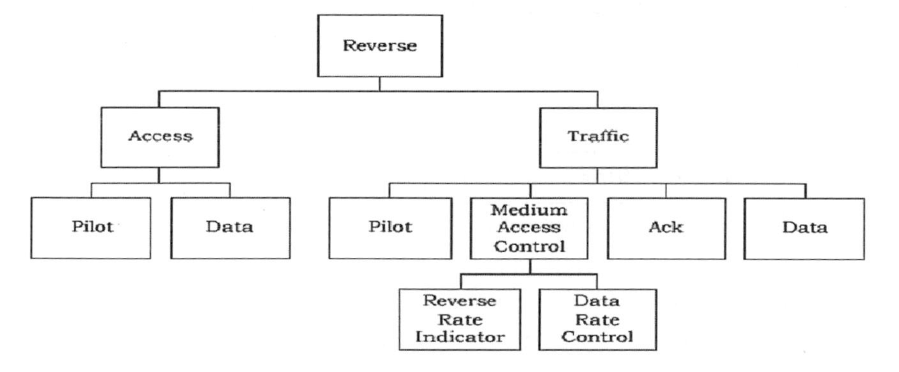


*Figure 9.* Reverse link channel structure

channels, and Traffic channels include the pilot channel, the MAC channel, the ACK channel, and the Data channel. The traffic MAC channel contains the Reverse Rate Indicator channel and the Data Rate Control channel.

The access channel enables the terminal to access the system or to respond to a message from the system. And the pilot channel is used for the coherent detection. The ACK channel of the traffic channel is related to the packet transmission on the forward link, and is used by the terminal to inform the system whether the data packet transmitted on the forward traffic channel has been received or not while supporting a H-ARQ operation. The Reverse Link Rate Indicator Channel (RRI) is used to inform the base station the data rate being transmitted by the terminal to help demodulation. The RRI is included as a preamble for reverse link frames. The DRC channel is used to determine the transmission rate in the forward link.

The reverse link data channel configures a packet based on the rate that is determined by the reverse rate decision algorithm and transmits the packet to the system. At this time, the power of the data channel is decided based on the reverse pilot power and pre-determined gain offset, which is assigned to each data rate as a different value.

- Transmission Scheme and Characteristics

With the 1xEV-DO system, each user takes turn in time to transmit their payload using the maximum base station power and time division multiplexing technique. It means that the 1xEV-DO system consists of a combination of CDMA and TDMA (Time Division Multiplexing Access). The basic transmission unit of the 1xEV-DO physical channel, as shown **Figure 10**, is a 1.67 ms long slot, and one slot consists of two half slots (with 1024 chips each) that have the same structure. Although the slot is the basic transmission unit of the forward link, practically 1xEV-DO systems supports packet repetition through the multi-slot. The repetition of the packet bits in a sub-packet is achieved through channel coding to further obtain coding gain. To allow time for the terminal to process each sub-packet and to feed back the information to the base station, each sub-packet is transmitted in disjointed time with 3 intervals in between. This is known as the 4 slots packet interlacing.

In 1xEV-DO, one traffic channel is divided based on time to separate the channel transmission node. When being served in EV-DO systems, a terminal receives the

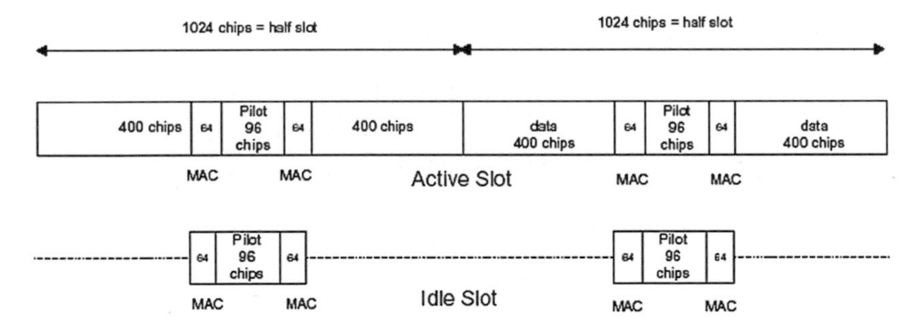


*Figure 10.* 1xEV-DO Forward link slot structure

full power of the cell transmitter. This is because the 1xEV-DO system employs a shared forward link and can serve a user at any instant to maximize the overall data throughput to a given sector. Therefore, unlike the CDMA 2000 1X, or the 1xEV-DO, every channel always can utilize maximum power.

In the 1xEV-DO, only one traffic channel is supported so that resource allocation must be scheduled in a multi-user environment. There are no predetermined time slots; the time the user is on the forward traffic channel depends on channel condition. When multiple users are waiting to be transmitted, RF conditions may be different. The base station can schedule the user with more favorable RF conditions to be transmitted first while the RF conditions from other users may improve before they are scheduled. This method of packet scheduling provides a gain known as the multi-user diversity gain. In EV-DO system, the proportional fairness scheduler is the default method for packet scheduling. In **Figure 11**, in the 1xEV-DO, the terminal decides the forward link data rate and requests, the decided rate to the system. The terminal measures the Carrier to Interface (C/I) of the pilot in its active set and selects the pilot PN with the best C/I as the best sector. Then, the terminal requests the data rate value (DRC value) corresponding to the measured C/I to the best sector using DRC cover assigned by the base station when the pilot is in transit to the active set of the terminal.

The forward link handoff in the 1xEV-DO system is considered a virtual hard handoff as shown in **Figure 12**. Because, although, the forward link for data

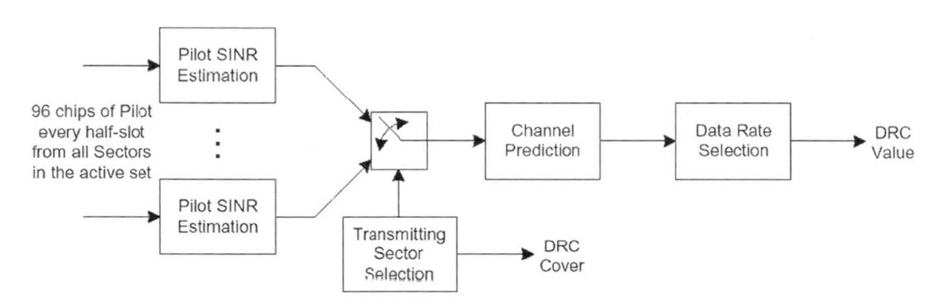


*Figure 11.* Best sector selection procedure

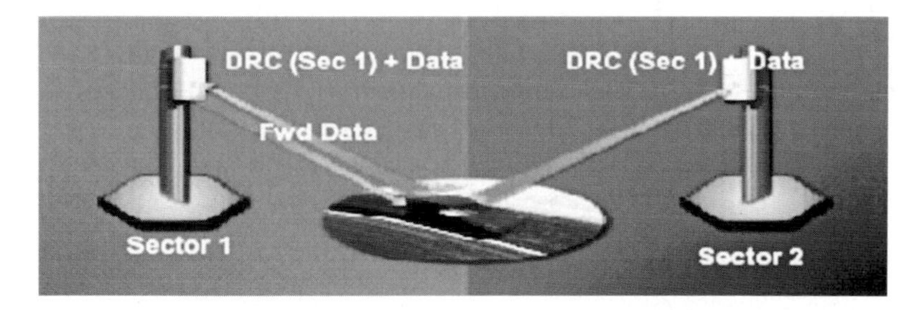


*Figure 12.* Data handoff (Virtual hard handoff)

transmission is connected only to one sector unlike the reverse link, which all sectors in the active pilot are connected, the best serving sector pilot is changed just by the terminal pointing without other handoff procedures. In **Figure 13**, the terminal is currently served by sector 1. But if C/I of sector 2 is higher than that of sector 1, the terminal shall point the DRC cover to sector 2.

The EV-DO forward link offers a range of different data rates. The data rates use each different modulation method as shown in **Table 2**. Modulation types used on the forward link are QPSK, 8PSK, and 16QAM. QPSK modulation is used to achieve 38.4Kbps through 1.2288Mbps data rates (with the exception of 921.6kbps), and 8PSK for 82.6 kbps and 1.8432Mbpsm and 16-QAM for 1.2288Mbps and 2.4576Mbps. The code rates used in the forward link are 1/5 and 1/3 and the maximum data bits per encoder packet range from 1024 to 4096 bits.

As shown in **Figure 14**, the data packet passes the turbo encoder, channel interleaver, Modulator, repletion, Walsh covering time-division-multiplexed with the MAC, Pilot, and the Packet Preamble and transmitted to the terminal after PN spreading.

The 1xEV-DO reverse link structure consists of fixed size physical layer packets (26.67ms frame unit). The reverse link uses a pilot-aided, coherently demodulated scheme. Traditionally IS-95/CDMA2000 1X power control mechanisms and soft handoffs are supported on the reverse link. The data rates used in the reverse link are in five ranges from 9.6Kbps to 153.6kbps, and the terminal deciding the data rate considering available power, the system load indicated by RRI, and the payload

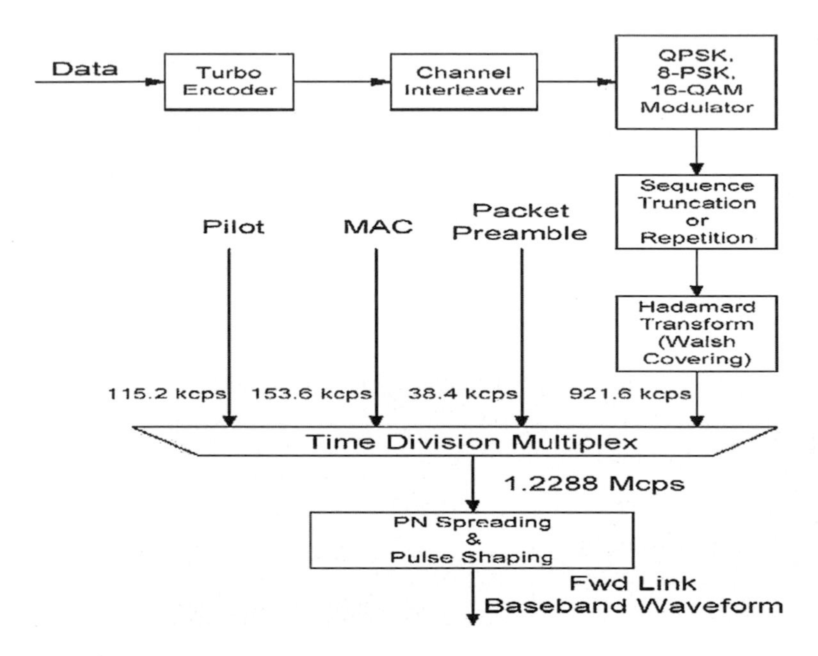


*Figure 13.* Transmission scheme of the forward link

*Table 2.* 1xEV-DO Forward Link Modulation

| | Physical Layer Parameters | | | | | | | | | | | |
|---|---|---|---|---|---|---|---|---|---|---|---|---|
| Data Rates (kbps) | 38.4 | 76.8 | 153.6 | 307.2 | 307.2 | 614.4 | 614.4 | 921.6 | 1228.8 | 1222.8 | 1843.2 | 2457.6 |
| Modulation Type | QPSK | QPSK | QPSK | QPSK | QPSK | QPSK | QPSK | 8PSK | QPSK | 16QAM | 8PSK | 16QAM |
| Bits per Encoder Packet | 1024 | 1024 | 1024 | 1024 | 2048 | 1024 | 2048 | 3072 | 2048 | 4096 | 3072 | 4096 |
| Code Rate | 1/5 | 1/5 | 1/5 | 1/5 | 1/3 | 1/3 | 1/3 | 1/3 | 1/3 | 1/3 | 1/3 | 1/3 |
| Encoder Packet Duration (ms) | 26.67 | 13.33 | 6.67 | 3.33 | 6.67 | 1.67 | 3.33 | 3.33 | 1.67 | 3.33 | 1.67 | 1.67 |
| Number of Slots | 16 | 8 | 4 | 2 | 4 | 1 | 2 | 2 | 1 | 2 | 1 | 1 |

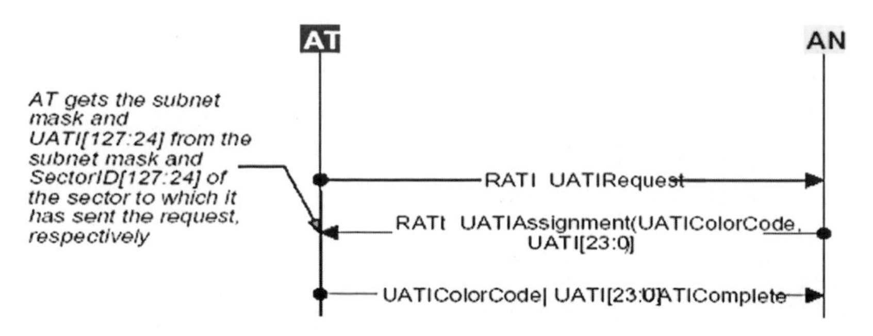


*Figure 14.* UATI Assignment flow

size to transmit. Main functions of the reverse link are transmission reverse data and the measured quality of the forward link, and the Ack function to support the hybrid ARQ for the forward link packet.

As shown **Figure 15** in **Table 3**, in the reverse link, there is just one modulation type- BPSK and two code rates; 4/1, 1/2. The bits per encoder packet range from 256 bits to 4096 bits. The basic transmission unit in the reverse link is a frame (26.67ms), so it is comparably long in length compared to a forward link transmission unit (1.67ms slot).

- Call Processing

– UATI Assignment Procedure

To process a call in the 1xEV-DO, the terminal must receive an address called the Unicast Access Terminal Identifier (UATI.). The UATI is an address that identifies the terminal, and the terminal attempts a call using the UATI. The terminal uses a Random access Terminal Identifier (RATI) to allocate the UATI. As we see in **Figure 14**, AT sends UATI Request message using RATI to get UATI and gets the subnet mask and other information to need for address assignment. Here, subnet means the area which UATI is assigned, so when terminal is moved another subnet area, the UATI should be updated as a new UATI.

– Connection and Session Procedure

After receiving the UATI, the terminal establishes a connection and a session. When the terminal attempts a call, the system transmits Pilot PN, MAC index, and a DRC

*Table 3.* 1xEV-DO Reverse link modulation

| | Physical Layer Parameters | | | | |
|---|---|---|---|---|---|
| Data Rates (kbps) | 9.6 | 19.2 | 38.4 | 76.8 | 153.6 |
| Modulation Type | BPSK | BPSK | BPSK | BPSK | BPSK |
| Bits per Encoder Packet | 256 | 512 | 1024 | 2048 | 4096 |
| Code Rate | 1/4 | 1/4 | 1/4 | 1/4 | 1/2 |
| Encoder Packet Duration (ms) | 26.67 | 26.67 | 26.67 | 26.67 | 26.67 |
| Number of Slots | 16 | 16 | 16 | 16 | 16 |

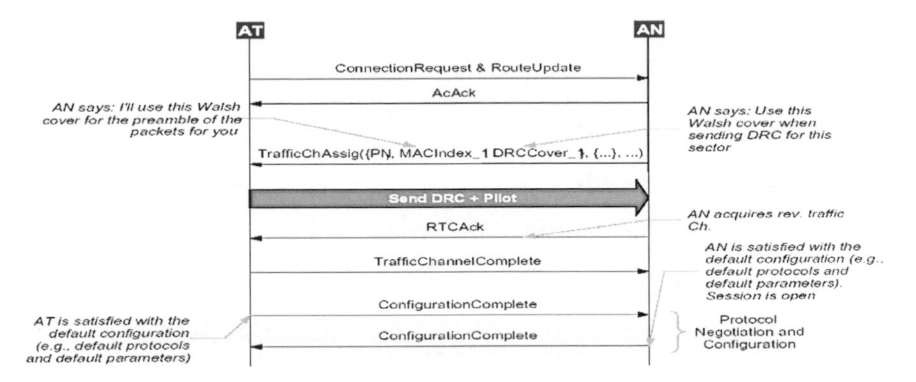


*Figure 15.* 1xEV-DO Connection procedure

cover to the terminal. After acquiring the traffic channel, the system sends the pilot and the DRC to the system. After a connection is established, the terminal and system negotiate to establish a session if needed. As known in **Figure 15**, AT shall send configuration message for session negotiation. If AN and AT are satisfied with the default configuration, AT and An shall send a configurationcomplete message to each other.

– Authentication Procedure
To receive the packet service in the 1xEV-DO, the terminal or the user must be authenticated, and the Challenge Handshaking Authentication Protocol (CHAP) is an example of such. In tThe CHAP procedure, the Network Access Identifier (NAI) is used to authenticate the user and the terminal. In case a mobile IP is used, the mobile IP authentication will be used between the terminal and the home agent instead of the CHAP. When the mobile IP is used, the home agent functions as the RADIUS server, and the foreign agent functions as the RADIUS client. The authentication procedure is described in detail in below **Figure 16**.

– Hybrid Operation
The 1xEV-DO network does not support voice communication so it uses the CDMA2000 1X network for voice communications. The hybrid terminal used in this case selects the CDMA2000 1X network first, acquires and registers the system, and searches and registers the 1xEV-DO network as shown in **Figure 17**. In case two systems are acquired, the hybrid terminal performs dual system scanning.

The 1xEV-DO terminal must interwork with the existing CDMA2000 1X network to use voice communication services. In its Idle state, the 1xEV-DO terminal must communicate with the CDMA2000 1X network, and when the user selects the packet data service, the terminal must connect the 1xEV-DO network.

• Protocol Layers
In the 1xEV-DO network, a total of seven layers are used below the Point-to-Point (PPP) and each layer defines the related protocol. IS-856 protocol stack shown in

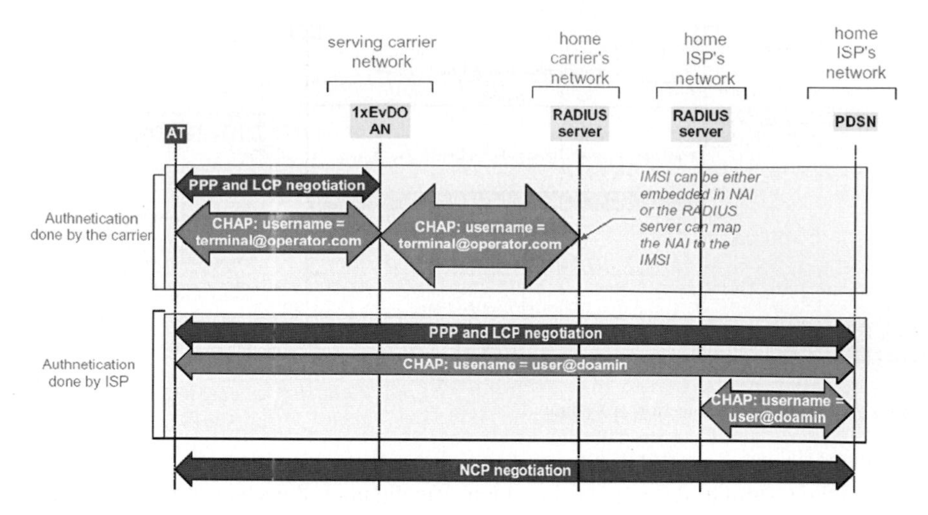


*Figure 16.* Authentication procedure

**Figure 18** is divided up 7 layers. IS-856 stack is under TCP/IP/PPP protocol stack and supports RLP (Radio Link Protocol).

– Physical Layer

The Physical layer modulates, codes, and interleaves the data from the upper layer to transmit it to the air link.

– MAC Layer

The MAC layer defines the procedure used to receive and transmit over the Physical Layer, and performs various operations such as the transmission of MAC layer packets, mapping for packet transmission, multiplexing, demultiplexing, priority handling, identification, and channel supervision. The MAC Layer is a key component optimizing the efficiency of air links and allowing access to the

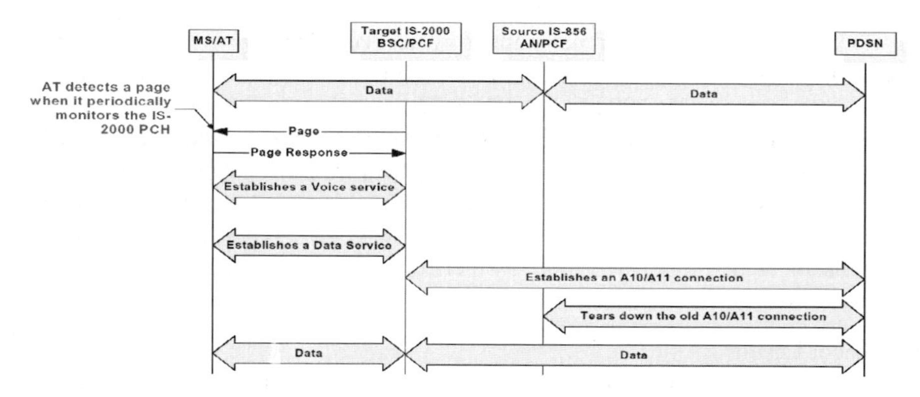


*Figure 17.* Hybrid operation procedure

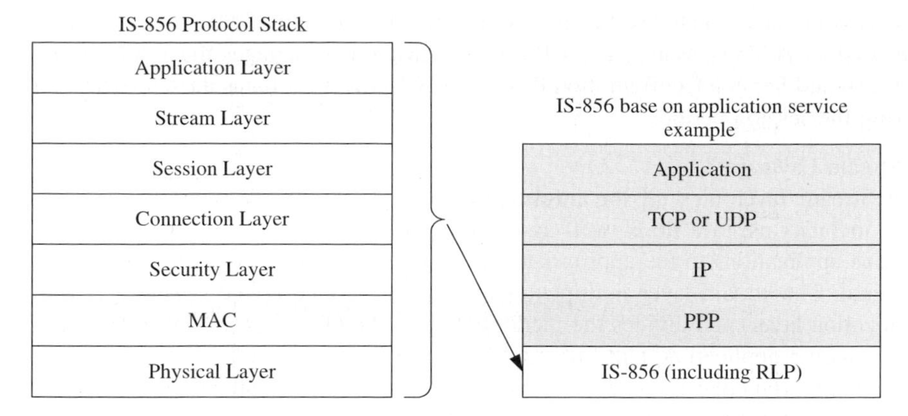


*Figure 18.* IS-856 Protocol stack

system. It is comprised of four protocols; Control Channel MAC Protocol, Access Channel MAC Protocol, Forward Traffic Channel MAC Protocol, and Reverse Traffic Channel MAC Protocol. The four protocols play a part transmitting data and system information over the air link.

– Security Layer

The security layer ensures security of the connection between the system and the terminal. The security layer provides four functions? key exchange, authentication, security, and encryption.

The upper connection layer packets are mapped on the MAC layer through passing the encryption protocol, the authentication protocol, and the security protocol.

– Connection Layer

The Connection layer is comprised of a group of protocols that are optimized for packet data. Combined they efficiently manage the 1xEV-DO air link, reserve resources, and prioritize each user's traffic. They are designed to enhance the user's experience while at the same time bring efficiency to the system. The connection layer consists of the protocols that control connections in the radio link. Protocols of the connection layer include air link management protocol, initialization state protocol, idle state protocol, connected state protocol, route reverse date protocol, overhead message protocol, and packet consolidation protocol.

– Session Layer

The Session layer provides a support system for the lower layers in the protocol stack. It manages the state between the base station and the terminal, and is responsible for address management, configuration negotiation, and protocol negotiation.

The session layer includes Session Management Protocol (SMP) that manages the session, Address Management Protocol (AMP) that manages the UATI of the terminal and Session Configuration Protocol (SCP) that negotiates the configuration during the session period.

– Stream Layer
The Stream layer tags all the information that is transmitted over the air link. This includes user traffic as well as signaling traffic. The stream layer maps the various applications to the appropriate stream and multiplexes the streams for one terminal. The stream layer multiplexes the data generated in the upper layer (the application layer) and divides the media based on the QoS. Stream 0 is assigned to the signaling application, and Stream 1 is used to manage the packet application. The stream protocol can process a maximum of four streams, and control the format and processing of the configuration message.

– Application Layer
The Application layer is a suite of protocols that ensure reliability and low reassure rate over the air link. The Application layer has two sub-layers, which are Default Signaling Application, which provides the best effort and reliable transmission of signaling messages, and the Default Packet Application that provides reliable and efficient transmission of the user's data.

Default Protocols of 1xEV-DO are shown in **Figure 19**.

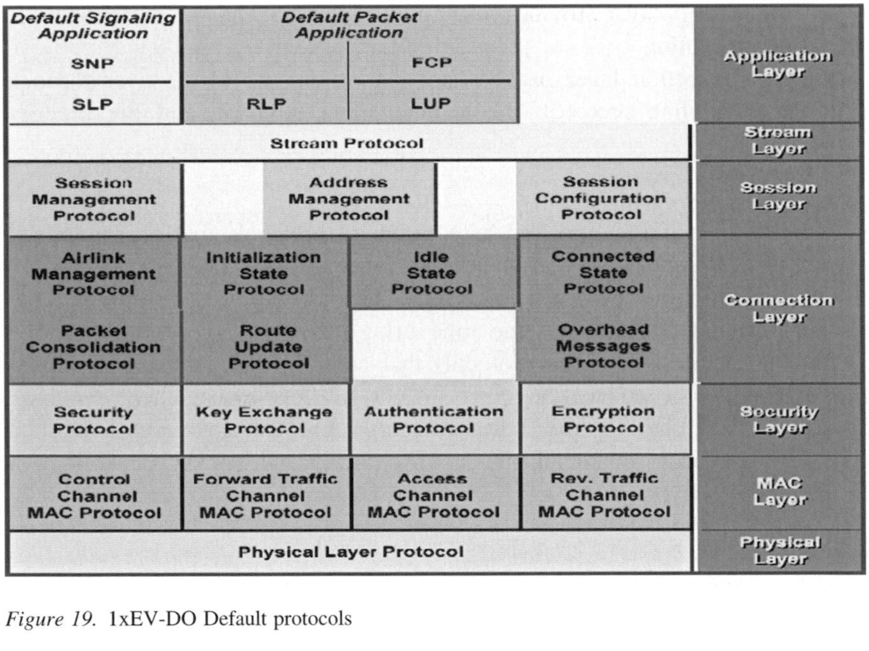

*Figure 19.* 1xEV-DO Default protocols

# 3. ENGINEERING & OPERATION

## 3.1 CDMA2000 1x

• Coverage & LBA (Link Budget Analysis)

The coverage of the CDMA2000 1x network is decided by Ec/I0 of the received signal. Ec/I0 refers to a ratio between the power of the pilot channel received by the terminal and the general signal power received in the band example of Forward LBA could be found in **Table 4**.

$$(1) \qquad E_c/I_1 = \frac{\xi_p P_1^e G_c G_m 1/L}{N_o W + I_o W + I_{oc} W} = \frac{\xi_p P_1^c 1/L_t}{N_o W + I_o W \left(1 + \frac{I_{oc}}{I_o}\right)}$$

| Parameters | Descriptions |
|---|---|
| $\xi_p$ | Portion of the cell power allocated to the pilot channel |
| $P_1$ | Cell PA Output |
| $G_c$ | Cell Ant. Gain |
| $G_m$ | Mobile Ant. Gain |
| $I_{oc}$ | Other Cell Interference Power Spectral Density |
| $I_o$ | Same Cell Interference Power Spectral Density |
| $N_o w$ | Thermal Noise at the Mobile LNA Input |

• Capacity

The physical capacity of the CDMA2000 1x system is decided by the Walsh code count and the number of Channel Elements (CEs.). A total of 64 Walsh codes are used, and a maximum of 61 Walsh codes except 3 Walsh codes for the pilot channel, the sync channel, and the paging channel can be used for voice traffic.

In addition, the capacity of the radio channel is influenced by transmission power of the base station and terminal. The higher the traffic and noise, the greater the output power in the base station and terminal. When the transmission power of the base station and the terminal reach the threshold, call dropping will occur although there are Walsh codes and CEs available. Limited system capacity due to lack of the Walsh code, the CE and the power may appear in a different manner depending on user distribution in the sector and the use of the data service.

• Scheduling Strategy

The CDMA2000 1x scheduler is used for data transmission through the SCH. This scheduler is launched every 260 msec, and determines related parameters such as User, SCH Start Time, SCH Duration, SCH Data Rate, and SCH Tx Power. The BTS Resource Control (BRC) module requests the scheduler to allocate the SCH

*Table 4*. Forward LBA

| | 1x Forward Link | | | | | |
|---|---|---|---|---|---|---|
| Morphology / Cell Model | Urban / Macro | | | | | |
| Service type | Voice | | Data | | | |
| Spreading rate | 1 | 1 | 1 | 1 | 1 | 1 |
| Chip rate [bps] | 1228800 | 1228800 | 1228800 | 1228800 | 1228800 | 1228800 |
| Bit rate [bps] | 9600 | 9600 | 76800 | 76800 | 153600 | 153600 |
| Test invironment | Pedestrian | Vehicular | Pedestrian | Vehicular | Pedestrian | Vehicular |
| Handoff state(with SHO=1, without SHO=O) | 1 | 1 | 0 | 0 | 0 | 0 |
| Transmitter (Base Station) | | | | | | |
| BS maximum transmitter power [dBm] | 43 | 43 | 43 | 43 | 43 | 43 |
| Maximum TCH fraction of total power. Ec/lor [dB] | −16.7 | −16.4 | −9.8 | −7.4 | −6.8 | −4.4 |
| BS Max.transmitter power per Traffic channel [dBm] | 26.3 | 26.6 | 33.2 | 35.6 | 36.2 | 38.6 |
| Cable, Connector and Combiner losses [dB] | 2 | 2 | 2 | 2 | 2 | 2 |
| BS transmitter antenna gain [dBi] | 17 | 17 | 17 | 17 | 17 | 17 |
| BS transmitter EIRP per Traffic channel [dBM] | 41.3 | 41.6 | 48.2 | 50.6 | 51.2 | 53.6 |
| Receiver (Mobile) | | | | | | |
| Thermal noise density [dBm/Hz] | −174 | −174 | −174 | −174 | −174 | −174 |
| Receiver noise figure [(dB)] | 8 | 8 | 8 | 8 | 8 | 8 |
| ^lor/ltc | 2 | 2 | 1 | 1 | 1 | 1 |
| Ltc/loc | 4 | 4 | 0.8 | 0.8 | 0.8 | 0.8 |

| | | | | | | |
|---|---|---|---|---|---|---|
| Same cell interference density, Isc [dBm/Hz] | −160.0 | −160.0 | −164.3 | −164.3 | −164.3 | −164.3 |
| Other cell interference density, Ioc [dBm/Hz] | −166.0 | −166.0 | −160.3 | −160.3 | −160.3 | −160.3 |
| Total effective noise + interference density [dBm/Hz] | −158.3 | −158.3 | −158.1 | −158.1 | −158.1 | −158.1 |
| Information rate at full rate [dB-Hz] | 39.8 | 39.8 | 48.9 | 48.9 | 51.9 | 51.9 |
| Required Geometry, Ior/(No+Ioc) [dB] | 6 | 6 | −2 | −2 | −2 | −2 |
| Required Eb/(No+Io) [dB] | 7.4 | 7.7 | 0.2 | 2.6 | 0.2 | 2.6 |
| Receiver sensitivity [dBm] | −111.1 | −110.8 | −109.0 | −106.6 | −106.0 | −103.6 |
| MS receiver gain [dBi] | 0 | 0 | 0 | 0 | 0 | 0 |
| Body Loss [dB] | 3 | 3 | 0 | 0 | 0 | 0 |
| Max. path loss [dB] | 149.37 | 149.37 | 157.17 | 157.17 | 157.17 | 157.17 |
| Handoff Gain [dB] | 3 | 3 | 3 | 3 | 3 | 3 |
| Penetration loss [dB] | 15 | 8 | 15 | 8 | 15 | 8 |
| Log-normal fading margin [dB] | 10.3 | 10.3 | 10.3 | 10.3 | 10.3 | 10.3 |
| Maximum Allowable Path Loss [dB] | 127.1 | 134.1 | 134.9 | 141.9 | 134.9 | 141.9 |
| Cell Range [km] | 0.92 | 1.46 | 1.53 | 2.44 | 1.53 | 2.44 |

every 260msec when data remains to be transmitted in the transmission buffer of the user. Then, the scheduler allocates the SCH according to the defined scheduler considering the number of users requesting for SCH. The user selection algorithm to define priorities is as follows:

$$\text{Priority\_Score} = A*\text{Starvation\_Number} + B*\text{Queue\_Length}+\Gamma$$
$$*\text{Expected\_Data\_Rate}$$

The Starvation_Number increases by 1 when it does not receive an SCH after making a request. The higher the A, the greater fairness among users.

The Queue_Length is the number of Mux PDUs in the transmission buffer of the channel card. The higher the B, the greater the link efficiency is. The Expected_Data_Rate can be transmitted when the corresponding user is allocated with the SCH. The higher the Γ, the greater the entire throughput.

- Quality Management

Unlike the overhead channels (pilot channel, sync channel, and paging channel) that have a fixed value, the transmission power of the base station of the forward traffic channel in the CDMA2000 1x network is not fixed and changes according to variable vocoder and multipath fading.

This is because protocol is mainly for voice communication, and the base station and terminal control power to manage voice quality. The set point of the terminal changes according to the forward link Frame Error Rate (FER). The terminal compares the received Eb/ Nt with the set point defined in the terminal and sends the Power Control bit (PCB) "1" or "0" to the base station. When the base station receives "1," it reduces the digital gain by Power_Control_Step, or when it receives "0," it increases the digital gain to maintain the target FER or the terminal.

The terminal transmission power in the reverse traffic channel is also determined in the same way. The Eb/ N0 measured at the demodulator of the base station is compared with Eb/ N0 predicted by the outer loop lower control. If the measured Eb/ N0 is higher, the base station will send "1" PCB to the terminal to lower the transmission power, or when it is lower, the base station will send "0" PCB to the terminal to increase the transmission power to maintain target FER.

- Resource Assignment (Only for Samsung System)

For communication between the system and the terminal, necessary resources must be allocated to the terminal. The FA, the transmission power, the channel element, the Walsh code, and E1 link are allocated in order. When the resources are more than the threshold, it can be allocated. If there is no resource available, the call connection request will be denied. The system will control call admission based on the cell load, channel element resources per FA/sector, and transmission power.

## 3.2 1xEV-DO

• Coverage & LBA

The coverage of the 1xEV-DO network refers to a zone where the service is provided at a certain throughput or higher throughput, or a zone with a certain C/I level or higher level. C/I refers to the ratio between signal and interception. "C" indicates the transmission power from the base station, and "I" indicates all interception signal power except carrier power C among the received power example of forward LBA could be found in **Table 5**.

$$\left.\frac{C}{I}\right|_{DO} = \frac{P_{TX,M} \cdot g_{M,00}(d)}{\sum_{i=1}^{36} \sum_{j=0}^{2} P_{TX,M} \cdot g_{Mij}(d) + \sum_{j=1}^{2} P_{TX,M} \cdot g_{M,0j}(d) + N_0 \cdot W}$$

| Parameters | Descriptions |
|---|---|
| $P_{Tx,M}$ | Transmission power of the cell |
| $g_{N,0}(d)$ | Transmission loss between the AT and the AP |
| $g_{M,y}(d)$ | Transmission loss between Ith AP and the AT in Jth sector |
| Now | Thermal noise |

• Do System Capacity

The forward link capacity of the 1xEV-DO network is determined by the MACIndex count that physically identifies the users. **Table 6** shows the MAC Channel and preamble use versus MACIndex, and total 59 MACIndex values can be used as traffic channels:

In the channel card of the 1xEV-DO, there are 96 reverse channels. Therefore, when there are three sectors, each sector will have 32 channels.

Unlike the CDMA 2000 1x network, in the 1xEV-DO network, it is the sector throughput, not the number of subscribers, which must be managed for the operation of the system. Depending on the location of the user and the number of users, the actual sector throughput may drop below 2.4 Mbps, which is the maximum physical throughput. Major factors that affect the throughput include the early termination and multi-user diversity gain.

Early termination occurs when the terminal requests a DRC that uses multi slots with repeated symbols, see **Table 7**. The terminal responds by sending an ACK through the ACK channel when correct demodulation is made before receiving all slots that the base station sends. When the base station receives the ACK, it will terminate the transmission without repetition. In short, transmission will be completed within a shorter slot time and the throughput will increase.

Multi-user diversity gain refers to an increase in sector throughput caused by characteristics of the proportional fair scheduler. The scheduler allocates the slots to the user in a relative better condition. Therefore, when there are multiple users, it is highly likely that uses in better conditions will be selected, hence increasing

*Table 5.* Forward LBA

| 1x EV-DO Forward Link | | | |
|---|---|---|---|
| Parameters | Value | Symbol | Equation |
| Average Throughput [bps] | 64300 | | |
| Bandwidth [Hz] | 1228800 | | |
| Bandwith [dB-Hz] | 60.895 | a | |
| Tranmitter (Base station) | | | |
| BTS Tx Power [Watts] | 20 | | |
| As above in dBm | 43.01 | b | |
| BTS Antenna Gain [dBI] | 17 | c | |
| BTS Cable Loss [dB] | 2 | d | |
| BTS EIRP [dBm] | 58.01 | e | b+c–d |
| Receiver (Mobile) | | | |
| MS Rx Antenna Gain [dBi] | 0 | f | |
| Body Loss [dB] | 3 | g | |
| Noise Figure (dB) | 8 | h | |
| Thermal Noise [dBm/Hz] | −166 | I | (–174)+h |
| Target PER [%] | 2 | | |
| (lor/No)req per Antenna [dB] | 4 | j | |
| Multi-user Diversity Gain [dB] | 2.25 | | |
| Rx Diversity Gain [dB] | – | | |
| MS Receiver Sensitivity [dBm] | −101.105 | k | 1+j+a |
| Log-normal Std. Deviation [dB] | 8 | | |
| Log-normal Fade Margin [dB] | 10.3 | l | |
| Handoff Gain [dB] | 4.1 | m | |
| Building Penetration Loss [dB] | 15 | n | |
| Maximum Pass Loss [dB] | 134.92 | o | e-k+f-g-l+m-n |
| Cell Range [km] | 1.54 | | |

the sector throughput. However, in case of the multi-user diversity gain, the sector throughput does not increase but is saturated when the number of subscribers reaches a certain level as shown in **Figure 20**:

• Scheduler

The 1xEV-DO system uses a proportional fair scheduler. The scheduler allocates the Nth slot to the user with the largest DRCi(N)/ Ri(N). DRCi(N)is the DRC that

*Table 6.* MAC Channel and preamble use Versus MACIndex

| MACIndex | Use |
|---|---|
| 0 And 1 | Not Used |
| 2 | 76.8kbps Control Channel |
| 3 | 38.4kbps Control Channel |
| 4 | RA Channel |
| 5 ~ 63 | Forward Traffic Channel, RPC Channel |

*Table 7*. Number of slots and symbol repetition count for each DRC

| | Values per Physical Layer Packet | | | | | | Approximate Coding | |
|---|---|---|---|---|---|---|---|---|
| Data Rate (kbps) | Number of Slots | Number of Bits | Number of Modulation Symbols Provided | Number of Modulation Symbols Needed | Number of Full Sequence Transmissions | Number of Modulation Symbols in Last Partial Transmission | Code Rate | Repetition Factor |
| 38.4 | 16 | 1,024 | 2.560 | 24,576 | 9 | 1,536 | 1/5 | 9.6 |
| 76.8 | 8 | 1,024 | 2,560 | 12,288 | 4 | 2,048 | 1/5 | 4.8 |
| 153.6 | 4 | 1,024 | 2,560 | 6,144 | 2 | 1,024 | 1/5 | 2.4 |
| 307.2 | 2 | 1,024 | 2,560 | 3,072 | 1 | 512 | 1/5 | 1.2 |
| 614.4 | 1 | 1,024 | 1,536 | 1,536 | 1 | 0 | 1/3 | 1 |
| 307.2 | 4 | 2,048 | 3,072 | 6,272 | 2 | 128 | 1/3 | 2.04 |
| 614.4 | 2 | 2,048 | 3,072 | 3,136 | 1 | 64 | 1/3 | 1.02 |
| 1,228.8 | 1 | 2,048 | 3,072 | 1,536 | 0 | 1,536 | 2/3 | 1 |
| 921.6 | 2 | 3,072 | 3,072 | 3,136 | 1 | 64 | 1/3 | 1.02 |
| 1,843.2 | 1 | 3,072 | 3,072 | 1,536 | 0 | 1,536 | 2/3 | 1 |
| 1,228.8 | 2 | 4,096 | 3,072 | 3,136 | 1 | 64 | 1/3 | 1.02 |
| 2,457.6 | 1 | 4,096 | 3,072 | 1,536 | 0 | 1,536 | 2/3 | 1 |

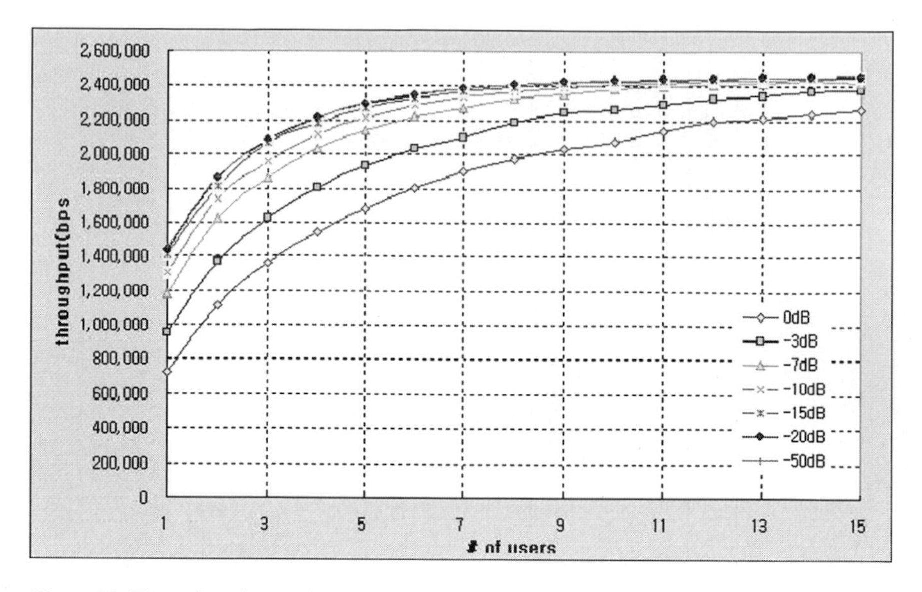


*Figure 20.* Throughput by number of users and idle slot gain

User i requests to the reverse link, and Ri(N) is the average rate of data that User i received during Tc, the time constant in the scheduler. Ri(N) can be updated as follows:

$$\text{Ri(N)} = (1 - 1/\text{Tc})^{*}\text{Ri}(\text{N} - 1) + (1/\text{Tc})^{*}$$
$$\text{(Served Rate In Slot N-1 To User i)}$$

As the default of Ri(N) is "0", the terminal that attempts to use the service for the first time in the cell will have priority. The user for whom the data transmission has not been allocated in the current slot will have a served rate of 0, and even the average rate of the user who does not have data to transmit in the buffer will be updated. This leads to giving a higher priority to the user who has not recently received the data.

- Quality Management

The 1xEV-DO controls the rate to manage traffic quality. The terminal measures the C/I and requests the maximum data rate to meet PER 1% using the DRC channel in the reverse link. Then, the system allocates the data rate each user requested. If the PER of received data is higher or lower than 1%, the terminal adjusts the DRC rate to maintain PER 1%.

The reverse rate control is for quality management of the reverse traffic channel, and the base station controls the reverse rate of the terminal based on probability. In case reverse traffic increases, the system load will also rise. However, if the reverse traffic crosses the threshold, the base station will set the Reverse Activity Bit (RAB) as "1" and sends the RAB to the terminal through the RA channel. After receiving "1" as the RAB, the terminal lowers the reverse rate based on the rate transition probability. If the system load is smaller than the threshold, the RAB will be "0" and the terminal will increase its transmission rate based on rate transition probability.

## 4.

### 4.1 Network Structure and Functional Elements

The data core network in the CDMA2000 network is configured as shown in **Figure 21**:

- Packet Data Serving Node (PDSN)

The PDSN allocates the IP addresses to the terminal through the PPP protocol, routes the packets between the terminal and the Internet network, and provides services according to user's authority given by the AAA. Based on the data collected through the signaling of the RP interface and packet use by the user, the PDSN creates the charging data and sends it to the charging equipment.

- Authentication, Authorization, Accounting (AAA) Server

When the user makes a new data call, the AAA server allocates the IP address and authenticates the user. Also, the AAA sends user authority data to the PDSN so that the PDSN can judge which users have access to which services. The AAA server

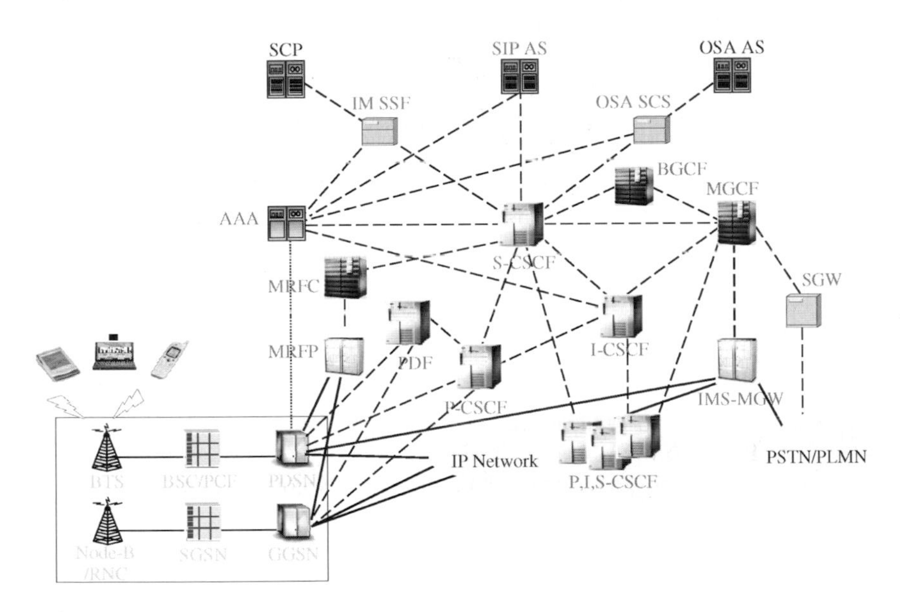


*Figure 21.* Core network structure

receives the charging data created when the user uses the packet data service from the PDSN, and executes charging features.

- Home Agent (HA)

When the user requests mobile IP service, the PDSN sends mobile IP user data to the HA. After receiving this data, the HA allocates the IP address to the terminal, and uses two addresses to transmit packets using the PDSN as a foreign agent. Tunneling protocol is used between the PDSN and the HA.

## 4.2 IMS

IMS stands for the IP Multimedia CN Subsystem, and is a core network domain that provides various IP-based multimedia services that are controlled by the SIP. The term “IMS” was first introduced when the 3GPP adopted the All-IP concept in Release 5, and since then, the IMS has been evolving through R5 and R7. Recently, the 3GPP2 and the fixed NGN network also adopted the IMS and are standardizing the related technologies, see **Figure 22**. The IMS adopted IETF standard protocols such as SIP and Diameter, and defined various capabilities necessary for the communication services to support global roaming and interworking.

## 4.3 Standardization Trend

The IMS, a standard for the IP multimedia service in the GPRS network was first defined in R5 specification of the 3GPP. In the same period, the 3GPP2 was also standardizing a technology similar to the IMS of the 3GPP. Later, to harmonize these two technologies, the 3GPP2 adopted the IMS of the 3GPP, not the All-IP standard, as the MMM standard. This has not yet changed. In addition, the 3GPP developed the IMS standard into R6 IMS through complementing the specification to give access independence to R5 IMS.

In this period, the TISPAN and ITU-T also recognized a need to introduce IMS technology to a fixed network. To meet this need, a work item called Fixed Broadband Access to IMS (FBI) is underway to integrate the IMS in a fixed-mobile convergence network as part of works for establishing the R7.

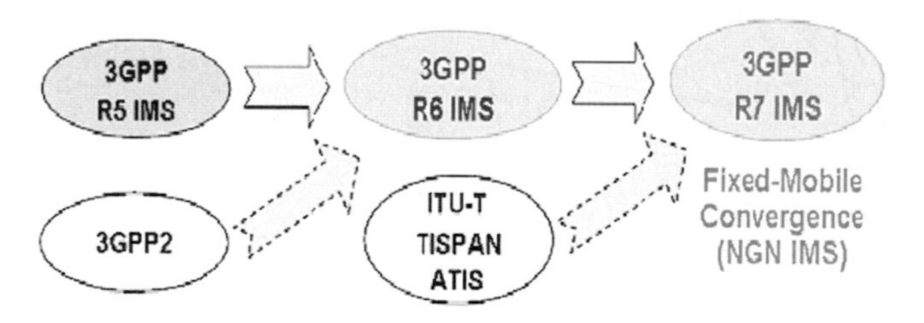


*Figure 22.* IMS Standard development

## 4.4 Terminal

The terminal is the endpoint that provides users with the services through the wireless communication network. The terminal has been developed in a way to meet various demands of users supporting multiple access networks and high data rate multimedia data services.

• Terminal H/W Architecture

The hardware of the terminal mainly consists of three parts: the modem that processes the call processing, the application that provides various value-added functions and services, and I/O peripherals such as keypad and display as shown in **Figure 23**.

The modem part is one of the core hardware elements of the terminal and deals with connections to the access network and manages data reception and transmission. Major modem parts includes the modem chip for baseband signal processing, the RF chip for the reception and the transmission of RF signals, the power management chip, and the antenna.

The application part is to handle various supplementary functions and multimedia services other than basic call processing. The application part includes various application processors and memory devices.

The I/O peripherals are exposed to the user. The I/O peripherals include the LCD (Liquid Crystal Display), the camera, the speakers, the keypad, the external memory reader, and many other I/O devices.

• Terminal S/W Structure

The software of the terminal consists of the operating system (OS) and the applications as shown in **Figure 24**. The terminal S/W is based on Dual-Mode Subscriber Software (DMSS), an MSM S/W, and the REX OS for multitasking. The terminal S/W also includes a vendor-specific platform to support unique features of each terminal over the operating system. The DMSS controls the model chip and enables interworking between the modem chip and various application processors providing a basic frame work for terminal S/W operations. The vendor-specific platform

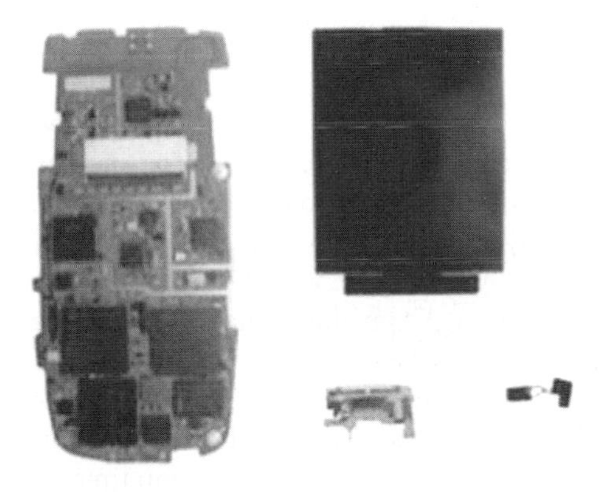

*Figure 23.* Example of terminal H/W component

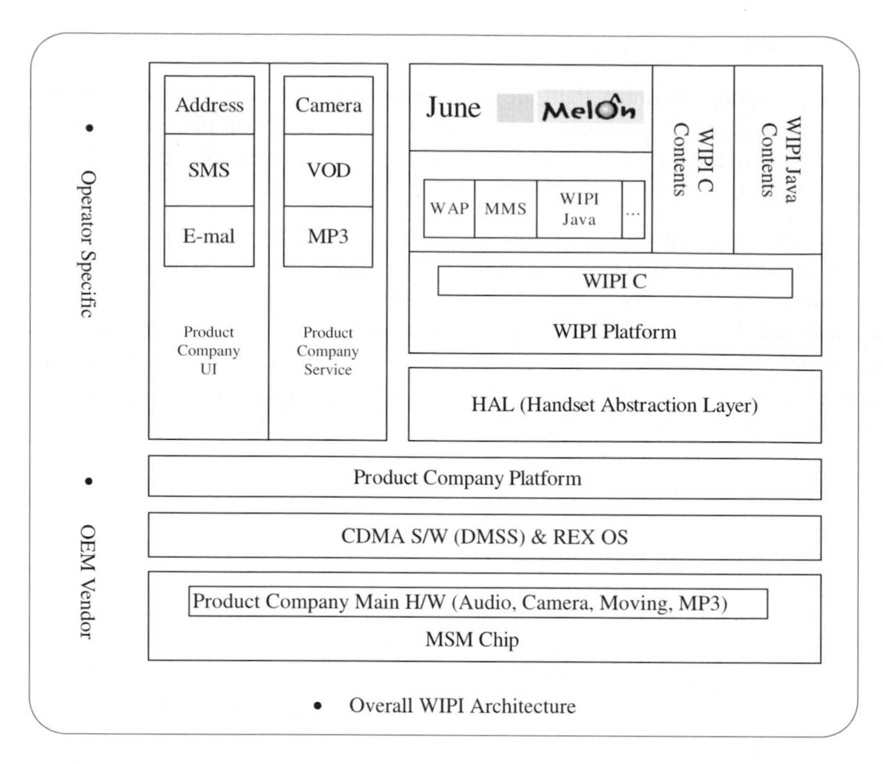


*Figure 24.* Example of terminal S/W architecture

processes features of each terminal such as address book, SMS and MP3 player and interworks with a higher layer platform. On the Hardware Abstraction Layer (HAL) above the vendor-specific platform, the application platform is installed to run various applications.

Korea adopted the Wireless Internet Platform for Interoperability (WIPI) platform as the standard, and all supplementary service applications are developed in C or Java language based on the WIPI.

- Terminal Evolution

The terminal evolution trend is mainly divided into two major streams: the support for the high data rate multi-access network and device convergence. The modem chipset is being developed in a way to support not only basic communication features of the IS-95A network in the early days but also various features of newly introduced access networks such as CDMA2000 1x, 1xEV-DO, and WCDMA to catch up with the access technology evolution. From the device convergence point of view, the terminal has been developed to have various designs and to reduce overall weight and size. Together with adopting System On Chip (SOC), the terminal is now developing into an integrated multimedia device that can support not only the basic voice calls but also the advanced and sophisticated multimedia service features.

# 5. FUTURE DEVELOPMENT

## 5.1 CDMA2000 TRM (Technology Road Map)

- IS-95

The IS-95A (Interim Standard 95A) is the first standardized CDMA technology. As the first digital mobile communication technology, the IS-95A was standardized in 1993 and upgraded later to the IS-95B. The IS-95 is recognized as the second-generation mobile communication technology by the International Telecommunication Union (ITU), see **Figure 25**.

- CDMA2000 1x

The CDMA2000 1x is the next-generation technology of the IS-95. 1x is an abbreviation of 1x Radio Transmission Technology, (1xRTT) which means a system that uses one 1.25MHz channel. The CDMA2000 1x system supports up to 3xRTT, and 1xRTT is the most basic system. The ITU classified the CDMA2000 1x technology as 3G technology (in November 1999).

- CDMA2000 1xEV-DO (1x Evolution-Data Optimized)

The 1xEV-DO is a dedicated standard for the data service and has been developed based on the CDMA2000 1x technology to provide mobile data service at a high speed. The ITU classified the 1xEV-DO as a 3G technology. Following 1xEV-DO Release 0 (CDMA2000 High Rate Packet Data Air Interface, IS-856) and 1xEV-DO Rev. A (TIA-856-A), 1xEV-DO Rev. B was standardized.

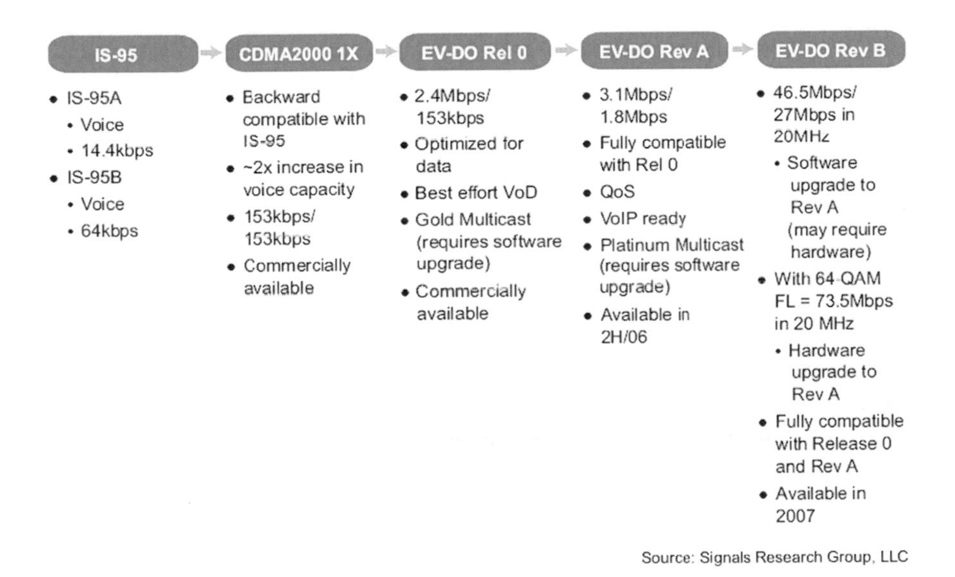


*Figure 25.* CDMA Technology TRM

## 5.2 1xEV-DO Revision A

1xEV-DO Revision A is the first upgraded version of 1xEV-DO Release 0 in the CDMA 2000 technology roadmap. It was first suggested to develop 1xEV-DO Release 0 into 1x EV-DV to support both voice and data, but this plan was suspended. Currently, 1xEV-DO Release 0 is being developed into 1xEV-DO Revision A dedicated to the data system.

The performance of the forward link in Revision A has increased by 20% due to the introduction of improved packet structure and equalizer compared to EV-DO Release 0, see the performance comparison in **Figure 26**. The improved packet structure and the equalizer increased the C/I of the signal received by the terminal so that a 1xEV-DO Revision A can support 3.1Mbps of the peak data rate.

However, the most significant characteristic of 1xEV-DO Revision A is greatly enhanced performance of the revers link compared to Release 0, see the performance comparison in **Figure 27**. While Release 0 supports 153.6kbps of the peak data rate, Revision A supports 1.8Mbps. This improvement was due to the introduction and application of new technologies such as higher modulation schemes, Rx civersity, pilot interference cancellation, and packet prioritization.

## 5.3 1xEV-DO Revision B

The next technology of Revision A that the 3GPP2 is now preparing is Revision B which is expected to completely standardize 1Q of 2006.

Revision B will support interference cancellation for both pilot and traffic signals and introduce 64QAM modulation to provide 3.7, 4.3, and 4.9Mbps of high peak data rates and to improve the efficiency of frequency use. Revision B will also adopt new technologies such as flexible carrier assignment and flexible duplexing to greatly improve the performance of the forward link so that it is expected that the the multimedia service can be provided through mobile Internet.

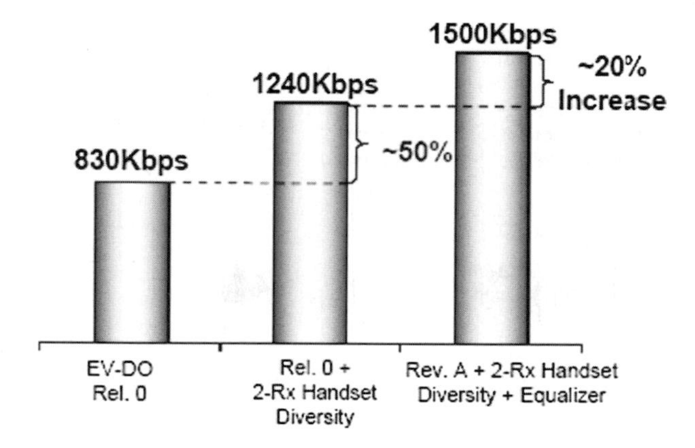


*Figure 26.* Forward link performance comparison (Aggregate throughput)

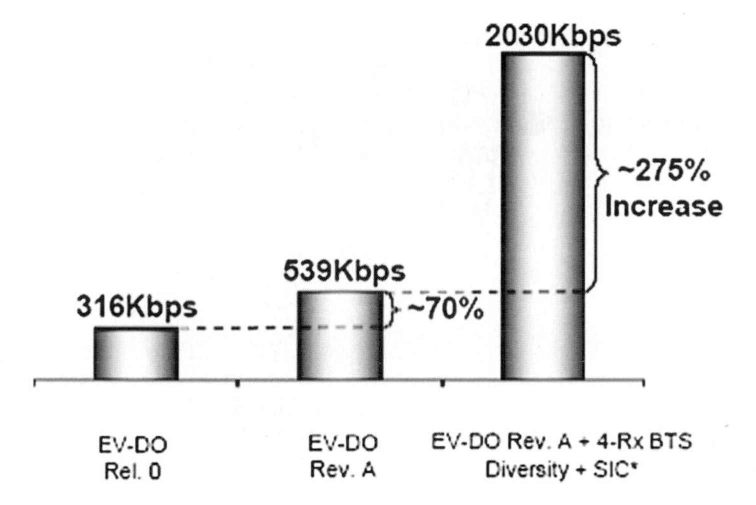


*Figure 27.* Reverse link performance comparison (Aggregate throughput)

Revision B introduced the scalable bandwidth technology and significantly enhanced the maximum throughput of each user in the forward and reverse link. **Figure 28** shows the comparison of the throughput of Revision A and Revision B.

- Flexible Carrier Assignment

The scalable bandwidth technology adopted by Revision B is to increase the throughput by using multiple carriers at the same time, as shown in **Figure 29**.

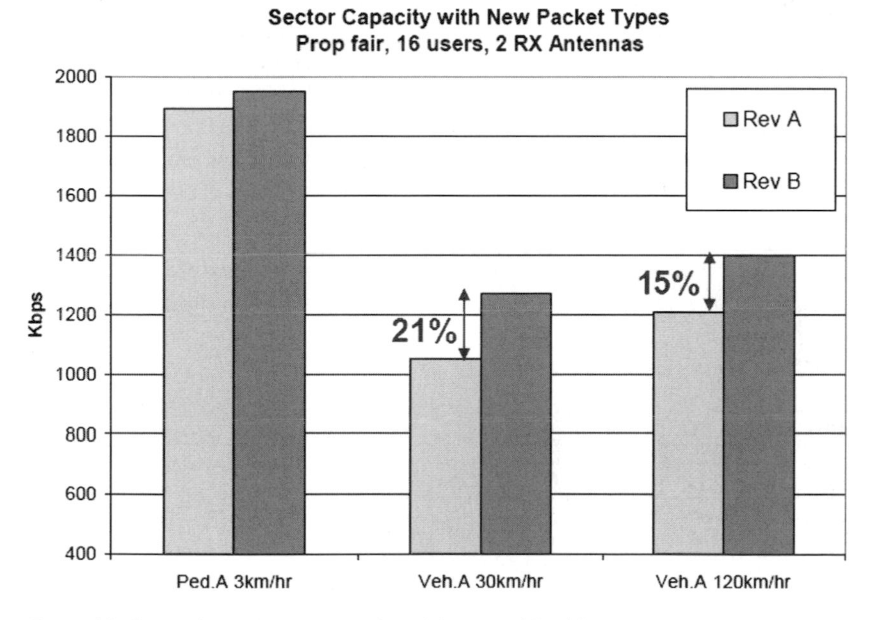


*Figure 28.* Comparison of throughput of Revision A and Revision B

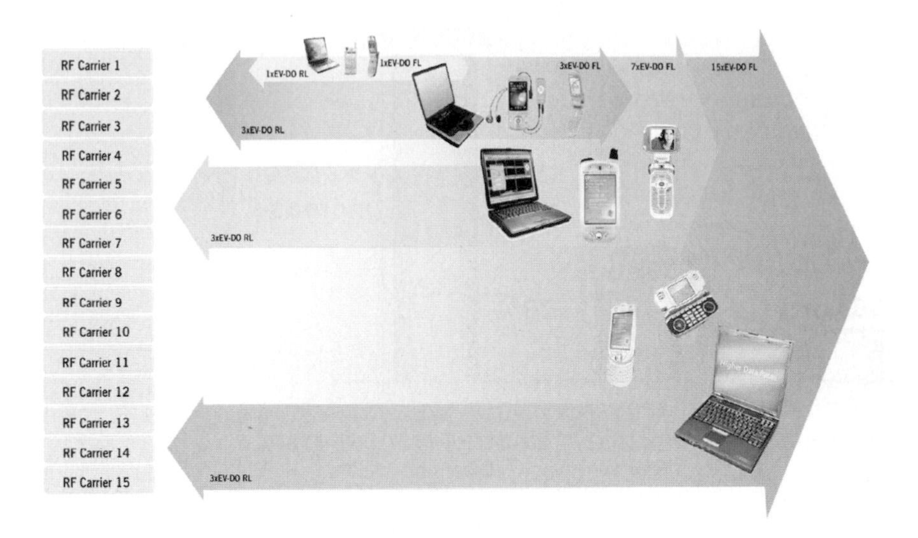


*Figure 29.* Flexible carrier assignment

Minimum one and maximum 15 carriers can be used at the same time. In case three 1.25MHz carriers are used, 14.7Mbps of the peak data rate will be possible, or in case 15 carriers are used, 73.5Mbps of the data rate will be possible. Therefore, one to three carriers can be used for the mobile terminal, and four or more carriers can be used to transmit 3D game data or high-resolution video data, which means more various types of services can be provided for users.

In the 1xEV-DO Revision B system, the spreading factor of the system is not changed and each 1.25MHz carrier is used as one unit. Therefore, unlike HSDPA (The carrier bandwidth : 5MHz), Wibro (The carrier bandwidth : 9MHz), and other technologies which use the carrier of broad bandwidth, the frequencies do not need to be adjacent so that frequency operation is more flexible, see an example of frequency allocation in **Figure 30**.

- Flexible Duplexing

The flexible duplexing feature has been adopted to Revision B. Flexible duplexing is an advanced form of scalable bandwidth technology. It uses an additional unpaired

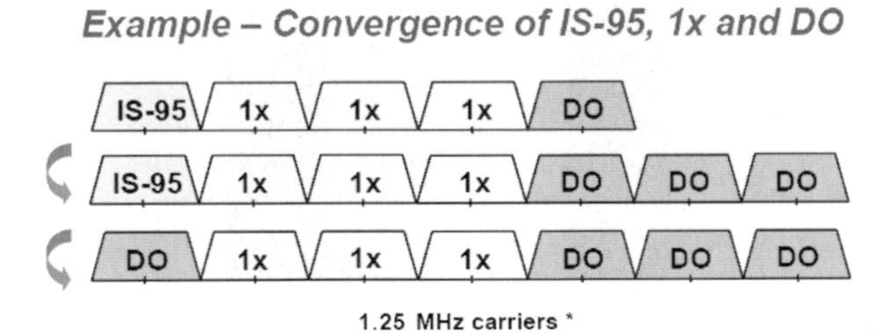


*Figure 30.* Example of frequency allocation

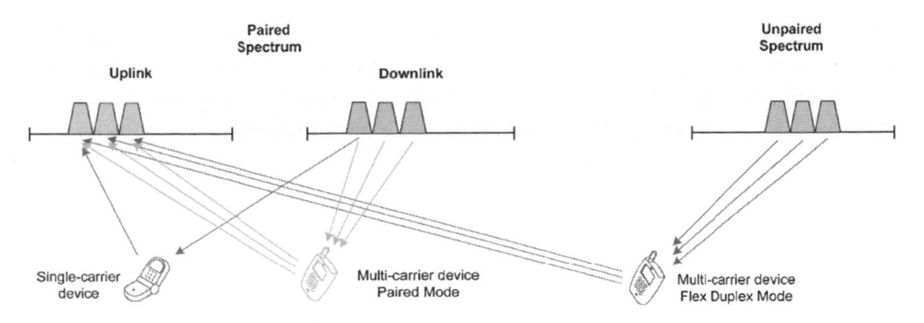


*Figure 31.* Flexible duplexing example

spectrum (for TDD service frequency) to increase the transmission capacity of the forward link, see an example in **Figure 31**.

- Upgrade From Revision A To B

As Revision A and Revision B are highly compatible, through a simple S/W upgrade, Revision A can evolve into Revision B. However, simple S/W upgrades does not provide all features of Revision B, and some features require H/W upgrades.

## 5.4 Major Rival Technologies

Revision B is expected to compete with standards that pursue High Data Rate (HDR) service. Major rivals will be HSDPA (or HSUPA) and Mobile WiMAX, and is expected to be commercialized between the end of 2006 and early 2008 like Revision B, see the comparison with major rival technologies in **Table 8**.

## 5.5 HSDPA (High Speed Downlink Packet Access)

The HSDPA is a fast forward link packet data service that is based on the WCDMA Release 5 standard. It includes a TDD mode that uses 5MHZ of broadband and

*Table 8.* Comparison with major rival technologies

| Charateristics Title | 1xEV-DO (Rev B) | WCDMA (R4) | HSDPA (R5) | WiBro |
|---|---|---|---|---|
| Bandwidth/FA | 1.25 MHz | 5 MHz | 5 MHz | 10 MHz |
| Service Type | Data only | Voice+Data | Voice+Data | Data Only |
| Peak Data Rate (F.L.) | 4.9 Mbps | 2 Mbps | 14.4 Mbps | 18 Mbps |
| Peak Data Rate (R.L.) | 1.8 Mbps | 2 Mbps | 2.3 Mbps | 6 Mbps |
| Mobility | | 250 Km/h | | ~60 Km/h |
| Throughput (DL, 1 FA/1 Sec) | 700 kbps | 838 kbps | 1.213 Mbps | 5.3 Mbps |
| QoS | Rarely Guaranteed | Guaranteed | Guaranteed | Partially Guaranteed |

an FDD mode that supports 1.25MHz of a narrow band. In the TDD mode, the HSDPA supports 14Mbps of the peak data rate for the forward link and 2Mbps for the reverse link. The HSDPA is expected to have advantages over other services in the global roaming, because 85% of the mobile communication service providers adopt WCDMA compatible standards.

### 5.6 Mobile WiMAX

Mobile WiMAX is a mobile communication service that complies with 802.16e, and is the most similar to Korea's Wireless Broadband Internet (WiBro) that uses a 2.3 GHz band and is scheduled to be serviced early 2006. Unlike Revision B, an FDD-based service, the WiBro is based on the TDD and can change the time ratio between the forward and the reverse link. Therefore, the WiBro is more suitable for data services with large forward link traffic volume. Unlike existing data service technologies that adopted CDMA-based modulation technique, the WiBro adopted the OFDM-based modulation technique to provide more reliable performance in urban areas with high multipath fading and to more easily adopt broadband services.

## 6. ABBREVIATION

**Access Channel**. A Reverse CDMA Channel used by mobile stations for communicating to the base station. The Access Channel is used for short signaling message exchanges, such as call originations, responses to pages, and registrations. The Access Channel is a slotted random access channel.

**Access Network (AN).** The network equipment providing data connectivity between a packet switched data network (typically the Internet) and the access terminals. An access network is equivalent to a base station.

**Access Terminal (AT).** A device providing data connectivity to a user. An access terminal may be connected to a computing device such as a laptop personal computer or it may be a self-contained data device such as a personal digital assistant. An access terminal is equivalent to a mobile station.

**Active Set.** The set of pilots associated with the CDMA Channels containing Forward Traffic Channels assigned to a particular mobile station.

**AMPS.** Advanced Mobile Phone Service.

**ARQ.** Automatic Repeat Request. Technique for providing reliable delivery of signals between communicating stations which involves autonomous retransmission of the signals and transmission of acknowledgments until implicit or explicit confirmation of delivery is received.

**Authentication.** A procedure used by a base station to validate a mobile station's identity.

**Base Station.** A fixed station used for communicating with mobile stations. Depending upon the context, the term base station may refer to a cell, a sector within a cell, an MSC, or other part of the wireless system. See also MSC.

**bps.**Bits per second.
**BPSK.**Biphase shift keying.

**Broadcast Control Channel.** A code channel in a Forward CDMA Channel used for transmission of control information and pages from a base station to a mobile station.

**CDMA.** Code Division Multiple Access.

**CDMA Channel.** The set of channels transmitted between the base station and the mobile stations within a given CDMA frequency assignment.

**Chip Rate.** Equivalent to the spreading rate of the channel. It is either 1.2288 Mcps or 3.6864 Mcps.

**Convolutional Code.** A type of error-correcting code. A code symbol can be considered as the convolution of the input data sequence with the impulse response of a generator function.

**CRC.** Cyclic Redundancy Code. A class of linear error detecting codes which generate parity check bits by finding the remainder of a polynomial division.

**Ec/I0.** The ratio in dB between the pilot energy accumulated over one PN chip period (Ec) to the total power spectral density (I0) in the received bandwidth.

**FDMA**. Frequency Division Multiplexing Access

**Forward CDMA Channel.** A CDMA Channel from a base station to mobile stations. The Forward CDMA Channel contains one or more code channels that are transmitted on a CDMA frequency assignment using a particular pilot PN offset.

**Forward Fundamental Channel.** A portion of a Forward Traffic Channel which carries a combination of higher-level data and power control information.

**Forward Common Control Channel.** A control channel used for the transmission of digital control information from a base station to one or more mobile stations.

**Forward Dedicated Control Channel.** A portion of a Radio Configuration 3 through 9 Forward Traffic Channel used for the transmission of higher-level data, control information, and power control information from a base station to a mobile station.

**Forward Power Control Subchannel.** A subchannel on the Forward Fundamental Channel or Forward Dedicated Control Channel used by the base station to control the power of a mobile station when operating on the Reverse Traffic Channel.

**Forward Supplemental Channel.** A portion of a Radio Configuration 3 through 9 Forward . Traffic Channel which operates in conjunction with a Forward Fundamental Channel or a Forward Dedicated Control Channel in that Forward Traffic Channel to provide higher data rate services, and on which higher-level data is transmitted.

**IS-95**. Industry Standard – 95, or EIA/TIA-95

**Long Code.** A PN sequence with period 242 – 1 that is used for scrambling on the Forward CDMA Channel and spreading on the Reverse CDMA Channel. The long code uniquely identifies a mobile station on both the Reverse Traffic Channel and the Forward Traffic Channel. The long code provides limited privacy. The long code also separates multiple Access Channels and Enhanced Access Channels on the same CDMA Channel. See also Public Long Code and Private Long Code.

**Mobile Station.** A station in the Public Cellular Radio Telecommunications Service intended to be used while in motion or during halts at unspecified points. Mobile stations include portable units (e.g., hand-held personal units) and units installed in vehicles.

**Multiplex Option** . Used to specify the multiplex sublayer operation for a physical channel. Each Multiplex Option specifies the available data rates for the physical channel (FCH, DCCH or max SCH rate).

**Orthogonal Transmit Diversity (OTD).** A forward link transmission method which distributes forward link channel symbols among multiple antennas and spreads the symbols with a unique Walsh or quasi-orthogonal function associated with each antenna.

**Paging Channel.** A code channel in a Forward CDMA Channel used for transmission of control information and pages from a base station to a mobile station.

**Protocol Data Unit (PDU).** Encapsulated data communicated between peer layers on the mobile station and the base station.

**Registration.** The process by which a mobile station identifies its location and parameters to a base station.

**Request.** A layer 3 message generated by either the mobile station or the base station to retrieve information, ask for service, or command an action.

**Response.** A layer 3 message generated as a result of another message, typically a request.

**Reverse CDMA Channel.** The CDMA Channel from the mobile station to the base station. From the base station's perspective, the Reverse CDMA Channel is the sum of all mobile station transmissions on a CDMA frequency assignment.

**Reverse Common Control Channel.** A portion of a Reverse CDMA Channel used for the transmission of digital control information from one or more mobile stations to a base station.

**Reverse Dedicated Control Channel.** A portion of a Radio Configuration 3 through 6. Reverse Traffic Channel used for the transmission of higher-level data and control information from a mobile station to a base station.
**Reverse Fundamental Channel.** A portion of a Reverse Traffic Channel which carries higher-level data and control information from a mobile station to a base station.

**Reverse Pilot Channel.** An unmodulated, direct-sequence spread spectrum signal transmitted continuously by a CDMA mobile station. A reverse pilot channel provides a phase reference for coherent demodulation and may provide a means for signal strength measurement.

**Reverse Power Control Subchannel.** A subchannel on the Reverse Pilot Channel used by the mobile station to control the power of a base station when operating on the Forward Traffic Channel with Radio Configurations 3 through 9.

**RLP** Radio Link Protocol. Connection-oriented, negative-acknowledgment-based data delivery protocol.

**Service Option.** A service capability of the system. Service options may be applications such as voice, data, or facsimile.

**Soft Handoff.** This handoff is characterized by commencing communications with a new base station on the same CDMA Frequency Assignment before terminating communications with an old base station.

**TIA.** Telecommunications Industry Association.

**QoS** Quality of Service. Metrics that affect the quality of a data service that is delivered to an end user (e.g., throughput, guaranteed bit rate, delay, etc.).

**UATI**. Unicast Access Terminal Identifier.

**LBA** Link Budget Analysis

**SCH** Supplemental Channel

**BRC** BTS Resource Control

**Mux PDU** Multiplex Sublayer Protocol Data Unit

**FER** Frame Error Rate

**PCB** Power Control Bit

**FA** Frequency Assignment

**MAC** Medium Access Control

**RA** Reverse Activity

**RPC** Reverse Power Control

**DRC** Data Rate Control

**RAB** Reverse Activity Bit

**PER** Packet Error Rate

**DMSS** Dual-Mode Subscriber Software

**HAL** Hardware Abstraction Layer

## 7. REFERENCES

[1] C.S0001-A, Introduction to cdma2000 Standards for Spread Spectrum Systems, July 2000.
[2] C.S0002-A, Physical Layer Standard for cdma2000 Spread Spectrum Systems, July 2000.
[3] C.S 003-A, Medium Access Control (MAC) Standard for cdma2000 Spread Spectrum System, July 2000.
[4] C.S0004-A, Signaling Link Access Control (LAC) Standard for cdma2000 Spread Spectrum Systems, July 2000.
[5] C.S0005-A, Upper Layer (Layer 3) Signaling Standard for cdma2000 Spread Spectrum Systems, July 2000.
[6] C.S0006-A, Analog Signaling Standard for cdma2000 Spread Spectrum Systems, July 2000.
[7] C.S0024-0 V4.0, cdma2000 High Rate Packet Data Air interface Specification, October 2002.
[8] TIA/EIA/IS-657, Packet Data Services Option Standard for Wideband Spread Spectrum Systems, July 1996.
[9] C.S0017-0, Data Service Options for Wide Spread Spectrum Systems, April 1999.
[10] C.S0017-0-1, Data Service Options for Wideband Spread Spectrum Systems-Addendum 1, January 2000.
[11] P.S0001-A, Wireless IP Network Standard.
[12] C.P9011, Recommended Minimum Performance Standards for cdma2000 High Rate Packet Data Access.
[13] TIA/EIA/TSB-707-A, Data Service Option for Widespread Spectrum systems.
[14] "IMT-2000 이동통신 원리" 진한도서, 김현욱외 3 명, 2001.5
[15] TIA/EIA IS-2000-A, "cdma2000 standards for Spread Spectrum Systems"
[16] TIA/EIA IS-856-1, "cdma2000 High Rate Data Air Interface Specification"
[17] EPBD-000907, "SCBS-418L BTS System Manual"
[18] "SCH Burst Operation (SCH Scheduler)"
[19] 80-H0410-1 Rev.X3, "Reverse Link Medium Access Control Algorithm for IS-856"

CHAPTER 6

# EVOLUTION OF THE WCDMA RADIO ACCESS TECHNOLOGY

ERIK DAHLMAN[1] AND MAMORU SAWAHASHI[2]

[1] *Ericsson AB, Sweden*

[2] *Musashi Institute of Technology, Japan*

**Abstract:** This chapter describes the evolution of WCDMA radio access technologies such as HSDPA (High-speed Downlink Packet Access), HSUPA (High-speed Uplink Packet Access), which is an enhanced uplink scheme, and MBMS (Multimedia Broadcast/Multicast Services) (i.e., Release 6 MBMS) Future WCDMA evolution including continuous connectivity and MIMO (Multiple-Input-Multiple-Output) channel transmission in WCDMA is also described.

**Keywords:** WCDMA, HSDPA, HSUPA (Enhance Uplink), Scheduling, Link Adaptation, Hybrid ARQ, MBMS, MIMO

## 1. INTRODUCTION

Already in its first release, the 3rd generation WCDMA standard developed by 3GPP allows for mobile-communication systems to provide packet-data services with a performance far exceeding what can be provided with mobile-communication systems based on earlier, 2nd generation standards such as GSM and PDC. However, user requirements and expectations in terms of packet-data services as well as other mobile-communication services are continuously expanding. As a consequence, 3G radio-access technologies such as WCDMA must evolve in order to match these requirements and expectations and stay competitive against other radio-access technologies. This section provides a description of the steps that have been taken on this WCDMA evolution and also provides an overview of the further WCDMA evolution steps that are currently being considered by 3GPP.

The first step of the WCDMA evolution consisted of the introduction of HSDPA or *High-Speed Downlink Packet Access* in 3GPP release 5 finalized during 2002. With HSDPA, the WCDMA support for packet-data services was significantly improved, especially in the downlink network-to-mobile-terminal direction.



*Y. Park and F. Adachi (eds.), Enhanced Radio Access Technologies for Next Generation Mobile Communication*, 191–216.

The introduction of HSDPA in 3GPP release 5 was followed by the introduction of *Enhanced Uplink*, also sometimes referred to as HSUPA or *High-Speed Uplink Packet Access,* in 3GPP release 6, finalized early 2005. Enhanced Uplink further improves the WCDMA support for packet-data services, as the name suggests with focus on improvements in the uplink mobile-terminal-to-network direction.

In parallel to the improved support for packet-data services provided by the introduction of the Enhanced Uplink feature, 3GPP release 6 also introduced improved support for broadcast/multicast services in the WCDMA standard by the introduction of the MBMS or *Multimedia Broadcast/Multicast Service* functionality.

Together these evolutionary steps have significantly enhanced WCDMA in terms of system performance and service provision. However, in order to continue to stay competitive also in the future the WCDMA radio-access technology must and will continue to evolve. As an example, at the time of writing 3GPP is finalizing the introduction of features for improved *Continuous Packet Connectivity* into the WCDMA standard. In parallel, 3GPP is also working on an introduction of MIMO or *Multiple Input Multiple Output* antenna processing for HSDPA.

In parallel to this step-by-step evolution of the WCDMA radio-access technology, there is also work ongoing within 3GPP on a more extensive long-term evolution of the 3GPP radio access technologies referred to as the *3GPP Long-Term Evolution* or *3GPP LTE.*

## 2. HSDPA – HIGH-SPEED DOWNLINK PACKET ACCESS

HSDPA or *High Speed Downlink Packet Access* was introduced in 3GPP release 5 with an aim to significantly improve the WCDMA support for packet-data services, more specifically targeting

- significantly improved downlink system capacity for packet-data services,
- possibility for significantly reduced delay/latency within the radio-access network, and
- possibility for significantly higher downlink data rates

To achieve these targets, HSDPA introduced the following performance-enhancing techniques into the WCDMA standard

- *Shared-channel transmission*, i.e. the possibility to dynamically share the downlink code resource between different users
- Possibility for *higher-order modulation* as a tool to provide higher data rates but also higher system efficiency
- Support for *fast channel-dependent scheduling* and *fast rate control* as tools to adapt to and utilize fast variations in the instantaneous channel conditions
- *Fast (hybrid) ARQ* with *soft combining* at the receiver side as a tool to reduce latency and improve system efficiency

HSDPA also introduced a shorter downlink *Transmission Time Interval* or TTI in order to allow for reduced radio-interface round trip time as well as to enable adaptation to fast variations in the channel conditions and fast ARQ.

## 2.1 Shared-channel Transmission

For downlink packet-data transfer in a mobile-communication system, *shared-channel transmission* is preferably used. Downlink shared-channel transmission implies that a certain amount of the downlink radio resources (channelization codes and transmit power in case of WCDMA) is seen as a common resource that is *dynamically* shared between users. Shared-channel transmission has two main benefits, both related to the efficient utilization of the available radio resources:

- As seen on a per-user basis, packet-data traffic typically has very bursty characteristics. Allocating a set of resources semi-statically for downlink transmission to a certain packet-data user may thus lead to in-efficient resource utilization. In contrast, dynamic resource allocation allows for radio resources to be instantaneously allocated/re-allocated for transmission to packet-data users on a per-need basis. This will allow for significantly improved resource utilization and a corresponding improved system efficiency.
- Dynamic resource allocation does not only imply that resources can be allocated to a user that momentarily needs them but also to a user that can, due to specific instantaneous downlink channel conditions, use the allocated resources most efficiently. The possibility for dynamic resource allocation by means of shared-channel transmission thus allows for so-called *channel-dependent scheduling*, see further Section 1.4.

It should be pointed out that dynamic power allocation by means of fast (closed-loop) power control was supported already in the first releases of the 3WCDMA standard. The key feature introduced into WCDMA as part of the introduction of HSDPA is the support also for dynamic sharing of the *channelization-code* resource.

To support shared-channel transmission, HSDPA introduced a new transport channel, the HS-DSCH or *High-Speed Downlink Shared Channel*, to which packet-data traffic can be mapped.

### (1) HS-DSCH code- and time-domain structure

The code resource used for HS-DSCH transmission consists of a set of channelization codes at spreading factor 16, see **Figure 1**. Note that the first (left most) channelization code at spreading factor 16 can not be allocated for HS-DSCH transmission as the corresponding node within the code tree e.g. includes the pre-allocated primary common pilot channel CPICH as well as the channel carrying the system broadcast information. Thus, a maximum of 15 channelization codes can be allocated for HS-DSCH transmission. If a substantial part of the overall code space is needed for other downlink channels, e.g. for services that are not to be carried on HS-DSCH, the number of channelization codes that can be used for HS-DSCH transmission is reduced. As an example, **Figure 1** assumes that twelve channelization codes are allocated for HS-DSCH transmission.

The HS-DSCH code resource can be dynamically allocated on a 2 ms TTI or *Transmission-Time-Interval* basis. The use of a 2 ms TTI for HS-DSCH transmission, which is significantly shorter than the 10 ms minimum TTI of earlier WCDMA releases, allows for a significantly reduced radio-interface delay. A short TTI is also important in terms of supporting tight adaptation to fast variations

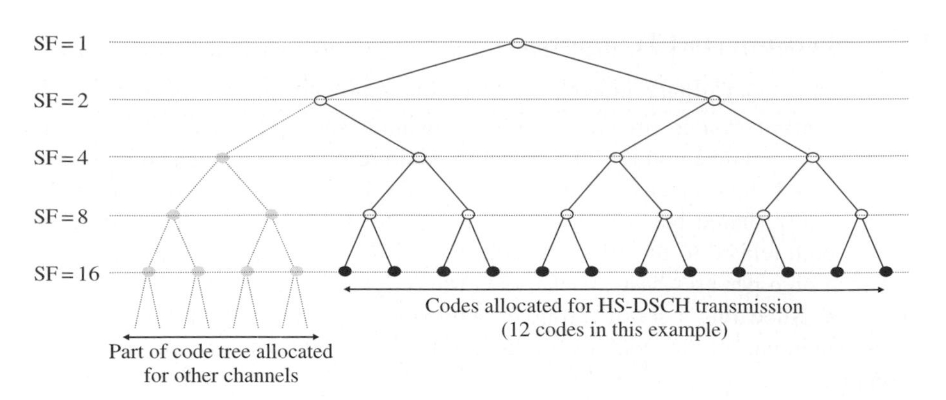


*Figure 1.* HS-DSCH code-domain structure assuming 12 channelization codes allocated for HS-DSCH transmission

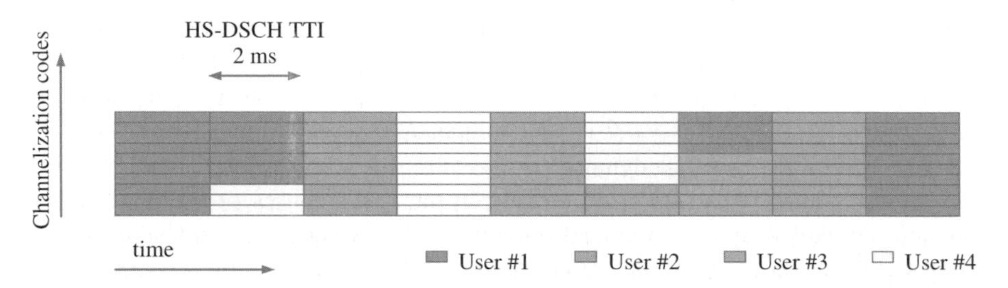


*Figure 2.* HS-DSCH time-domain structure, assuming 12 channelization codes allocated for HS-DSCH transmission

in the instantaneous channel conditions, i.e. channel-dependent rate control and scheduling, as described in Sections 1.3 and 1.4, as well as being an enabler for fast retransmissions, see Section 1.5.

As illustrated in **Figure 2** the full set of HS-DSCH channelization codes can be dynamically allocated for downlink transmission to a single user, see for example the first TTI of **Figure 2**. This obviously allows for an instantaneously very large resource allocation and the possibility for a corresponding very high instantaneous transmission data rate to a single user. Allocation of the entire HS-DSCH code resource for transmission to a single user at a time is often referred to as *Time Division Multiplexing* (TDM).

Alternatively, different sub-sets of the HS-DSCH channelization codes can be allocated to different users, see e.g. the second TTI of **Figure 2** Using different sub-sets of the HS-DSCH code resource for parallel transmission to different users is often referred to as *Code Division Multiplexing* (CDM). Although from a channel-dependent-scheduling point-of-view, see Section 1.4, TDM operation is fundamentally more efficient, there are several reasons why also CDM-based allocation of the HS-DSCH code resource is supported for HSDPA:

- Low-end mobile terminals may not be capable of receiving and demodulating the full set of HS-DSCH channelization codes (up to 15 codes of spreading factor 16 according to above) or support decoding of the corresponding potentially very high instantaneous data rates. As an example, there may be HSDPA-capable terminals that only support reception of a maximum of five HS-DSCH channelization codes. For efficient channelization-code usage, the possibility for sharing of the HS-DSCH code resource also in the code domain (CDM) is then required.
- Even if all mobile terminals are able to receive the full set of HS-DSCH channelization codes, the per-user payload available at the base station for transmission to a specific user may not be sufficient to efficiently fill up the full HS-DSCH code resource. Thus, once again, for efficient channelization-code usage, the possibility for CDM is required.

### (2) HS-DSCH power-domain structure

In addition to being allocated a certain part of the overall channelization-code resource, HS-DSCH transmission should also be allocated a part of the total cell power. There are two main alternatives for the HS-DSCH power allocation, see also **Figure 3**.

- *Semi-static power allocation*, i.e. a fixed part of the overall cell power is allocated for HS-DSCH transmission. The remaining power is shared between other channels such as common-control channels with constant power and power-controlled dedicated channels.
- *Dynamic power allocation*, i.e. for each TTI HS-DSCH transmission can use the remaining power after power has been dynamically allocated to common-control channels and power-controlled dedicated channels. Consequently, with dynamic power allocation, a temporarily reduced need for power for the power-controlled

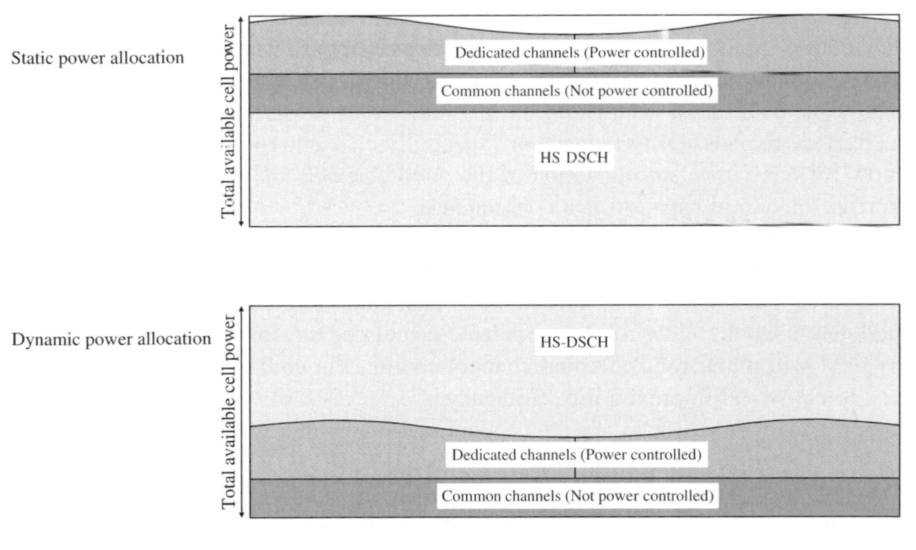


*Figure 3.* HS-DSCH power allocation, semi-static vs. dynamic

dedicated channels may immediately be used to provide increased HS-DSCH capacity. Thus dynamic power allocation is more efficient and should be employed in order to achieve maximum utilization of the overall cell transmit power.

The amount of power that can typically be used for HS-DSCH transmission depends on how much power is needed for other channels. In practice it can be expected that, as a maximum, in the order of 70–80% of the total cell power can be used for HS-DSCH transmission, leaving sufficient power for necessary (common and/or dedicated) control channels. If there are additional downlink channels, e.g. for circuit-switched speech, the power available for HS-DSCH transmission will obviously be lower.

## 2.2 Support for Higher-order Modulation

The first releases of the 3G standards, including WCDMA, only supported QPSK modulation for downlink transmission, thus allowing for two bits of information to be transmitted per modulation symbol. In order to support higher data rates within a given bandwidth, HSDPA introduced additional support for downlink 16QAM modulation, allowing for up to four bits per modulation symbol, i.e. twice the spectral efficiency of QPSK.

Together with the support for shared-channel transmission, the support for 16QAM modulation allows for significantly higher downlink data rates to a single user, compared to earlier 3GPP releases as well as other mobile-communication technologies. Assuming the largest possible HS-DSCH channelization-code allocation (15 codes), HSDPA allows for peak data rates up to 14 Mbps in the current releases. This is expected to increase further, in a first step to around 28 Mbps, by the introduction of e.g. *multi-layer transmission* for HSDPA, see Section 4.2.

It is important to understand that the introduction of 16QAM does not only allow for higher downlink peak data rates. In a cellular system only supporting QPSK modulation, a mobile terminal may, in some cases, experience such good downlink channel conditions (high signal-to-noise and signal-to-interference ratios) that the achievable data rate is bandwidth limited rather than noise/interference limited. In such cases, the possibility to use more spectrally efficient 16QAM modulation will allow for more efficient utilization of the good channel conditions with an overall increase in system capacity as a consequence.

Furthermore, certain data rates, although possible to support with QPSK modulation, may be more *efficiently* supported with higher order modulation (16QAM). This will e.g. be the case for data rates for which the use of QPSK modulation would allow for very limited channel coding. In such cases, the use of 16QAM will allow for additional channel coding, the coding gain of which may lead to an overall improved link efficiency.

## 2.3 Fast Link Adaptation / Rate Control

In a cellular system, the radio-channel conditions experienced by different network-to-user-terminal links will typically vary significantly, both in time and between different positions within the cell. In general there are several reasons for these variations and differences in the instantaneous channel conditions:

- The channel conditions will differ significantly between different mobile-terminal positions due to distance-dependent path loss and shadowing.
- The instantaneous channel conditions will vary rapidly due to multi-path fading. The rate of these variations depends on the speed of the mobile terminal. However, typically there will be significant variations during a fraction of a second.
- The channel conditions will vary due to variations in the interference level. The interference level will depend on the position of the user terminal within the cell with typically higher interference level close to the cell border. The interference level will also depend on the instantaneous transmission activity of neighbor cells.

Downlink power control can be used to compensate for differences and variations in the instantaneous downlink channel conditions. In principle, downlink power control allocates a proportionally larger part of the total available cell power to communication links with instantaneously bad channel conditions. This can be seen as one example of so-called *link adaptation*, i.e. the adjustement of transmission parameters, in this case the transmission power, to compensate for differences and variations in the instantaneous channel conditions.

In general, the goal of link adaptation is to ensure sufficient received energy per information bit for all communication links, despite variations and differences in the channel conditions. Power control achieves this by adjusting the transmission power while keeping the data rate of each communication link constant. Keeping the data rate constant, regardless of the instantaneous channel conditions, is obviously often desirable and may even be required for some services. However, for services that do not require a specific data rate, such as many packet-data services, the energy per information bit can also be controlled by adjusting the data rate while keeping the transmission power constant. This kind of link adaptation can be referred to as *rate control* or *rate adjustment*. Especially in combination with shared-channel transmission and channel-dependent scheduling, see Section 1.4, rate control is a more efficient approach to link adaptation, compared to power control.

In case of HS-DSCH transmission, the instantaneous data rate can be controlled by selecting between different modulation schemes (QPSK vs. 16QAM) as well as selecting different channel-coding rates. Together this is referred to as dynamic selection of the HS-DSCH *transport format*. Selecting higher-order modulation (16QAM) and/or a higher channel-coding rate, allows for higher data rates over the radio interface. However, this is only applicable in case of instantaneously good channel conditions. In case of not-so-good channel conditions more robust QPSK modulation together with a lower channel-coding rate should be used.

Obviously, the downlink channel conditions are never known exactly at the base-station. Instead, some estimates of the instantaneous channel conditions must be used for the rate control. In case of HSDPA, the mobile terminal is continuously estimating the instantaneous downlink channel conditions based on measurements on the common pilot channel. Based on the estimated channel conditions the mobile terminal then estimates a suitable HS-DSCH transport format (modulation scheme and coding rate) and reports this to the network in form of a *Channel Quality Indicator* CQI. The CQI reporting can be done as often as once every TTI, i.e. once every 2 ms, allowing for tracking also of fast variations in the instantaneous channel conditions.

The base station can then use the reported CQI for deciding on the actual transport format to use for HS-DSCH transmission to this specific mobile terminal. It should be pointed out though that the base station does not *need* to follow the reported CQI. Rather, the reported CQI should be seen as a *recommended* transport format or data rate.

It should also be pointed out that, despite the term *Channel Quality* Indicator, the CQI reported by the mobile terminal is not an objective measure of the channel quality such as the received signal-to-noise/interference ratio. Instead the reported CQI is, as already mentioned, expressed as a transport format, more specifically the transport format that the mobile terminal estimates would be a suitable transport format for the network to use, given the estimated channel conditions. The reason for reporting a (recommended) transport format instead of the actual estimated channel quality is that different mobile terminals may implement more or less advanced receiver structures, including multiple receive antennas and different types of equalization. Thus for the same channel quality, different mobile terminals may be able to receive HS-DSCH with different data rates (different transport formats). By specifying the CQI reporting as a supportable transport format rather than a mobile-terminal-independent channel-quality measure, a user may benefit from mobile terminal with a more advanced receiver by being served with higher data rates.

## 2.4 Channel-dependent Scheduling

As described above, link adaptation by means of rate control, i.e. the adjustment of the data rate to match the instantaneous channel conditions, can be used together with shared-channel transmission to improve overall system performance, assuming that variations and differences in the instantaneous data rate are acceptable. Another important question is how to select the mobile terminal(s) to which to transmit data on the shared channel in a given time interval. This is also referred to as the *scheduling strategy*.

In case of so-called *Round-Robin* (RR) scheduling, radio resources are basically allocated to communication links on a sequential basis and especially without any considerations to the downlink channel conditions experienced by different mobile terminals. However, further improvements in system performance can be achieved by taking the instantaneous channel conditions into account not only for link adaptation but also in the scheduling process. This is also referred to as *channel-dependent scheduling*.

To achieve maximum system throughput, data should as much as possible be scheduled on communication links with the best instantaneous channel conditions, also known as *max-C/I scheduling*. Combined with link adaptation based on rate control, max-C/I scheduling allows for the highest possible data rate at each scheduling instant and thus maximizes the overall cell throughput. However, this is achieved at the expense of potentially large differences in the throughput experienced by different users, i.e. large differences in service quality between users. In the extreme case, no downlink data what-so-ever may be scheduled to users with bad channel conditions, e.g. users at the cell border.

As Round-Robin scheduling does not take the instantaneous channel conditions into account in the scheduling process it will lead to lower overall system performance but more equal service quality between different communication links, compared to max-C/I scheduling. Round-Robin scheduling can be seen as *fair scheduling* in the sense that the same amount of radio resources (the same amount of time) is given to each communication link. However, Round-Robin scheduling is not fair in the sense that it provides the same service quality to all communication links. In that case more radio resources (more time) must be given to communication links with bad channel conditions.

When discussing and comparing different scheduling algorithms it is important to distinguish between different types of variations in the service quality:

- Fast variations in the service quality corresponding to e.g. fast multi-path fading and fast variations in the interference level. For many packet-data applications, relatively large short-term variations in service quality are often acceptable or not even noticeable to the user.
- More long-term differences in the service quality between different communication links corresponding to e.g. differences in the distance to the cell site and shadow fading. In many cases there is a need to limit such long-term differences in service quality.

A practical scheduler should thus operate somewhere in-between the max C/I scheduler and the RR scheduler, i.e. try to utilize fast variations in channel conditions as much as possible while still satisfying some degree of fairness between users.

The *Proportional Fair* scheduling algorithm is an example of a scheduling algorithm that has been designed with an aim to achieve this. In a given scheduling instant, a proportional-fair scheduler schedules data on a communication link for which the data rate is highest relative to the average data rate of that communication link, with averaging based on some time constant $T_{PF}$.

To ensure efficient usage of the short-term channel variations and, at the same time, limit the long-term differences in service quality to an acceptable level, the time constant $T_{PF}$ should be set longer than the time constant for the short-term variations. At the same time $T_{PF}$ should be sufficiently short so that quality variations within the interval $T_{PF}$ are not strongly noticed by a user. Typically, $T_{PF}$ can be set to be in the order of one second.

It should be noted that there is little difference between different scheduling algorithms at low system load, i.e. when only one or, in some cases, a few users have data waiting for transmission at the base station at each scheduling instant. The differences between different scheduling algorithms are primarily visible at high load.

It should also be noted that the scheduling algorithm is a base-station-implementation issue and nothing that is typically specified in any standard. What is needed in a standard to support channel-dependent scheduling is channel-quality measurements/reports and the signaling needed for dynamic resource allocation. For HSDPA, the same CQI measurements/reports as is used for the fast rate control can be used to also support fast channel-dependent scheduling at the base station.

## 2.5 Fast Hybrid ARQ with Soft Combining

Strictly speaking, the term *Hybrid ARQ* simply implies that a communication link relies on a combination (hybrid) of forward-error correction ("channel coding") and error-detection/retransmissions. The key properties of the Hybrid ARQ scheme adopted for HSDPA are instead the following:

- It allows for very fast error detection and sub-sequent retransmissions, with a retransmission round-trip time as low as 12 ms, thus the term *fast* Hybrid ARQ. This implies that the HSDPA retransmission protocol can operate with a significantly higher error rate and retransmission probability, without the radio link suffering from very large overall delays. Operating on a higher retransmission probability in general implies improved system efficiency.
- It allows for soft combining of different retransmissions at the receiver side, thus the term Hybrid ARQ *with soft combining*, see below. This also leads to a significant improvement in the efficiency of the retransmission protocol.

In order to minimize the signaling associated with a retransmission protocol, something which is especially important if fast retransmissions and a low round-trip time is to be achieved, a *Stop-and-Wait* ARQ protocol is often preferred. However, a Stop-and-Wait ARQ protocol suffers from bad link efficiency due to the constraint that no new transmissions can be carried out until an acknowledgement of a previous transmission has been received. Thus, in order to have a simple ARQ protocol with a minimum amount of signaling and still allow for full link efficiency, a so-*called N-channel stop-and-wait* ARQ protocol was adopted for HS-DSCH transmission. As outlined in **Figure 4**, an N-channel stop-and-wait protocol consists of N staggered stop-and-wait protocols. By selecting N equal to or larger than the ARQ protocol round-trip time, the radio link can be fully utilized by transmissions to a single mobile terminal.

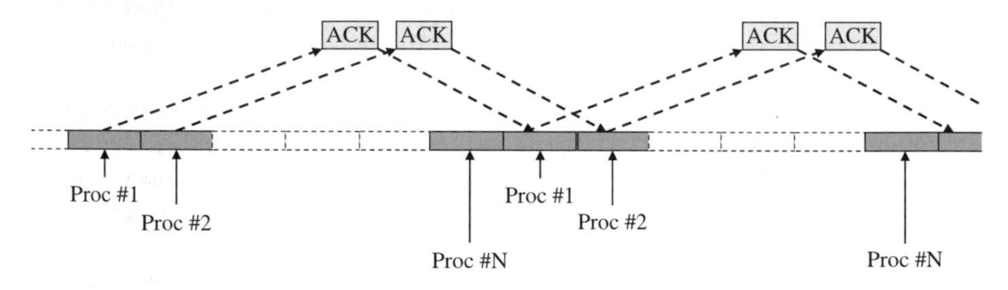


*Figure 4.* N-channel stop-and wait ARQ protocol

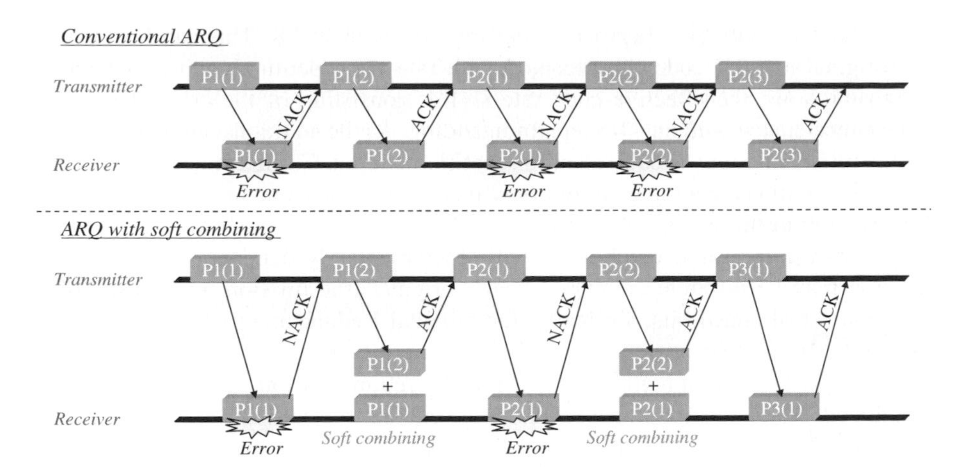


*Figure 5.* Conventional ARQ vs. ARQ with soft combining at the receiver side

The basic principle of Hybrid ARQ with soft combining is illustrated in **Figure 5**. With a conventional ARQ scheme, received data blocks that can not be correctly decoded are discarded by the receiver and received retransmitted data blocks are separately decoded. In contrast, in case of Hybrid ARQ with soft combining, received data blocks that can not be correctly decoded are not discarded. Instead the corresponding received signal is buffered and soft combined with later received retransmissions. Decoding is then applied to the combined signal.

The use of Hybrid ARQ with soft combining increases the total received energy per information bit and thus the effective received $E_b/I_0$ for each retransmission. As an example, assuming that the original transmission and a retransmission are received with an $E_b/I_0$ equal to $\gamma_1$ and $\gamma_2$ respectively, the effective or accumulated $E_b/I_0$ after the retransmission will be $\gamma_{tot} = \gamma_1 + \gamma_2$, compared to an $E_b/I_0$ equal to $\gamma_2$ in case of conventional ARQ (without soft combining). Clearly, with soft combining, the probability for correct decoding after the retransmission and subsequent soft combining will be improved, compared to conventional ARQ.

There are many different schemes for Hybrid ARQ with soft combining. However, there are two main approaches, frequently referred to as *Chase combining* and *Incremental Redundancy*, respectively. These two approaches differ in the structure of the retransmissions and in the way by which the soft combining is carried out at the receiver side.

In case of Chase combining each retransmission is an identical copy of the original transmission, i.e. each retransmission consists of exactly the same set of coded bits as the original transmission. As a consequence, soft combining can be done on the received signals before demodulation.

As each retransmission is an identical copy of the original transmission, the retransmissions with Chase combining can be seen as additional repetition coding. As an example, assume that the original transmission is coded with a rate-3/4 code.

The effective code rate after one retransmission is then 3/8. This can be seen as the original rate-3/4 code concatenated with rate-1/2 repetition coding. After two retransmissions, the effective code rate is 1/4, consisting of the original rate-3/4 code concatenated with rate-1/3 repetition coding. As the additional coding provided by the retransmissions is equivalent to repetition coding, Chase combining does not give any additional coding gain 'but only increases the accumulated received $E_b/I_0$ for each retransmission.

In contrast to Chase combining, with Incremental Redundancy each retransmission may be different compared to the original transmission, i.e. consist of a different set of coded bits. Typically, Incremental Redundancy is based on a low-rate code. In the first transmission only a sub-set of the coded bits is transmitted, effectively leading to a high-rate code. For the retransmissions, a different sub-set of the coded bits, i.e. additional redundancy, can be transmitted.

In addition to a gain in accumulated received $E_b/I_0$, Incremental Redundancy may thus lead to a true reduction in the channel-coding rate and thus to an additional coding gain for each retransmission. Thus, at least from a link-level-performance point-of-view, Incremental Redundancy should be superior to Chase combining. This can also be seen from **Figure 6**, which compares the error-rate performance of Chase combining and Incremental Redundancy after different number of retransmissions and for different initial coding rates. As expected, the gain with Incremental Redundancy is larger in case of higher a initial coding rate (right-hand figure). With a high initial coding rate, the initial coding gain is relatively small and the additional redundancy provided by Incremental Redundancy is more important. The gain with IR will also be larger in case of higher-order modulation.

As each retransmission will normally be different compared to the original transmission, i.e. consist of a difference set of coded bits, each IR retransmission must be separately demodulated and buffered at the receiver side. Soft combining is then done implicitly as part of the decoding process.

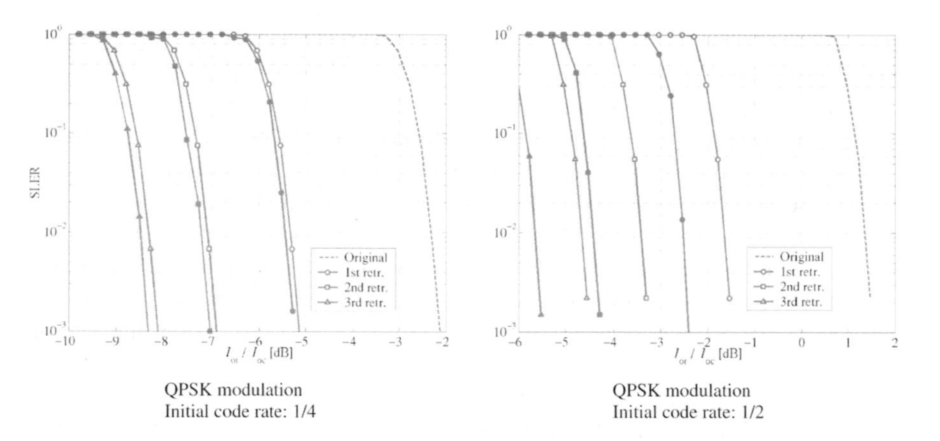


*Figure 6.* Gains of IR vs. Chase Combining. IR curves to the left in each figure

## 3. HSUPA – HIGH SPEED UPLINK PACKET ACCESS ("ENHANCED UPLINK")

Enhanced Uplink, also sometimes referred to as HSUPA or *High Speed Uplink Packet Access,* was introduced in 3GPP release 6 (finalized in 2005) to complement HSDPA and further improve the WCDMA packet-data support, with focus on the uplink, mobile-terminal-to-network, direction. Jointly, HSDPA and Enhanced Uplink are often referred to as HSPA or *High Speed Packet Access.*

The aim of Enhanced Uplink is to further improve the WCDMA support for packet-data services, targeting

- significantly improved uplink system capacity
- further reduced delay/latency with focus on the uplink
- possibility for significantly higher uplink data rates

To achieve these targets, Enhanced Uplink introduces improved *base-station-controlled uplink* scheduling allowing for more efficient utilization of the uplink radio resources. Base-station-controlled uplink scheduling also enables the possibility to provide significantly higher instantaneous uplink data rates to a single user, without the risk for system instability. With Enhanced Uplink, peak uplink data rates beyond 5.7 Mbps can be provided in the uplink in case of good channel conditions.

In addition, Enhanced Uplink also introduces support for fast Hybrid ARQ with soft combining also for the uplink. Similar to the downlink, fast Hybrid ARQ with soft combining for the uplink provides both improved system efficiency and possibility for significantly reduced delay. It should be noted that a reduced uplink delay is beneficial also for downlink data transfer due to its positive impact on the overall radio-interface round-trip time. Thus the introduction of Enhanced Uplink also implies a further improvement in the WCDMA downlink packet-data performance.

These techniques are introduced into the WCDMA standard as part of a new transport-channel type, the *Enhanced Dedicated Channel* or E-DCH. In addition to a 10 ms TTI, the E-DCH also supports a TTI of 2 ms, reducing the radio-interface delays, allowing for fast adaptation of the transmission parameters, and enabling fast retransmissions.

Unlike the downlink direction, the WCDMA uplink is inherently non-orthogonal even within the cell. Fast power control is therefore needed for the uplink also in the case of E-DCH transmission, in order to handle the so-called "near-far problem" and to ensure coexistence on the same carrier with terminals and services not relying on the E-DCH for uplink traffic. The E-DCH is transmitted with a power offset relative to the WCDMA power-controlled uplink control channel, the *DPCCH.* By adjusting the maximum allowed E-DCH/DPCCH power offset, the uplink scheduler at the base station can control the E-DCH data rate, see further below.

Enhanced uplink also retains the uplink *macro diversity* (*"soft handover")* supported in earlier WCDMA releases. In practice, the support for uplink macro diversity implies two things:

(1) Uplink data transmissions can be received by multiple cells, more specifically the cells in the so-called *Active Set* of the mobile terminal

(2) Mobile terminals can be jointly power controlled by multiple cells, more specifically by all the cells in the Active Set

There are two reasons for supporting uplink macro diversity also for E-DCH:

- Receiving transmitted data at multiple cell sites provides a macro-diversity gain which offers the possibility for improved coverage and cell-edge data rates also for E-DCH
- Power control from multiple cells is beneficial in terms of limiting the amount of interference generated in neighbor cells.

One cell within the Active Set of a mobile terminal is defined as the *E-DCH serving cell*. The E-DCH service cell is the cell that has the main responsibility for scheduling of the uplink transmissions from the mobile terminal.

As discussed in Section 1.2, HSDPA introduced the support for higher-order modulation in case of downlink (HS-DSCH) transmission. As described, higher-order modulation for the downlink is useful in situations where the data rates, without the possibility for higher-order modulation, would be bandwidth limited rather than power/SIR limited.

However, on the uplink the situation is somewhat different with regards to higher-order modulation

- Due to the use of mutually non-orthogonal codes for different mobile terminals in WCDMA, there is no need to share channelization codes between mobile terminals on the uplink. Thus, there is less probability for the uplink to be "bandwidth" limited, compared to the downlink.
- Due to power limitations, very high SNR occurs less frequently for the uplink compared to the downlink. This further reduces the probability for the uplink to be bandwidth limited rather than power limited.

For these reasons and in order to reduce the mobile-terminal complexity, higher-order modulation was not introduced as part of the Enhanced Uplink. Once again, note that even without the support for higher-order modulation, uplink data rates beyond 5.7 Mbps can be supported with Enhanced Uplink.

### 3.1 Fast Base-station-controlled Scheduling

Similar to HS-DSCH, Enhanced Uplink introduces fast base-station-controlled scheduling also for the uplink. However, due to fundamental differences between the downlink and uplink transmission directions, the basic scheduling principles are quite different between the downlink and the uplink.

- For the downlink, the cell transmit power and the set of channelization codes are the shared radio resources. The task of the downlink scheduler at the base station is to ensure as efficient utilization as possible of these resources, e.g. by means of channel-dependent scheduling, while also taking e.g. quality-of-service requirements into account.
- For the uplink, the shared resource is instead the amount of tolerable interference at the cell site. The fundamental task of the uplink scheduler is to control the uplink transmissions from the different mobile terminals so that the overall uplink

interference is as close as possible to the maximum tolerable interference level without exceeding it. In this way, maximum system efficiency can be achieved. To achieve this, the scheduler controls what mobile terminals are allowed to transmit at a given time instant as well as with what rate each terminal is allowed to transmit. By moving the uplink scheduling functionality from the *Radio Network Controller (*RNC) to the base station, faster reaction to interference variations is possible. This allows for operation closer to the interference limit and thus allows for more efficient uplink resource utilization.

Channel-dependent scheduling, which typically is used for HSDPA, is possible also for the uplink. However the benefits with uplink channel-dependent scheduling are different and typically smaller, compared to the downlink. As fast closed-loop power control is used for the uplink, including E-DCH, a mobile terminal experiencing good instantaneous channel conditions will still be received with approximately the same power, and thus be able to transmit with similar data rates, as a terminal with more unfavorable channel conditions. This is in contrast to HSDPA and the downlink direction where, at least in principle, a constant transmission power is used and the data rates are adapted to the channel conditions, resulting in a possibility for higher data rates for users with good channel conditions, compared to users experiencing not-as-good channel conditions.

However, for the uplink, a difference in the channel conditions between two mobile terminals will lead to a difference in the *transmission* power of the two terminals and hence a difference in the amount of interference the two terminals will cause to neighbor cells. Thus, the gain in system performance due to uplink channel-dependent scheduling is more indirect, i.e. a reduction in inter-cell interference, compared to the more direct gains in the downlink HSDPA case.

The E-DCH scheduling framework is based on *scheduling requests* sent by the mobile terminal to the network to request uplink transmission resources and corresponding *scheduling grants* provided by the base-station scheduler to control the mobile-terminal transmission activity.

The scheduling requests sent by the mobile terminals contain information about the amount of available transmit power at the mobile terminal and the amount of data, including the traffic priority of that data, available for transmission.

The scheduling grants control the maximum allowed E-DCH transmit power or, more exactly, the maximum allowed E-DCH-to-DPCCH power ratio. More specifically, each mobile terminal maintains a *serving grant* which directly determines the maximum E-DCH/DPCCH power ratio, with a larger grant implying that the terminal can use a higher relative E-DCH power. Due to the use of fast closed-loop power control which, in principle, ensures a constant received DPCCH power at the base station, a larger serving grant indirectly implies a higher received E-DCH power allowing for a higher E-DCH data rate. At the same time, a higher relative E-DCH power implies that the terminal will cause more interference and thus use a larger part of the overall uplink radio resource.

The reasons for expressing the limitation imposed by the serving grant as a power ratio and not as a data rate or transport format are twofold:

- The fundamental quantity the scheduler is controlling is the interference caused to the system. This interference is directly proportional to the transmission power.
- It allows the E-TFC selection algorithm to autonomously select transport formats targeting different number of transmission attempts (and consequently different data rates and delays) for different MAC-d flows as long as the total E-DCH transmission power is within the limits set by the grant. This is further discussed in Section 2.2.

The serving grant can be updated by the network in two different ways

- By means of *Absolute Grants*, setting an absolute value for the serving grant at the mobile terminal.
- By means of *Relative Grants*, providing a relative, step-wise update of the serving grant of the mobile terminal

Absolute grants can be received by a mobile terminal only from the E-DCH serving cell and can either be set on a per-mobile-terminal basis ("*Dedicated scheduling*") or jointly for a group of mobile terminals ("*Common scheduling*"). Common scheduling is especially useful at low uplink loads as it allows for relatively large serving grants to be provided to multiple mobile terminals. When a mobile terminal has data to transmit it may then immediately transmit with a high data rate, without first going through a request phase.

Dedicated scheduling provides tighter control of the uplink load and is more suitable at high system loads. In case of dedicated scheduling, the base-station scheduler determines what user(s) are allowed to transmit and set the serving grant(s) specifically for the intended user(s). In this case, only one or a few users at a time are allowed to transmit any substantial amount of uplink data.

Relative grants can be sent from both the serving cell and non-serving cells. However, although the term 'relative grant' is used in both cases, there is a significant difference between relative grants received from the serving cell and from non-serving cells.

Relative grants received from the serving cell are targeting a specific mobile terminal and can take one out of three possible values: 'UP', 'HOLD', or 'DOWN'. An 'up' ('down') command instructs the mobile terminal to increase (decrease) the serving grant, i.e., to increase (decrease) the allowed E-DPDCH/DPCCH power ratio compared to the last used power ratio. The 'hold' instructs the mobile terminal not to change the upper limit. A schematic illustration of the operation due to relative grants received from the serving cell is given in **Figure 7**.

To implement the increase (decrease) of the serving grant, the mobile terminal maintains a table of possible E-DCH/DPCCH power ratios as illustrated in **Figure 8**. The up/down commands corresponds to an increase/decrease of the power ratio in the table by one step compared to the power ratio used in the previous TTI in the same hybrid ARQ process. There is also a possibility to have a larger increase (but not decrease) for small values of the serving grant. This is achieved by configuring two thresholds in the E-DCH/DPCCH power ratio table, below which the mobile terminal may increase the serving grant by three and two steps, respectively, instead of only a single step. The use of the table and the two indices allow the network to

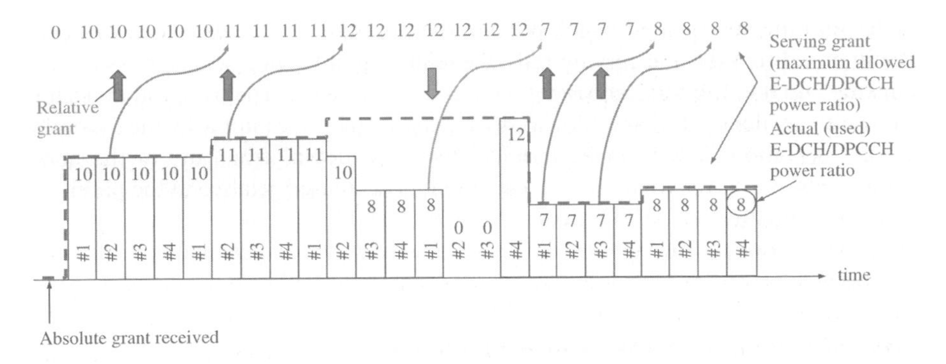


*Figure 7.* Schematic illustration of relative grant usage

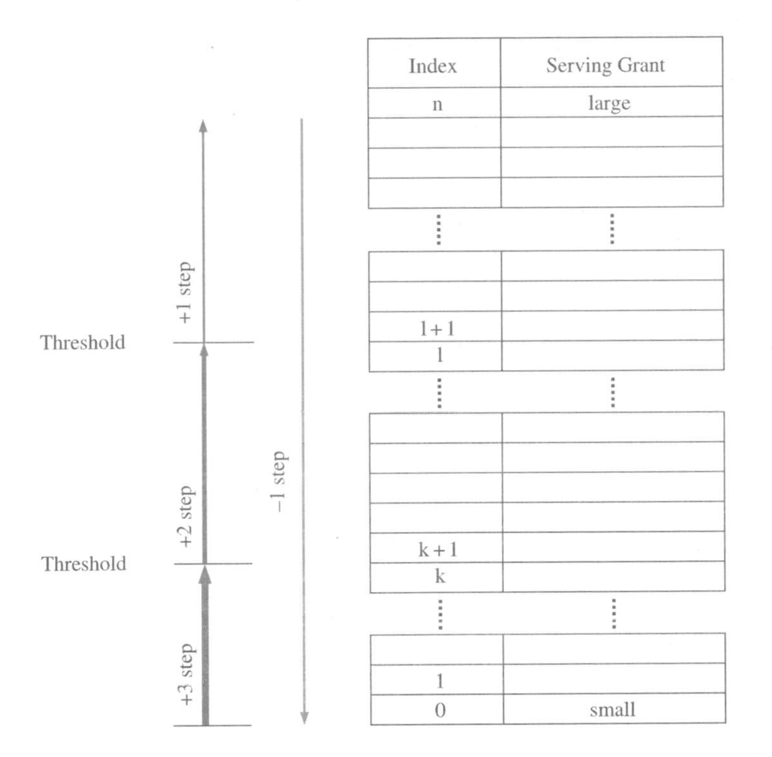


*Figure 8.* Example of grant table

increase the serving grant efficiently without extensive repetition of relative grants for small data rates (small serving grants) and at the same time avoiding large changes in the power offset for large serving grants.

Relative grants from non-serving cells provide the possibility for the non-serving cells in the Active Set to control the *inter-cell interference,* in contrast to the grants from the serving cell which provide the possibility to control the *intra-cell interference*. From the non-serving cells, the relative grant is in essence an "overload indicator", used to limit the amount of inter-cell interference. The overload indicator can take two values: 'dtx' and 'down', where the former does not affect the mobile terminal operation. If the mobile terminal receives 'down' from any of the non-serving cells in the Active Set, the serving grant is decreased relative to the previous TTI in the same hybrid ARQ process.

In soft handover, the serving cell thus has the *main* responsibility for the scheduling operation but the non-serving cells can request all its non-served users to lower their E-DCH data rate by transmitting an overload indicator in the downlink. This mechanism ensures a stable network operation.

Fast scheduling allows for a more relaxed connection admission strategy. A larger number of bursty high-rate packet-data users can be admitted to the system as the scheduling mechanism can handle the situation when multiple users need to transmit in parallel. Without fast scheduling, the admission control would have to be more conservative and reserve a margin in the system in case of multiple users transmitting simultaneously.

## 3.2 Hybrid ARQ for E-DCH

The E-DCH hybrid ARQ scheme is similar to that supported for HS-DSCH on the downlink, see Section 1.5. For each transport block received in the uplink, a single bit is transmitted from the base station to the mobile terminal after a well-defined time duration from the reception to indicate successful decoding (ACK) or to request a retransmission of the erroneously received transport block (NACK). In a soft handover situation, if an ACK is received from at least one of the base stations in the Active Set, the mobile terminal considers the data to be successfully received by the network.

Hybrid ARQ with soft combining can be exploited not only to provide robustness against unpredictable interference and reduce delay, but also to improve the link efficiency in order to, in the end, improve capacity and/or coverage. As a straight-forward example, consider a target data rate of x Mbps. This can obviously be achieved with a link data rate in the order of $x$ Mbps with the power control set to target a low error probability in the first transmission attempt. However, as an alternatively, the same effective data rate can be achieved with a link data rate in the order of $n$ times x Mbps at an unchanged transmission power. Clearly, the error rate at the first retransmission will be much higher in this case. However, if the hybrid ARQ scheme can ensure that the information is recovered at the receiver side after, on average, less than n retransmission, there is an overall gain is system efficiency, i.e. the same effective data rate have been achieved using overall less radio resources. Obviously, the same principle can be applied also for HS-DSCH in the downlink direction. The drawback is a somewhat larger radio-interface delay. Thus the Hybrid ARQ with soft combining can be used to trade-off efficiency vs. delay by adjusting the target settings (initial error rate) for the Hybrid ARQ scheme.

## 4. MBMS – MULTIMEDIA BROADCAST/MULTICAST SERVICES

MBMS or Multimedia Broadcast/Multicast Services, introduced in WCDMA release 6 in parallel to Enhanced Uplink, provides WCDMA with a powerful tool to offer true broadcast/multicast services over a mobile-communication network, in parallel to normal unicast services. With MBMS, the same content is transmitted to multiple users in a unidirectional fashion, typically over multiple cells to cover a large area in which the service is provided.

MBMS in 3GPP provides a full set of functionally to support broadcast/multicast services in a mobile-communication network, including both core-network and radio-access-network functionality.

**Figure 9** illustrates the overall MBMS structure in a 3GPP-based radio-access network. A new core-network node, the BM-SC or *Broadcast Multicast Service Center* is introduced as part of MBMS. The BM-SC is responsible for authorization and authentication of content providers, charging, and the overall configuration of the data flow through the core network. The BM-SC is also responsible for so-called *application-level coding*.

One of the main benefits of MBMS is a general resource saving in both the core network and the radio-access network as a single stream of data may serve multiple

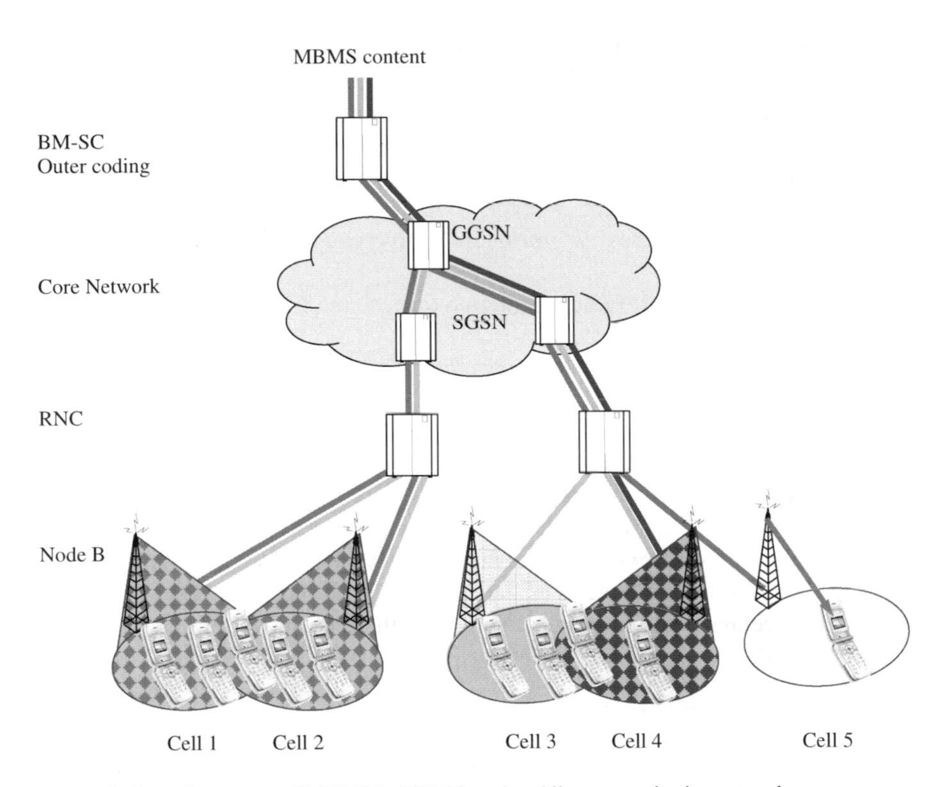


*Figure 9.* Overall structure of MBMS in 3GPP-based mobile-communication networks

users. This can clearly be seen from **Figure 9** where three different services are offered in different areas. From the BM-SC, data streams are fed, via intermediate core- and radio-network nodes, to each of the base stations involved in providing the MBMS services over the radio interface. As seen in the figure, the data stream intended for multiple users is, in general, not split until necessary. For example, there is only a single stream of data jointly sent to all the users in cell 3. This is in contrast to earlier releases of WCDMA where one stream per user had to be configured throughout both the core network and the radio-access network, even when identical information was to be provided to multiple users.

As indicated above, one of the main benefits with MBMS is resource savings in the network as multiple users can share a single stream of data. This is valid also from a radio-interface point-of-view where a single transmitted MBMS signal may be received by multiple users. Obviously such point-to-multipoint radio-transmissions within a cell imply very different requirements on the radio interface compared to downlink unicast transmissions based on e.g. HDSPA. As an example, user-specific adaptation of the radio parameters (link adaptation, channel-dependent scheduling, etc.) cannot be used for MBMS as the transmitted signal is intended for multiple users experiencing different instantaneous channel conditions. Instead transmission parameters such as transmit power and transport format must be selected based on what may be required by the worst-case mobile-terminal position, typically at the cell border. This also implies that different forms of diversity to suppress the impact of multi-path fading on the radio channel are highly important when providing broadcast/multicast services over a radio interface. Furthermore, different types of ARQ protocols are obviously also not suitable when the broadcast/multicast information is to be received by a large number of mobile terminals within the cell.

The two main techniques for providing diversity for MBMS services in WCDMA are

- time-diversity against fast fading through a long 80 ms TTI and application-level coding
- downlink macro-diversity, i.e., combining of transmissions received from multiple cells.

Fortunately, MBMS services are not delay sensitive and the use of a long TTI is not a problem from the end-user perspective. Additional means for providing diversity can also be applied in the network, e.g., open-loop transmit diversity. Receive diversity in the terminal also improves the MBMS reception performance, but as the 3GPP mobile-terminal requirements for 3GPP release 6 are set assuming single-antenna mobile terminals, it is hard to exploit this type of diversity in the planning of MBMS coverage.

### 4.1 MBMS Macro Diversity

An MBMS services is often provided simultaneously over a large number of cells. This provides the opportunity for *multi-cell reception* of the MBMS signal for mobile-terminals at or close to the border between cells, providing macro diversity and, as a consequence, substantially improved coverage for the MBMS service.

Combining transmissions of the same content from multiple cells provides a significant diversity gain, in the order of 4-6 dB reduction in transmission power, compared to single-cell reception only, as illustrated in **Figure 10**. Two combining strategies are supported for MBMS, *Soft Combining and Selection Combining.*

In case of soft combining the soft bits received from the different radio links are combined prior to (Turbo) decoding. In principle, the mobile terminal descrambles and RAKE combines the transmission from each cell individually, followed by soft combining of the different radio links. Note that WCDMA uses cell-specific scrambling of all data transmissions. Hence, the soft combining is performed by the appropriate mobile-terminal processing, which also is responsible for suppressing the interference caused by the transmission activity in the neighbor cells. To perform soft combining, the physical channels to be combined should be identical. For MBMS, this implies the same S-CCPCH content and structure should be used on the radio links which are soft combined.

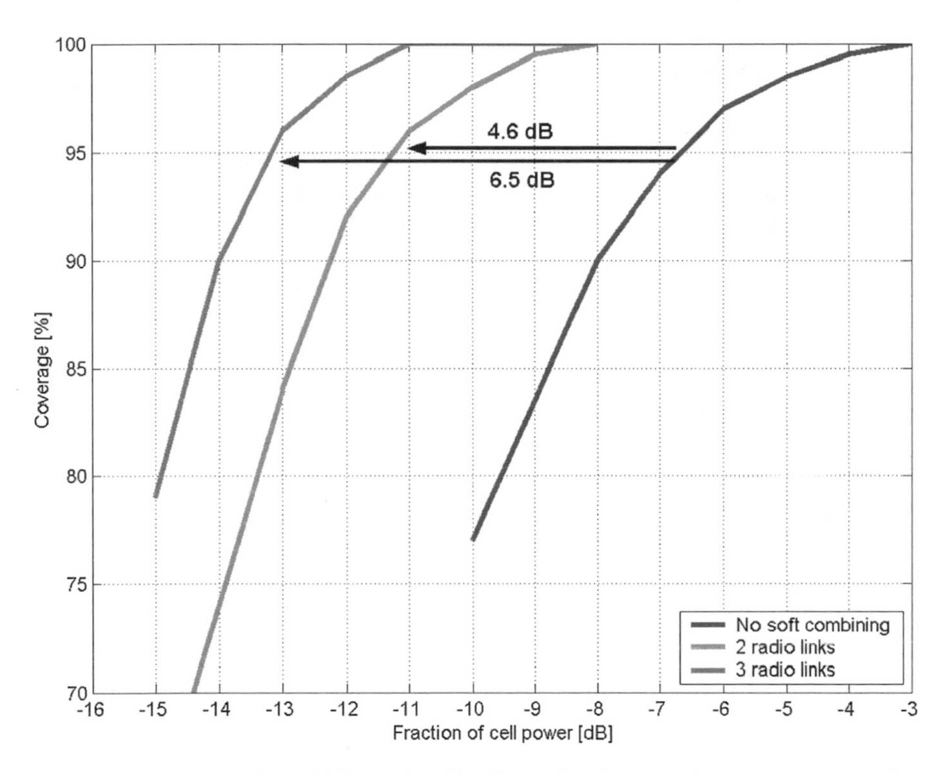


*Figure 10.* Gain with soft combining and multi-cell reception in terms of coverage vs. power for a 64 kbit/s MBMS service (Vehicular A, 3 km/h, 80 ms TTI, single receive antenna, no transmit diversity, 1% BLER)

Selection combining, on the other hand, decodes the signal received from each cell individually and for each TTI selects one (if any) of the correctly decoded data blocks for further processing by higher layers. From a performance perspective, soft combining is preferred over selection combining as it provides not only diversity gains but also a power gain as the received power from multiple cells is exploited. Relative to selection combining, the gain with soft combining is in the order of 2–3 dB.

The reason for supporting two different combining strategies for MBMS is to handle different levels of asynchronism in the network. For soft combining, the soft bits from each radio link have to be buffered until the whole TTI is received from all involved radio links and the soft combining can start while, for selection combining, each radio link is decoded separately and it is sufficient to buffer the decoded information bits from each link. Hence, for a large degree of asynchronism, selection combining requires less buffering in the mobile terminal at the cost of an increase in turbo decoding processing. The mobile terminal is informed about the level of synchronism and can, based upon this information and its internal implementation, decide to use any combination scheme as long as it fulfills the minimum performance requirements mandated by the specifications.

## 4.2 Application-level Coding

Many end-user applications require very low error probabilities, e.g., in the order of $10^{-6}$. Providing such low error probabilities on the radio-link level can be very costly from a transmit-power point-of-view. In point-to-point communications, for example for HS-DSCH and E-DCH, some form of ARQ mechanism is therefore typically used to reduce the residual error rate to the required level. However, as previously mentioned, it is not straightforward to apply an ARQ protocol for broadcast transmissions. For MBMS, application-level forward error-correcting coding has instead been adopted as a tool to reduce the overall error rates to the required level as shown in **Figure 11**.

The MBMS application-level coding resides in the BM-SC and is thus, strictly speaking, not part of the radio-access network, With application-level coding, the system can operate at a transport-channel block error rate in the order of 1%–10% instead of fractions of a percent, which significantly lowers transmit power requirement. As the application-layer coding resides in the BM-SC, it is also

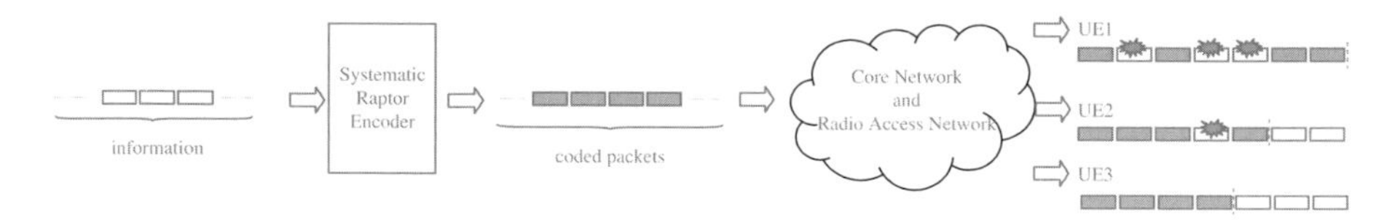


*Figure 11.* Illustration of application-level coding, depending on their different ratio conditions, the number of coded packets required for the mobile terminals to be able to reconstruct the original information differs

effective against occasional packet losses in the transport network, e.g., due to temporary overload conditions.

So called *Systematic Raptor codes* have been selected for the application-level coding in MBMS. Raptor codes belongs to a class of so-called *Fountain* codes, i.e., as many encoding packets as needed can be generated on-the-fly from the source data. For the decoder to be able to reconstruct the information, it only needs to receive sufficiently many coded packets. It does not matter which coded packets it receive, in what order they are received, or if certain packets were lost.

In addition to provide additional protection against packet losses and to reduce the required transmission power, the use of application-level coding also simplifies the procedures for mobile-terminal measurements, e.g., for cell reselection in case of MBMS reception. For HSDPA, the scheduler can avoid scheduling data to a given mobile terminal in certain time intervals. This allows for the mobile terminal to use the receiver for measurement purposes, e.g., to tune to a different frequency and possible also to a different radio-access technology. In a broadcast setting, scheduling measurement gaps is cumbersome as different mobile terminals may have different requirements on the frequency and length of the measurement gaps. Furthermore, the mobile terminals need to be informed when the measurements gaps occur. Hence, a different strategy for measurements is adopted in MBMS. The mobile-terminal measurements are done autonomously, which could imply that a mobile terminal sometimes miss (part of) an MBMS TTI. In some situations, the inner Turbo code is still able to decode the transport channel data, but if this is not the case, the outer application-level code will ensure that no information is lost.

## 5. FUTURE WCDMA EVOLUTION

Although the introduction of HSPA (HSDPA + Enhanced Uplink) provides substantial enhancements to the WCDMA support for packet-data services in terms of system as well as end-used performance, there continues to be a need to further enhance and evolve the WCDMA radio-access technology in order to ensure WCDMA competitiveness also in a longer time perspective. This section with briefly discuss two features currently being specified for WCDMA.

- *Continuous Packet Connectivity* (CPC) aiming at a further improvement in the WCDMA resource utilization and faster access to radio resources
- Downlink *multi-layer transmission* or “*MIMO*” aiming at a further increase in the downlink data rates supported by the WCDMA radio interface, at a first step up to 28 Mbps

### 5.1 Continuous Connectivity

In general, a WCDMA mobile terminal may be in one of different “states” as outlined in **Figure 12**, depending on the current “activity” of the mobile terminal. These states differ in terms of how fast the mobile terminal can get access to

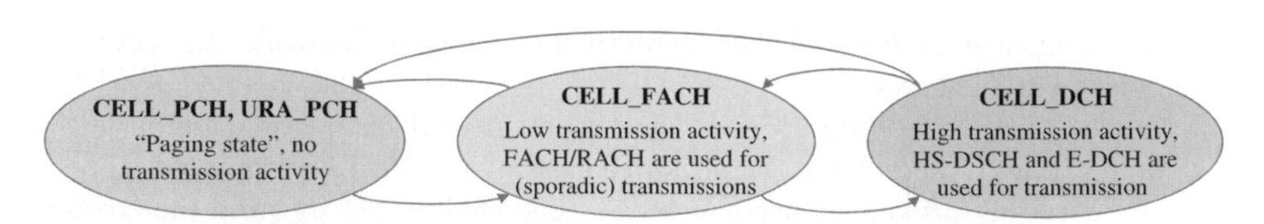


*Figure 12.* The WCDMA state model in conneceted mode

large radio resources for high speed downlink and/or uplink packet data transfer, with CELL_DCH providing the fastest (basically immediate) access to large radio resources.

However, at the same time as the CELL_DCH state provides the fastest access to large radio resources and thus, from this point-of-view, can provide the best user experience, the CELL_DCH state is also the most expensive state in terms of radio-resource usage. The reason is that, when in CELL_DCH state, a mobile terminal is continuously using a certain amount of radio resources on both uplink and downlink. These resources are used to keep the radio link between the mobile terminal and the network, including power control, up and running, to ensure a steady flow of uplink CQI reports for downlink scheduling, etc. Thus, in order to maximize the amount of radio resources available for actual packet-data transmission, the number of mobile terminals in CELL_DCH state should be limited. As a consequence, mobile terminals that have not transmitted or received any data for a certain time interval are "moved" to the, from a resource-utilization point-of-view, more efficient CELL_FACH state. However, CELL_FACH state also implies a certain delay when, eventually, there is data to transmit to or from the mobile terminal as the mobile terminal must then first move back to the CELL_DCH state. Thus there is a desire to improve the efficiency of the CELL_DCH state allowing for more mobile terminals to simultaneously be in CELL_DCH state without an unreasonably negative impact on resource availability. If this would be possible, a mobile terminal can also stay in CELL_DCH state for a longer time, reducing the time needed to get access to large radio resources and thus improve the overall user experience. This is the target for the specification of the *Continuous Packet Connectivity* to be part of WCDMA release 7.

### (1) Uplink Overhead Reduction

In the current WCDMA releases (up to release 6), a mobile terminal in CELL_DCH state continuously transmits an uplink control channel, the *uplink DPCCH*. This control channel is used to maintain uplink synchronization as well uplink power control. At the same time, this control channel causes uplink interference and thus uses a certain amount of uplink radio resources. In order to reduce that uplink interference from the DPCCH while still maintaining uplink synchronization and power control, Uplink DPCCH Gating is considered to be introduced for WCDMA. Uplink DPCCH gating implies that, instead of being transmitted continuously even when there is no uplink data transmission, the DPCCH is transmitted only intermittently.

Basically, if there is neither E-DCH nor HS-DSCH transmissions in the uplink, the UE automatically stops continuous DPCCH transmission and uses a DPCCH on/off transmission pattern. The gating pattern, configured in the UE and the Node B by the RNC, defines when not to transmit the DPCCH unless there is an E-DCH transmission in the uplink. When an E-DCH transmission is to take place, the DPCCH is also transmitted, regardless of the on/off pattern. The DPCCH is also transmitted when uplink acknowledgements, corresponding to downlink HS-DSCH transmissions, are to be transmitted.

(2) **Downlink Overhead Reduction**
In the downlink, each mobile terminal in CELL_DCH state uses a certain amount of downlink radio resource (channelization code and transmission power) due to the fact that each terminal has been allocated and uses an associated physical control channel, a *Dedicated Physical Channel* DPCH, mainly used for power control of uplink transmission. The *Fractional DPCH*, F-DPCH, introduced in Release 6 improves efficiency by allowing for multiple mobile terminals to share a common physical channel thus significantly reducing the overhead in terms channelization-code space.

However, another source of downlink overhead is the scheduling information transmitted on the *High Speed Shared Control Channel* or HS-SCCH. In case of medium-to-large payloads on the HS-DSCH, the relative HS-SCCH overhead is small. However, for services such as VoIP with frequent transmissions of small payloads, the HS-SCCH overhead compared to the actual HS-DSCH payload may not be insignificant. Therefore, an additional HS-SCCH structure with reduced overhead can be used for such services. By limiting the possible transmission formats on the HS-DSCH, the number of bits on the HS-SCCH using the new format can be kept small, thereby reducing the overhead in terms of downlink power.

## 5.2 MIMO/Multi-layer Transmission

Strictly speaking, MIMO or *Multiple Input Multiple Output antenna process* in its most general interpretation denotes the use of multiple antennas at both the transmitter and the receiver side of a radio link. However, the term MIMO is commonly used to denote the transmission of *multiple layers* or *streams* as a mean to increase the maximum data rate that can be provided over a given radio link. Hence, MIMO, or *multi-layer transmission*, should mainly be seen as a tool to improve the *end-user throughput* by acting as a "data rate booster" (Note that another MIMO scheme called MIMO diversity was specified as transmit diversity in 3GPP. Thus, we focus on MIMO SDM in the section). However, similar to higher-order modulation, multi-layer transmission will also provide a possibility for an increased system throughput especially in scenarios with, in general, high signal-to-noise/interference ratios.

MIMO schemes are designed to exploit certain properties in the radio propagation environment to attain high data rates through multiple layers. However, in order to achieve these high data rates, a correspondingly high received signal-to-noise and/or signal-to-interference ratios are required. Multi-layer transmission is therefore mainly applicable in smaller cells or close to the base station, i.e. in situations where high signal-to-noise/interference ratios are more often experienced. In situations where a sufficiently high signal-to-noise/interference ratios cannot be achieved, the multiple receive antennas, which MIMO-capable mobile terminals are equipped with, should instead be used for receive diversity in order to boost the received signal quality.

The MIMO scheme adopted for HSDPA-MIMO is referred to as *D-TxAA* or *Dual Transmit-diversity Adaptive Antennas*. D-TxAA is *multi-codeword* MIMO scheme with *rank adaptation* and *pre-coding*, in which per stream rate control is applied to each codeword. The scheme can be seen as a generalization of the closed-loop mode 1 transmit-diversity scheme, present already in the first WCDMA release.

Transmission of up to two streams is supported by HSDPA-MIMO, thus providing the possibility for peak data rates twice that of current HSDPA or approximately 28 Mbps. Each stream is subject to the same physical layer processing in terms of coding, spreading, and modulation as the corresponding single-layer HSDPA case. After coding, spreading, and modulation, linear pre-coding is used before the result is mapped to the two transmit antennas. There are several reasons for the pre-coding. Even if only a single stream is transmitted, it can be beneficial to exploit both transmit antennas. Therefore, the pre-coding in the single-stream case is identical to closed-loop transmit diversity mode 1. In essence, this can be seen as a simple form of beam-forming. Furthermore, the pre-coding attempts to pre-distort the signal such that the two streams are (close to) orthogonal at the receiver. This reduces the interference between the two streams and lessens the burden on the receiver MIMO processing.

## REFERENCES


[1] H. Holma and A. Toskala, "WCDMA for UMTS" John Wiley & Sons, Ltd., June 2000

[2] 3GPP TR25.903 v1.0.0, "*Continuous Connectivity for Packet Data Users*"

[3] 3GPP TR25.876, v1.8.0, "*Multiple Input Multiple Output in UTRA*"

[4] H. Ekström et al, "*Technical Solutions for the 3G Long Term Evolution*", IEEE Communications Magazine, vol. 44, No 3, March 2006

[5] A. Jolali et al, "Data Throughput of CDMA_HDR a High Efficiency-High Data Rate personal Communication Wireless System", Proc. 51st IEEE VTC 2000-Spring

[6] D. Chase, "*Code Combining – A Maximum-Likelihood Decoding Approach for Combining an Arbitrary Number of Noisy Packets*" IEEE Trans. Commun., vol. 33, pp. 385–393

[7] J. Hagennauer, "*Rate-Compatible Punctured Convolutional Codes (RCPC codes) and Their Applications*" IEEE Trans. Commun., vol. 36, pp. 389–400

[8] P. Frenger et al, "*Performance Comparison of HARQ with Chase Combining and Incremental Redundancy for HSDPA*," in Proc. IEEE VTC 2001-Fall

[9] A. Shokrollahi, "Raptor Codes", In Proceedings from International Symposium on Information Theory (ISIT 2004)

[10] 3GPP TSG-R1-040336: "Double TxAA for MIMO"

CHAPTER 7

# EVOLVED UTRA TECHNOLOGIES

MAMORU SAWAHASHI[1], ERIK DAHLMAN[2], AND KENICHI HIGUCHI[3]
[1] *Musashi Institute of Technology, Japan*
[2] *Ericsson AB, Sweden*
[3] *NTT DoCoMo, Japan*

**Abstract:** This chapter describes the radio access technologies and physical-layer channels for the Evolved UTRA (UMTS Terrestrial Radio Access, UMTS: Universal Mobile Telecommunications System) and UTRAN (UMTS Terrestrial Radio Access Network), which represent the long-term evolution of UMTS. Discussion on the Evolved UTRA is ongoing in the 3GPP (3rd Generation Partnership Project) with the target of completing the work item (WI) specifications by around September 2007

**Keywords:** WCDMA, Evolved UTRA, OFDM, Single-Carrier FDMA, Reference signal, Broadcast information, Paging information, Shared data channel, Scheduling, Link Adaptation, Hybrid ARQ, MBMS, L1/L2 control information, MIMO, Inter-cell interference coordination

## 1. INTRODUCTION

The 3GPP (3rd Generation Partnership Project) study item (SI) on Evolved UTRA (UMTS Terrestrial Radio Access, UMTS: Universal Mobile Telecommunications System) and UTRAN (UMTS Terrestrial Radio Access Network) was initiated in December 2004. Evolved UTRA and UTRAN (E-UTRA/E-UTRAN) represent the long-term evolution of 3G radio access intended to be highly competitive even in the future 4G era. Evolved UTRA will support emerging multimedia technology and ubiquitous traffic over cellular networks through the use of only packet domain, which has affinity to IP-based core networks. The system requirements and targets of Evolved UTRA were agreed by TSG RAN in June 2005 and documented in 3GPP Technical Report TR-25.913. After extensive studies on basic radio-access schemes and physical-layer technologies in TSG RAN WG1 , the use of OFDM-based and Single-Carrier (SC)-FDMA based radio access was decided for the E-UTRA downlink and uplink respectively. This section addresses the Evolved UTRA technologies with focus on physical-layer aspects.



*Y. Park and F. Adachi (eds.), Enhanced Radio Access Technologies for Next Generation Mobile Communication*, 217–276.

## 2. REQUIREMENTS IN EVOLVED UTRA AND UTRAN

The Evolved UTRA and UTRAN targets high-capacity high-speed radio access and radio-access networks with low latency in order to support full IP functionalities. With low-latency characteristics E-UTRA/E-UTRAN will be able to provide current circuit-switched mode services over the packet domain, such as Voice-over-IP (VoIP). The work plan for establishing SI and Work Item (WI) specifications was approved. The system requirements on E-UTRA were also agreed upon and specified. Major system requirements related to the physical layer are given below.

- In order to allow for deployments in differently-sized spectrum allocation, E-UTRA should support multiple transmission bandwidths on both uplink and downlink. The current assumption is that E-UTRA should support transmission bandwidths corresponding to spectrum allocations of size 1.25, 1.6, 2.5, 5.0, 10.0, 15.0, and 20.0 MHz.
- Peak data rates of at least 100 Mbps and 50 Mbps are to be supported for downlink and uplink respectively. These data rates are to be achievable with two-branch MIMO transmission.
- Compared to WCDMA release 6, gains in average user throughput of at least 3 – 4 times and 2 – 3 times are to be achieved for downlink and uplink respectively. Corresponding gains in cell-edge user throughput should be at least 2 – 3 times that of WCDMA release 6.
- Gains in spectrum efficiency (capacity), compared to WCDMA release 6, are to be at least 3 – 4 times and 2 – 3 times for downlink and uplink, respectively.
- The minimum achievable one-way transmission delay over the radio-access network should be less than 5 msec.
- The minimum transition time from idle mode to active mode should be less than 100 msec. Corresponding transition time from dormant mode to active mode should be less than 50 msec.

As addressed, a high user throughput and high capacity, i.e., spectrum efficiency, are the eternal requirements in cellular systems in order to offer high-speed multimedia services efficiently at low cost. In particular there is demand for improvement in the user throughput over the entire cell area including the cell boundary. Furthermore, it is no exaggeration to say that low latency (short delay) is a most important requirement in terms of providing real-time services such as VoIP. The flexibility for different spectrum arrangements is necessary from operators' point of view. Moreover, simple channel and protocol structures and optimization of control-signal formats are also highly desirable.

## 3. RADIO ACCESS SCHEMES

An important requirement on the Evolved UTRA is to support operation in both paired and unpaired spectrum. Thus, Evolved UTRA should support operation with both Frequency-Division Duplex (FDD) and Time Division Duplex (TDD).

Furthermore, in order to reduce terminal complexity, there should be maximum commonality between the transmission scheme used in paired and unpaired spectrum, i.e. between FDD and TDD. Thus, the same E-UTRA radio-access schemes has been adopted for both FDD and TDD: OFDM based radio access in the downlink and Single-carrier (SC)-FDMA based radio access in the uplink. The E-UTRA radio frame length is 10 msec, which is identical to that of UMTS (i.e., WCDMA).

## 3.1 OFDM Base Radio Access in Downlink

Especially for signal bandwidths wider than 5 MHz, increased multipath interference (MPI) impairs the achievable data rate and coverage. For this reason, Orthogonal Frequency Division Multiplexing (OFDM) based downlink radio access was adopted for E-UTRA due to the inherent immunity of OFDM to MPI. OFDM also provides access to the frequency domain for the scheduling. Moreover, OFDM has superior flexibility to accommodate different spectrum arrangements due to its small granularity in frequency domain. Finally OFDM provides specific benefits in case of downlink multi-cell transmission as will be further discussed below. OFDM signal is generated and separated by IFFT and FFT based implementations at the transmitter and receiver, respectively.

### *3.1.1 Radio parameters*

**Table 1** lists the current 3GPP assumptions regarding radio parameters for the E-UTRA OFDM based downlink radio access. As can be seen, the sub-carrier spacing is constant regardless of the transmission bandwidth. To allow for operation in differently sized spectrum allocations, the transmission bandwidth is instead varied by varying the number of OFDM sub-carriers. The major downlink parameters were decided as follows:

- **Radio frame length**

Considering the simultaneous use of UTRA and Evolved UTRA, i.e., dual-mode usage, and backward compatibility with WCDMA/HSDPA, the same radio-frame length as WCDMA is desirable. Therefore, a 10-msec radio frame was adopted for Evolved UTRA.

- **Sub-frame and TTI lengths**

The E-UTRA sub-frame length was set to 0.5 msec in both the downlink and uplink. The sub-frame length corresponds to the minimum Transmission Time Interval (TTI). As mentioned previously, the E-UTRA requirement for one-way radio-access-network (RAN) latency is 5 msec. The total RAN latency includes the transmission delay between the UE and Node B, the transmission delay in the backhaul network between the Node B and access router, and the processing delays of the UE, Node B, and access router[1]. In particular, the transmission delay in the

[1] The term "access router" refers to a higher-level node than Node B

*Table 1.* Radio Parameters in Downlink OFDM Based Radio Access

| Transmission BW | | 1.25 MHz | 2.5 MHz | 5 MHz | 10 MHz | 15 MHz | 20MHz |
|---|---|---|---|---|---|---|---|
| Sub-frame duration | | | | | 0.5 ms | | |
| Sub-carrier spacing | | | | | 15 kHz | | |
| Sampling frequency | | 1.92 MHz | 3.84 MHz | 7.68 MHz | 15.36 MHz | 23.04 MHz | 30.72 MHz |
| FFT size | | 128 | 256 | 512 | 1024 | 1536 | 2048 |
| Number of occupied sub-carriers | | 76 | 151 | 301 | 601 | 901 | 1201 |
| Number of OFDM symbols per sub frame (Short/Long CP) | | | | | 7/6 | | |
| CP length (μs/samples) | Short | (4.69/9) × 6, (5.21/10) × 1 | (4.69/18) × 6, (5.21/20) × 1 | (4.69/36) × 6, (5.21/40) × 1 | (4.69/72) × 6, (5.21/80) × 1 | (4.69/108) × 6, (5.21/120) × 1 | (4.69/144) × 6, (5.21/160) × 1 |
| | Long | (16.67/32) | (16.67/64) | (16.67/128) | (16.67/256) | (16.67/384) | (16.67/512) |

wireless channel between the UE and the Node B contributes a significant part to the total RAN latency. This delay depends on the Transmission Time Interval (TTI) length, which directly determines the control delay of channel-dependent scheduling, link adaptation, and hybrid ARQ. Thus, the TTI length was set to 0.5 msec in the study item (SI) specification. However, in the work item (WI) evaluation, the necessity for a long TTI was presented in order to enable repetition of the L1/L2 control signaling bits over multiple sub-frames particularly in the uplink with a strict transmission power restriction. Moreover, one option for the TTI length was claimed commonly in the downlink and uplink for the sake of implementation simplicity. As a result, the TTI length was decided to be 1.0 msec based on the tradeoff relation between a short control delay and the repetition of L1/L2 control signaling bits to achieve wide coverage provisioning. Since one TTI comprises two sub-frames, intra-TTI frequency hopping is beneficial to achieving a larger frequency diversity effect.

- **CP length**

At an OFDM receiver, the coded data symbol of each sub-carrier is retrieved by sampling the received signal over the duration of an OFDM symbol and performing fast Fourier transform (FFT) processing. Thus, in the case of multipath propagation, inter-symbol interference (ISI) occurs since the delayed part of the previous OFDM symbol falls within the FFT processing window. In addition, since the orthogonality between sub-carriers is destroyed due to the delayed paths, inter-sub-carrier interference (ICI) is generated. To avoid ISI and ICI in multipath propagation, *Cyclic Prefix* (CP) *insertion* is used. Cyclic-prefix insertion implies that the last part of the OFDM symbol, of length $T_g$, is copied and attached to the beginning of the OFDM symbol. As long as the delayed paths are received within a delay window of size $T_g$ , the orthogonality between sub-carriers is then retained, i.e., ISI and ICI do not occur. The CP length should be selected based on the maximum time delay of the multi-path channel, which depends on the inter-site distance (ISD), frequency reuse, i.e., influence of other-cell interference, transmission power, etc. Meanwhile, the shortest possible CP is desired from the viewpoint of achieving efficient radio resource usage as the CP implies a certain overhead and waste of bandwidth.

In E-UTRA, two CP lengths are specified for effectively supporting physical channels for system-dependent environments: short and long CPs.

- A short CP is used for Unicast services.
- A long CP is used for Multimedia Broadcast Multicast Services (MBMS) and for Unicast services in environments with an extraordinarily long time dispersion.

The required CP lengths were investigated in system-level simulations in multi-cell environments. In the evaluations, the received signal-to-interference plus noise power ratio (SINR) is calculated using the method in assuming Greenstein's power-delay-profile model, which well approximated the root mean square (r.m.s.) delay spread. It was reported that the required short CP is approximately 3 to 5 μsec from the viewpoint of accommodating delayed paths for the cell radius of less than 5 km, assuming the support of various environments with a low-to-high channel load in the surrounding cells. Furthermore, to provide high-data-rate MBMS, a long CP is necessary to obtain benefits from soft-combining for long delays from far cell sites and to compensate for the residual inter-Node B synchronized timing error. It was reported that a long CP is required to accommodate the delayed paths in MBMS of approximately 10 to 15 μsec in order to gain the soft-combining effect including the delay of RF filtering in the relay station using simple RF conversion. To support a long ISD of greater than 10 km, a longer CP is not necessary because the major impairments are background noise and other-cell interference, and not the MPI from the target cell under such long ISD conditions. Therefore, the short and long CPs become 4.69 / 5.21 μsec and 16.67 μsec, respectively. Furthermore, in the WI evaluation, the necessity for a CP longer than 16.67 sec was presented for MBMS in the region with a very long ISD using high transmission power. In addition to the short and long CPs, a very long CP, which is twice as long as the long CP (= 33.3 μsec), was agreed only for dedicated MBMS carrier application together with 7.5-kHz sub-carrier spacing.

- **Sub-carrier spacing**

As mentioned above, the CP length should be selected based on the multi-path scenarios that are to be supported according to the coverage requirement. By reducing the sub-carrier spacing, i.e., by increasing the number of sub-carriers, the CP insertion loss decreases, which is important from the viewpoint of efficient radio-resource utilization. Accordingly, a narrow sub-carrier spacing with a long OFDM symbol duration is necessary to suppress the CP insertion loss to a low level.

On the other hand, the sub-carrier spacing should be selected such that the influence of the Doppler effect and phase noise is limited. However, the influence of inter-carrier interference caused by phase noise is small when the sub-carrier spacing is wider than approximately 10 kHz. Accordingly, the minimum sub-carrier spacing is determined such that the influence of the Doppler effect becomes small for the maximum terminal speed to be supported by E-UTRA, which is approximately 350 km/h, corresponding to a maximum Doppler frequency of 840 Hz at the carrier frequency of 2.6 GHz. This view is based on the system design concept which indicates that although the E-UTRA system including radio parameters should be optimized in low mobility environments, the mobility up to the maximum speed in the system is supported while minimizing the performance degradation. It is given that the sub-carrier spacing should be greater than approximately 11 kHz to suppress the loss in the achievable throughput at 350 km/h from that at 30 km/h to within less than approximately 0.5 Mbps (2%) when 64QAM is employed at a 2.6-GHz carrier frequency. As a result, the sub-carrier spacing was decided to be 15 kHz both for short and long CPs. However, in the WI evaluation, the necessity for a very long CP, which is twice as long as the long CP (= 33.3 μsec) was presented for MBMS in the region with a very long ISD using high transmission power. Thus, the use of 7.5-kHz sub-carrier spacing was agreed only for dedicated MBMS carrier application to generate such a very long CP.

A physical resource block (PRB) is defined as the minimum radio-resource unit that can be assigned to a UE for data transmission. The bandwidth for the minimum resource block (RB), in which a distinct gain from frequency domain channel-dependent scheduling is obtained considering the control signaling overhead, is around 25 sub-carriers or approximately 375 kHz assuming the six-ray Typical Urban channel model. However, a narrower bandwidth for the minimum RB is considered such as 12 sub carriers or 180 kHz to support low-rate data, e.g., VoIP.

### *3.1.2 Modulation schemes*

Higher-order modulation scheme such as 64QAM has been adopted for Evolved UTRA to provide higher data rates as well as high system efficiency due to the robust feature against time dispersion of OFDM based radio access. Consequently, data modulation schemes supported in the downlink are QPSK, 16QAM, and 64QAM.

### *3.1.3 Channel coding and interleaving*

Turbo coding based on Release 6 is the working assumption for the Evolved UTRA SI.

### *3.1.4 Downlink data multiplexing*

The channel-coded, interleaved, and data-modulated Layer 3 information is mapped onto OFDM symbols in the frequency and time domains. PRBs consist of a number of consecutive sub-carriers for a number of consecutive OFDM symbols. The Node B scheduler allocates frequency and time radio resources to map coded data symbols for a certain UE. The Node B scheduler also determines the channel-coding rate and the modulation scheme based on the reported channel-quality indicator (CQI) and quality-of-service (QoS). In the OFDM-based downlink, two types of data multiplexing were adopted for coded data symbols as illustrated in **Figures 1(a)** and **1(b)**, respectively: *Localized* OFDM transmission and *distributed* OFDM transmission. In case of localized transmission, the coded data symbols are transmitted block wise in the frequency domain. Meanwhile, coded data symbols are transmitted on non-consecutive sub-carriers in case of distributed transmission. To describe the operations of localized and distributed OFDMA transmissions, the notation of the virtual resource block (VRB) was introduced. The VRB is characterized by type and size. The type indicates either localized or distributed transmission. The size is defined as the numbers of sub-carriers and OFDM symbols. Distributed VRBs are mapped onto the PRBs in a distributed manner. Localized

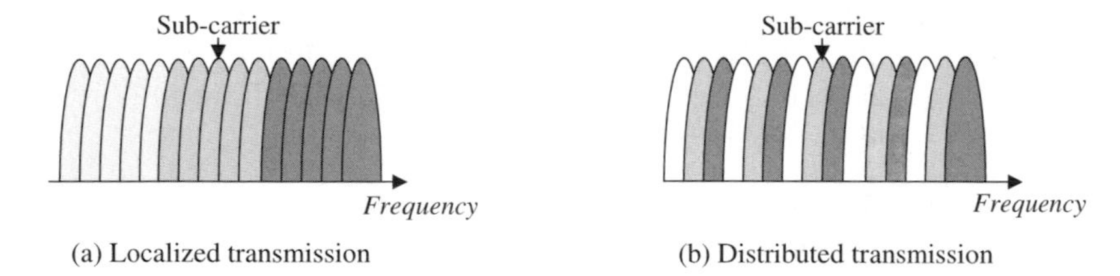


*Figure 1.* Localized and distributed OFDMA transmissions

VRBs are mapped onto the PRBs in a localized manner. The multiplexing of localized and distributed transmissions within one sub-frame is accomplished using frequency division multiplexing (FDM).

### 3.2 Single-Carrier Based Radio Access in Uplink

Achieving wide-area coverage is one of the most important requirements in a cellular system. This is particularly critical for the uplink where UE transmission power and power consumption are restricted due to battery limitations. Accordingly, Single-Carrier Frequency Division Multiple Access (SC-FDMA) was adopted as the E-UTRA uplink radio access scheme. Single-carrier transmission achieves a lower peak-to-average power ratio (PAPR) and thus improves power-amplifier efficiency for a given transmitter output power, compared to multi-carrier based radio access such as OFDM. Moreover, signal-waveform generation in the frequency domain was proposed based on Discrete Fourier transform (DFT)-Spread OFDM, see **Figure 2**. Similar to OFDM, SC-FDMA also has flexibility for different spectrum arrangements. The Evolved UTRA DFT-Spread OFDM uplink radio access has high commonality in terms of radio parameters with the OFDM-based downlink radio access, for example in terms of, sampling rate and sub-carrier spacing.

The input of the DFT is sampled using the sampling clock, which corresponds to the symbol rate of the incoming coded data symbol. Either a localized or distributed FDMA signal is generated in the frequency domain. Let $N_{DFT}$ and $N_{IFFT}$ be the size of the DFT and inverse fast Fourier transform (IFFT), respectively. Then, the sampling rate at the output of the IFFT, $R_{IFFT}$, is given by the following equation.

$$(1) \qquad R_{IFFT} = R \text{ x } N_{IFFT} \,/\, N_{DFT}.$$

After mapping either to a localized or distributed FDMA signal format, the signal is converted into a time-domain signal by IFFT processing. After IFFT processing, cyclic-prefix (CP) is applied similar to basic OFDM. Cyclic-prefix insertion for SC-FDMA is used to avoid inter-block interference for frequency-domain equalization. Finally, a time-window filter suppresses out-of-band emissions due to

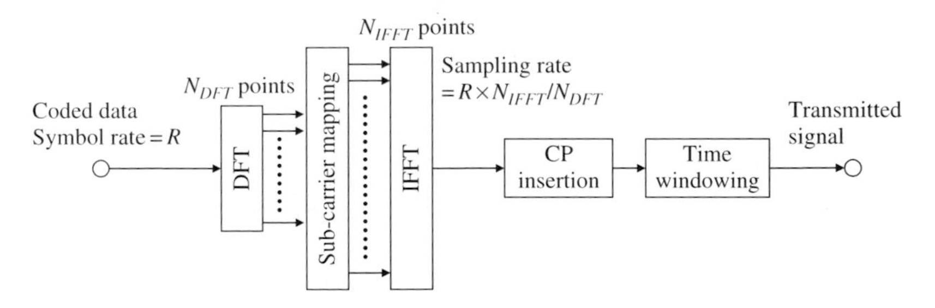


*Figure 2.* Transmitter block diagram of DFT-Spread OFDM

the discontinuity of contiguous blocks. The original DFT-Spread OFDM provides steep attenuation of the frequency domain power spectrum, which corresponds to a roll-off factor of zero in the raised cosine Nyquist filter. It was reported that the roll-off factor of zero achieves a higher user throughput than that for a roll-off factor value of greater than zero. On the other hand, the PAPR is reduced by increasing the roll-off factor value of the pulse-shaping filter. However, the effective channel rate, corresponding to the transmission bandwidth, is decreased. In other words, the channel coding rate becomes higher assuming the same information bit size, with a reduced coding gain as a consequence. The optimum roll-off factor value should thus be decided based on the trade-off relationship between the reduction in the PAPR and the decrease in the channel coding gain due to a decreasing channel coding rate. The results showed that the loss in the channel coding gain exceeds the gain from the PAPR reduction and thus roll-off of zero is preferred.

With DFT-Spread OFDM, sub-carrier mapping is possible in the frequency domain by inserting the output of the DFT and zeros at the input of the IFFT. **Figures 3(a)** and **3(b)** show the mapping schemes for the localized and distributed FDMA transmissions in DFT-spread OFDM, respectively. The sub-carrier mapping determines which part of the spectrum that is used for transmission by inserting a suitable number of zeros at the upper and/or lower end as indicated in the figures. As shown in the figures, between each DFT output sample, $L-1$ zeros are inserted. A mapping with $L=1$ corresponds to a localized transmission. In this case, the DFT output is mapped to consecutive sub-carriers to generate a normal localized FDMA signal. Meanwhile, when $L$ is set to greater than 1, a distributed transmission signal is generated with a comb-shaped spectrum.

### *3.2.1 Radio parameters*

**Figure 4** illustrates the basic TTI structure comprising two sub-frames for uplink transmission. In the SI specification, a sub-frame comprises two short blocks (SBs) and six long blocks (LBs). SBs are used for reference signals for coherent demodulation and channel-quality measurement at the Node B and/or control and data

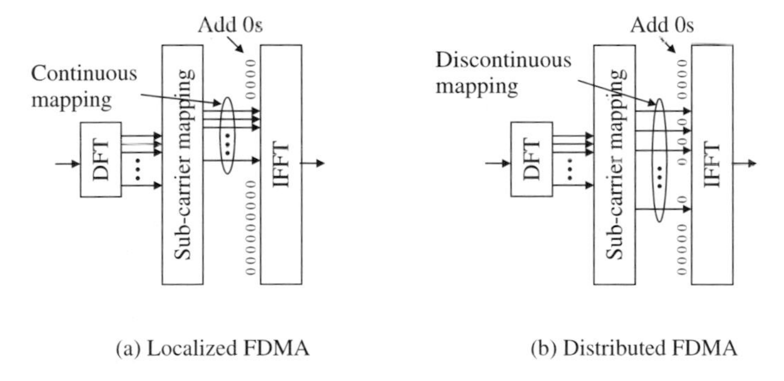


*Figure 3.* Mapping schemes for localized and distributed FDMA transmissions

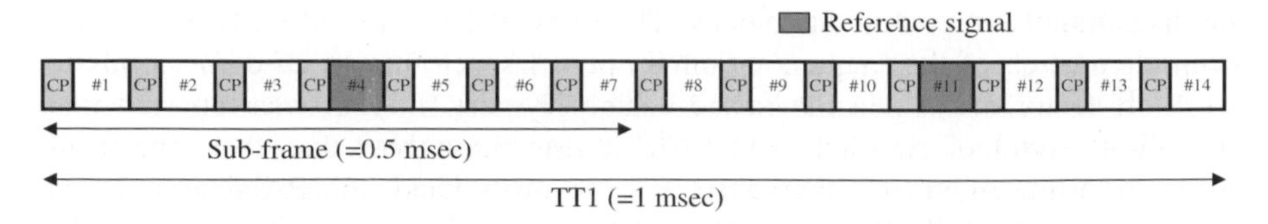


*Figure 4.* Basic sub-frame structure for uplink transmission

transmissions. Meanwhile, LBs are used for control and/or data transmissions. However, in the WI evaluation, the following two modifications were added for the uplink TTI structure.

First, the reference signal for channel-quality measurement is designed separately from the reference signal (RS) for coherent demodulation. Thus, it was decided that a LB at the beginning of each TTI (= 1.0 msec) would be used for the RS for channel-quality measurement. Second, to simplify the TTI structure, one LB instead of two SBs per sub-frame (i.e., two LBs instead of four SBs per TTI) is required in the design for the RS for coherent demodulation. The LB for the RS is multiplexed into the middle position of each sub-frame. The data part can include either contention-based data transmission or scheduled-based data transmission.

- **Sub-frame and TTI lengths**

In the SI evaluation, the TTI was set to the sub-frame duration, i.e., 0.5 msec. This setting was also established in the downlink.

However, in the WI evaluation, the necessity for concatenation, i.e., repetition, of multiple sub-frames into a longer TTI was presented to increase the area coverage for L1/L2 control signaling bits by multiplexing the information bits over a concatenated long TTI duration. The long TTI is particularly of interest for the uplink to extend the coverage area, since the achievable uplink data rate is often power limited rather than bandwidth limited. Moreover, one option for the TTI length was claimed commonly in the downlink and uplink for the sake of implementation simplicity. As a result, the TTI length was decided to be 1.0 msec based on the tradeoff relation between a short control delay and the repetition of L1/L2 control signaling in order to achieve wide coverage provisioning. Since one TTI comprises two sub-frames, intra-TTI frequency hopping is beneficial to achieving a larger frequency diversity effect particularly in the uplink with restricted transmission power at a UE.

- **CP length**

The required uplink CP length is decided from the time delays of multipaths to be supported and other impairment factors such as the path-timing-detection error, residual transmission-time-alignment error, and residual frequency drift between Node B and simultaneously accessing UEs. The E-UTRA uplink CP length is almost identical to that for the downlink because the influence from the impairment factors is slight. The CP lengths are slightly different in the respective transmission bandwidths.

- **Block size**

The LB size is constant regardless of the transmission bandwidth. The numbers of sub-carriers in the occupied bandwidth and samples per block are changed according to the transmission bandwidth. According to the reduction in the block size, the CP insertion loss becomes larger. Meanwhile, as the block size increases, the tracking ability for fast channel variation under high mobility conditions is degraded. Therefore, the optimum block size was investigated based on these tradeoff relationships. For instance, the LB comprises 512 samples in a 5-MHz transmission bandwidth case.

### *3.2.2 Modulation schemes*

Currently assumed E-UTRA uplink data-modulation schemes are BPSK, QPSK, 8PSK, and 16QAM. Higher-order modulation schemes such as 8PSK and especially 16QAM, are especially applicable to small-cell environments and low-load conditions. To achieve efficient modulation with a low PAPR, phase-shift type and offset type modulation schemes were investigated. However, $\pi/4$-shifted QPSK and 16QAM modulations and offset QPSK and 16QAM modulations have only a slight effect in decreasing the PAPR although $\pi/2$-shifted BPSK modulation reduces the PAPR to some extent compared to that for BPSK assuming the same conditions. Furthermore, it was reported that (8, 8)-star 16QAM reduces the required average received $E_b/N_0$ considering a cubic metric (CM) compared to square 16QAM in the case of a low channel coding rate such as 1/3. However, when the adaptive modulation and coding (AMC) scheme is used, the merit of (8, 8)-star 16QAM is not gained since the application region of the (8, 8)-star 16QAM with $R= 1/3$ is concealed by a lower modulation and coding scheme (MCS) such as QPSK and $R = 2/3$. As a result, square 16QAM is the current working assumption for 16QAM modulation scheme.

### *3.2.3 Channel coding and interleaving*

Similar to the downlink, Turbo coding based on WCDMA Release 6 is the working assumption for Evolved UTRA SI.

### *3.2.4 Uplink multiplexing*

The channel-coded, interleaved, and data-modulated Layer 3 information is mapped onto SC-FDMA symbols in the time and frequency domains. The overall SC-FDMA time/frequency resource symbols can be organized into a number of resource units (RUs). Each RU consists of a number (M) of consecutive or non-consecutive sub-carriers during the *N* long blocks within one sub-frame. In order to support localized and distributed transmission, two types of RUs are defined: Localized RU (LRU) and distributed RU (DRU). The LRU consists of *M* consecutive sub-carriers during *N* long blocks, which is a conventional single-carrier signal generated in the frequency domain. The DRU consists of *M* equally spaced non-consecutive sub-carriers with a comb-shaped spectrum during *N* long blocks.

# 4. DOWNLINK PHYSICAL CHANNEL STRUCTURES

In this section, we explain the features and structures of the E-UTRA downlink physical-channel structure related to:

- Reference signals
- Broadcast Channel (BCH)
- Paging Channel (PCH)
- Synchronization Channel (SCH)
- Multimedia Broadcast Multicast Services (MBMS)
- L1/L2 Control Channel
- Shared Data Channel

## 4.1 Downlink Reference Signal

E-UTRA downlink reference signal(s) is used for the following purposes[2].

- Downlink channel-quality measurements
- Downlink channel estimation for coherent demodulation/detection at the UE
- Cell search and initial acquisition

### *4.1.1 Mapping of reference signal symbols in time and frequency domains*

**Figure 5** illustrates the multiplexing of reference symbols in the downlink. The basic downlink reference-signal structure consists of known reference symbols transmitted at known positions within the OFDM time/frequency grid. Reference symbols are multiplexed into the first and third last OFDM symbol of every sub-frame with a frequency-domain separation of six sub-carriers. Furthermore, the reference symbols at the third last OFDM symbol are staggered in the frequency

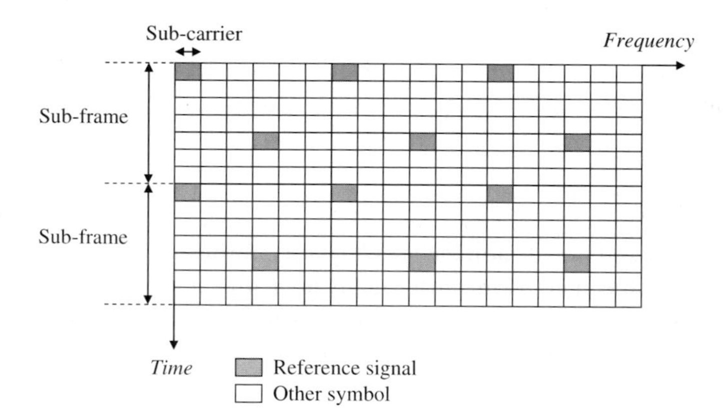


*Figure 5.* Multiplexing of downlink reference symbols in downlink

[2] A reference signal basically corresponds to the pilot channel of WCDMA

domain compared to the reference symbols of the first OFDM symbol. It was reported that by multiplexing reference signals into two OFDM symbols within a sub-frame, low-to-high mobility environments up to, e.g., 350 km/h can be supported without additional reference signals in the time domain.

### *4.1.2 Orthogonal reference signals*

In E-UTRA, it should be possible to provide orthogonal reference signals between cells of the same Node B as well as between different transmit antennas of the same cell. Orthogonal reference signals between transmit antennas within the same cell is e.g. needed to support downlink transmit diversity and MIMO transmission.

**(1) Orthogonal reference signals for different transmission antennas**
Orthogonal reference signals for different transmit antennas of the same cell/beam is established by means of FDM, possibly in combination with TDM. Thus, reference-signal multiplexing with different antenna-specific frequency (or time) shifts is used for each antenna. The main reason for relying on FDM/TDM-based orthogonality between transmit antennas of the same cell/beam is that it provides more accurate orthogonality compared to CDM-based orthogonality since no inter-code interference occurs in a frequency-selective fading channel. A high level of orthogonal accuracy is necessary to separate composite streams from different antennas in MIMO multiplexing and MIMO diversity schemes.

**(2) Orthogonal reference signals for different cells in the same Node B**
CDM-based reference-signal orthogonality is used between different cells/beams belonging to the same Node B in order to suppress the mutual interference particularly near the cell boundary. The merit of CDM-based orthogonality, compared to FDM-based orthogonality, between cells of the same Node B is a better tracking ability for the channel estimation, particularly UEs far from sector borders, since the density of the CDM-based orthogonal reference symbols in the frequency domain is higher than in case of FDM-based orthogonality.

**Figure 6** shows the principle of the intra-Node B orthogonal reference signal employing the combination of a Node B-specific scrambling code and cell-specific orthogonal sequence in the same Node B. As shown in **Figure 6**, we employ the same scrambled code among all cells belonging to the same Node B unlike in the WCDMA scrambled code assignment. Furthermore, a cell-specific orthogonal sequence is applied in order to distinguish cells (typically three or six) within the same Node B. Therefore, the resultant cell-specific scrambled code for the reference signal, $p_{n,m}$ ($n$ is the cell belonging to the same Node B and $m$is the index for the reference symbols), is generated through the combination, i.e., multiplication, of a Node B-specific scrambled code and cell-specific orthogonal sequence represented as

$$p_{n,m} = c_m \cdot s_{n,m \bmod SF}. \tag{2}$$

In this equation, $c_m$ denotes the Node B-specific scrambled code, and $s_{n,m}$ is the orthogonal sequence with the spreading factor of $SF$ employed in the $n$-th cell.

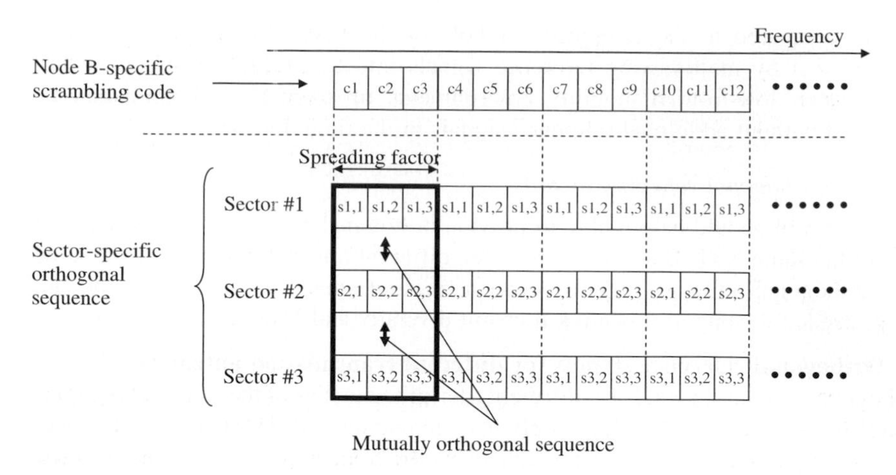


*Figure 6.* Principle of intra-Node B orthogonal reference signal structure

The cell-specific orthogonal sequence is generated by a Walsh-Hadamard sequence or phase rotation sequence. Here, we assume a cell-specific orthogonal sequence generated by phase rotation as indicated in the following equation assuming $N$sectors ($SF = N$) in the same Node B.

$$s_{n,m} = \exp\left(j\frac{2n\pi}{N}m\right). \tag{3}$$

Thus, in the three-cell configuration at each Node B, the phase rotation of 0, $2\pi/3$, and $4\pi/3$ is added to Sectored beams 1, 2, and 3, respectively. Using the orthogonal reference signal in **Figure 6**, intra-Node B orthogonality in the channel estimate is achieved by despreading CDM based reference symbols in the frequency or time domain. Note that the channel estimate at each sub-carrier is directly used without despreading for the UE without intra-Node B macro-diversity.

## 4.2 Broadcast Channel (BCH)

The broadcast channel (BCH) is used to broadcast system and cell-specific control information over the entire cell area. The broadcast control information includes information related to connection setup, cell selection, and re-selection, etc..

### *4.2.1 Broadcast Control Information*

Broadcast control information can be categorized into cell-specific information, Node B-specific information, and system-specific information. Furthermore, another level of categorization is primary information, which is necessary to be immediately available to UE after cell search and initial acquisition, and non-primary information. **Table 2** lists different kinds of broadcast control information together with the categorization according to above.

*Table 2.* Broad cast Control Information

| System control information elements | Classification (Area scope) | Primary or not |
|---|---|---|
| SFN (System Frame Number) | Node B specific or cell specific | Primary |
| PLMN (Public Land Mobile Network) identity | Node B specific or cell specific | Primary |
| Overall transmission bandwidth | Node B specific | Primary |
| Number of transmit antennas | Node B specific or cell specific | Primary |
| Scheduling and update information index (value tag) of system control information | Cell specific | Primary |
| NAS (Non Access Stratum) system information | Node B specific | Non-primary |
| UE (User Equipment) timers and counters | Node B specific | Non-primary |
| Cell selection and re-selection parameters | Node B specific or cell specific | Non-primary |
| Common physical channel configuration | Node B specific or cell specific | Non-primary |
| UL interference | Cell specific | Non-primary |
| Dynamic persistence level | Cell specific | Non-primary |
| Measurement control information | Cell specific | Non-primary |
| Time of day | PLMN specific | Non-primary |
| UE positioning related information | Cell specific | Non-primary |
| Stored RB (Radio Bearer) configuration | PLMN specific | Non-primary |
| PLMN Ids of neighboring cells | Cell specific | Non-primary |

### 4.2.2 *Multiplexing of BCH*

(1) **Primary broadcast information**

The primary broadcast information is transmitted using the BCH with a pre-determined radio resource, which is known to all UEs. The BCH is multiplexed into one or a few sub-frames during one radio frame.

The BCH is transmitted from the center part of the overall cell transmission band as shown in **Figure 7**, regardless of the overall cell transmission bandwidth, similar to the case of the synchronization channel (SCH), see below. Accordingly, no change in the carrier frequency is necessary after establishing the initial acquisition. In terms of the BCH transmission bandwidth, a wide transmission bandwidth such as 5 MHz can achieve superior link performance compared to e.g. a 1.25-MHz transmission bandwidth due to a larger frequency-diversity effect. On the other hand, a 1.25-MHz transmission bandwidth for the BCH has advantages in that the UE can decode the BCH of the target cell to perform handover without a change in the carrier frequency when the BCH is transmitted from the central part of the 20-MHz transmission bandwidth of the neighboring cell where the UE capability for the minimum reception bandwidth is 10 MHz (note that the assumption is that the UE capability for the minimum reception bandwidth is slightly extended).

(a) Time domain

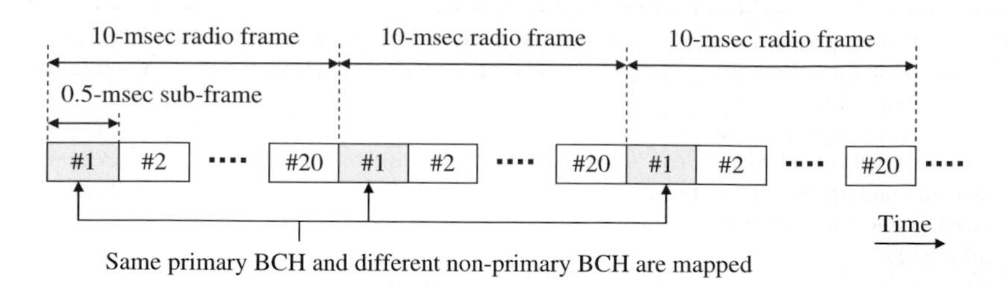


(b) Frequency domain

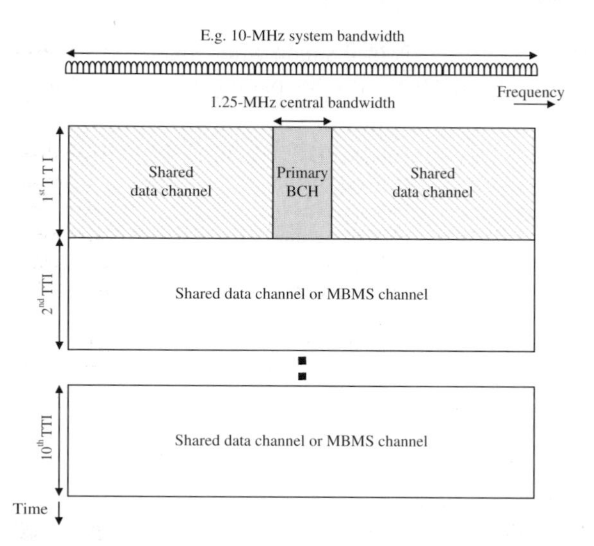


*Figure 7.* BCH Multiplexing

A constant 1.25-MHz transmission bandwidth for the BCH is also beneficial in order to achieve simple cell search since the UE does not need to detect the BCH bandwidth prior to decoding it.

(2) **Non-primary broadcast information**
Non-primary broadcast information is transmitted employing a scheduled-based shared data channel. A set of UE is informed of the RB assignment for non-primary broadcast information using the primary broadcast information in the BCH.

## 4.3 Paging Indicator and Paging Channel (PCH)

The paging channel (PCH) is used for network-initiated connection setup. Efficient reception of the PCH is necessary to obtain a high power saving effect.

### *4.3.1 Control Information in Paging indicator and PCH*

A paging indicator (PI) is used before receiving the PCH similar to WCDMA. The number of bits for the PI information is much less than that for the PCH. Thus, the time duration of the PI is much shorter than that for the PCH. Therefore, by using the PI, a much higher gain for power saving at a set of UE using intermittent reception is obtained compared to the case with direct PCH reception without the PI. PI information contains the following.

- Group ID: The group ID indicates the ID of the user group who are to receive the subsequent PCH.
- Mapping information: This information indicates the location of the RBs where the PCH to be decoded is multiplexed.

The PCH conveys the following information employing a scheduled-based shared data channel.

- User ID: The user ID indicates the ID of the user who is paged from the Node B.
- Cause ID: The cause ID indicates the cause for paging such as the traffic service type.

The flow of the decoding procedure for the paging information is given in **Figure 8**.

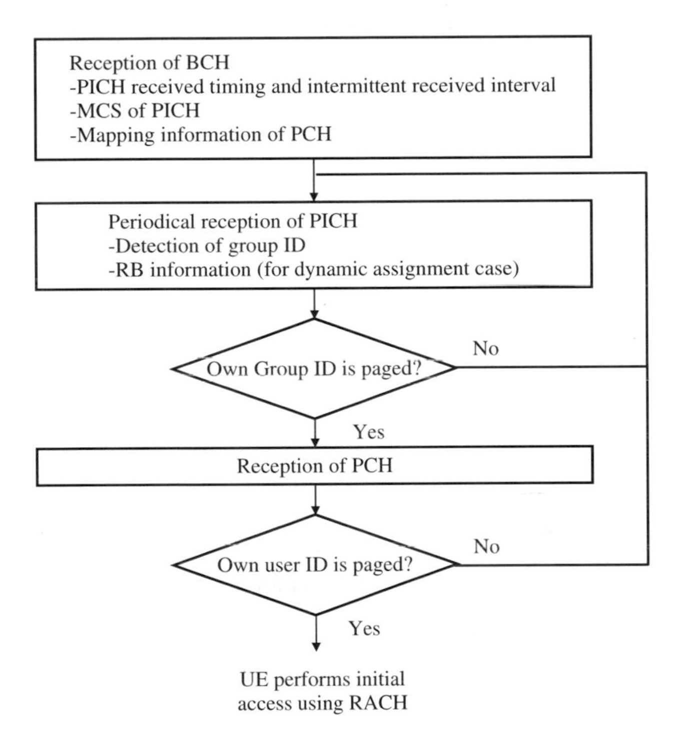


*Figure 8.* Flow of decoding procedure for paging information

### *4.3.2 Multiplexing of PI information and PCH*

**Figure 9** illustrates an example of the multiplexing of PI information and the PCH. The PI information is conveyed using the downlink L1/L2 control channel. In the figure, the PI information is multiplexed into the same OFDM symbol duration as the L1/L2 control channel using distributed transmission. Note that different cell-specific control information in the same Node B is sent on the L1/L2 control channel, whereas the cell-common PI information in the same Node B is transmitted, and coordinated transmission is applied to the PI information. By using separate coding between Cat. 1 information (control information related to scheduling (resource assignment)) and Cat. 2/3 information (control information related demodulation and decoding), transmission of the Cat. 2/3 information can be omitted to avoid an unnecessary increase in the overhead. This configuration also allows application of the synchronous PI information and PCH transmission schemes employing coordinated transmission among cells within the same Node B.

The PI information is transmitted from the system-dependent pre-assigned transmission frequency band. For example, in the figure assuming a 20-MHz system bandwidth, two 10-MHz frequency blocks of the L1/L2 control channel are defined, but only the central 5-MHz band is used as the pre-assigned transmission band for the PI information.

The PCH is transmitted within the pre-assigned transmission band similar to the case of the PI information. In the example in **Figure 9**, the system allocated bandwidth and system-dependent pre-assigned transmission band for the PCH are

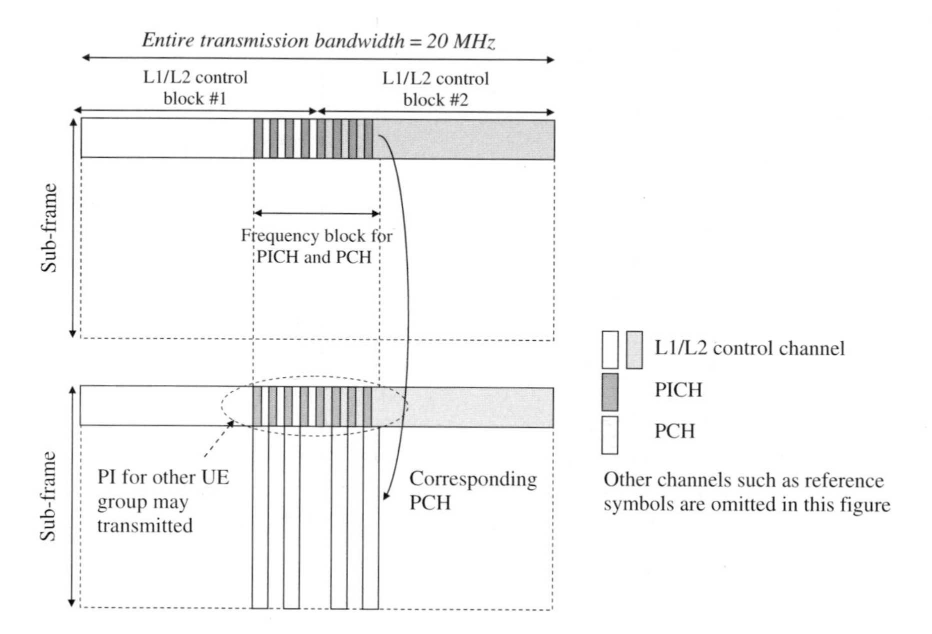


*Figure 9.* Multiplexing of PI information and PCH

20 and 5 MHz, respectively. Sets of UE are notified of the pre-assigned transmission band at each cell site using the broadcast information. It should be noted that assigning the central part of the system bandwidth as the pre-assigned transmission band for the PCH can be beneficial in simplifying the cell search procedure for the neighboring cells with the same carrier frequency, since the change in the center frequency at the UE can be avoided. By transmitting the PICH in advance using a pre-decided duration before the PCH, the decoding processing of the PCH can be simplified (see **Figure 9**).

### *4.3.3 Resource assignment for PCH*

There are two possibilities for RB assignment for the PCH within the pre-assigned frequency block: Semi-static assignment and dynamic assignment. When semi-static assignment is used, the RB positions for the PCH are fixed. The number of assigned RBs may be changed according to the amount of paging information. In this case, the UE is informed of the number of assigned RBs for the PCH using the information regarding the number of RBs. The assigned RBs within the pre-decided transmission band are pre-decided according to the number of assigned RBs. Therefore, the UE can know the positions of the RBs for the PCH by decoding only the RB index information. In order to achieve coordinated synchronous transmission within the same Node B, the position of the RBs for the PCH must be common to all sectors within the same Node B. Meanwhile, when dynamic assignment is used, the assigned RB position can be dynamically changed according to the frequency domain channel dependent scheduling results on the shared data channel. Typically, by prioritizing the frequency domain channel dependent scheduling of the shared data channel, the PCH is transmitted using the remaining RBs. This brings about increased channel dependent scheduling gain for the shared data channel. However, the number of control signaling bits for the PI information will be increased compared to the case with semi-static assignment since the UE must be informed of the detailed RB positions of the PCH by using the PI. Similar to the semi-static assignment, to achieve coordinated transmission among sectors within the same Node B, the position of the RBs for the PCH must be common to all sectors within the same Node B.

### *4.3.4 Synchronous transmission and soft-combining reception*

Since the PI and PCH convey sector-common information from all sectors in the same Node B, synchronous transmission associated with soft-combining among cells within the same Node B was proposed to achieve high quality transmission of the PI and PCH. **Figure 10** shows synchronous transmission employing delay diversity among cells in the same Node B, i.e., sectors, and soft-combining reception. As shown in **Figure 10**, the same paging information or PCH is transmitted among cells in the same Node B using coordinated delay diversity so that the time delays of the paths of all cells in the same Node B are aligned within the CP. Then, since soft-combining within the CP is used at the UE, high

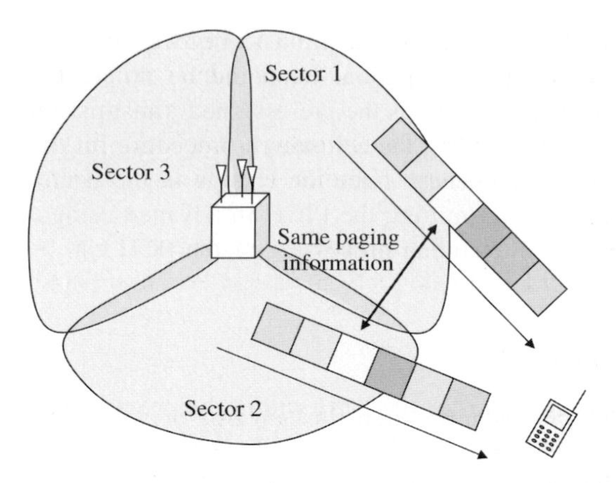


*Figure 10.* Principle of simultaneous transmission and soft-combining reception

quality reception is achieved for UEs located near the cell boundary. This coordinated transmission and soft-combining can be applied regardless of the usage of repetition (spreading) for the PCH. In synchronous transmission with soft-combining, two reference signal structures are considered: Cell-common reference signals in the same Node B and cell-specific orthogonal reference signals in the same Node B. It was reported that the cell-common reference signals in the same Node B achieved better packet error rate performance than the cell-specific orthogonal reference signals in the same Node B, even though an additional cell-specific orthogonal reference signal is necessary for demodulation of the L1/L2 control channel within the same sub-frame. This is because when cell-specific orthogonal reference signals in the same Node B are used, the influence of the background noise is greater than that with cell-common reference signals, since the received signal is demodulated independently at each cell and then soft-combined.

## 4.4 Downlink L1/L2 Control Channel

### *4.4.1 Control signaling bits in L1/L2 control channel*

The following L1/L2 control signaling bits are transmitted using the downlink L1/L2 control channel.

- Downlink scheduling information for the downlink shared data channel
  - UE identity: Identification of the assigned UE
  - RB assignment information: Location of the assigned RBs
  - MIMO related information: Employed MIMO mode (MIMO multiplexing or MIMO diversity, etc.) and the number of data streams (note that a portion of the information may be transmitted as downlink demodulation-related information)
- Control information for demodulation of the downlink shared data channel
  - MCS information

- Control information for decoding of the downlink shared data channel
  - Hybrid ARQ related information: hybrid ARQ process number and redundancy version including new data indicator
- Uplink scheduling information for the downlink shared data channel
  - UE identity and RB assignment information: Similar to downlink-related information
- Control information for demodulation of the uplink shared data channel
  - MCS information and MIMO related information: Similar to downlink-related information
- ACK/NACK bit in response to uplink transmission
- Other information
  - Transmission timing control bits for adaptive transmission timing alignment in the uplink
  - Transmission power control (TPC) command for uplink transmission
  - PI information (this information can be categorized into downlink scheduling information)

The UE first detects the scheduling-related information, and the demodulation and decoding-related information are subsequently detected. It should be noted that the number of control signaling bits for demodulation and decoding of the shared data channel may change according to the MIMO configuration when the per antenna rate control (PARC) is applied. However, since the MIMO configuration is sent as a part of the scheduling-related information in advance, the number of bits for demodulation and decoding of the shared data channel can be identified before the UE decodes these bits.

### *4.4.2 Multiplexing of L1/L2 control channel*

As shown **Figure 11**, there are two candidates for multiplexing of the L1/L2 control channel with other physical channels: Time domain multiplexing (TDM) and frequency domain multiplexing (FDM). Here, we compare TDM and FDM

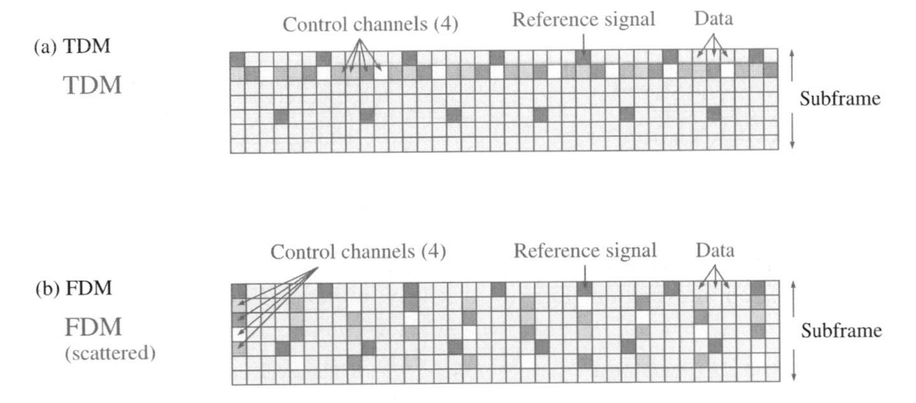


*Figure 11.* TDM and FDM Multiplexing of downlink L1/L2 control channel

multiplexing from the viewpoints of the possibility of power savings using the micro-sleep mode, processing delay, and a method for increasing the coverage. From the viewpoint of power saving TDM is potentially more advantageous than FDM, due to the possibility for micro-sleep. In addition, compared to FDM, TDM can somewhat reduce the processing delay due to the reception and demodulation time for the L1/L2 control channel. However, FDM can allow for power balancing between coded data symbols, reference symbols, and the L1/L2 control channel, which may improve coverage, see **Figure 12**. In this case, for UEs near the cell edge, more power can be allocated to the L1/L2 control information symbols by reducing the transmission power of the data symbols at the cost of decreased throughput. However, in the TDM structure, the total transmission power for the L1/L2 control channel can be increased to increase the coverage using the following methods. The first is using a long TTI at the cost of increasing the control delay. By repeating the same L1/L2 control information over multiple sub-frames, the received power of the L1/L2 control channel is increased. The second is to use a low coding rate including a large repetition factor within one sub-frame by reducing the number of coded data symbols in the shared data channel. A low coding rate including a large repetition factor in in case of TDM is fundamentally the same as power balancing in case of FDM although the lower coding rate method in TDM needs additional signaling to inform UE of the transport format of the L1/L2 control channel. It should be noted though that power balancing in case of FDM may require signaling of the transmission power ratio between the reference signal and the shared data channel in case of 16QAM or 64QAM modulation. Alternatively, blind estimation can be applied as is used for HSDPA.

It should also be mentioned that a lower coding rate for the L1/L2 control channel requires a change in the transport format of the shared data channel since the number of symbols available to the shared data channel is changed according to the coding rate of the L1/L2 control channel. This brings about some degree of

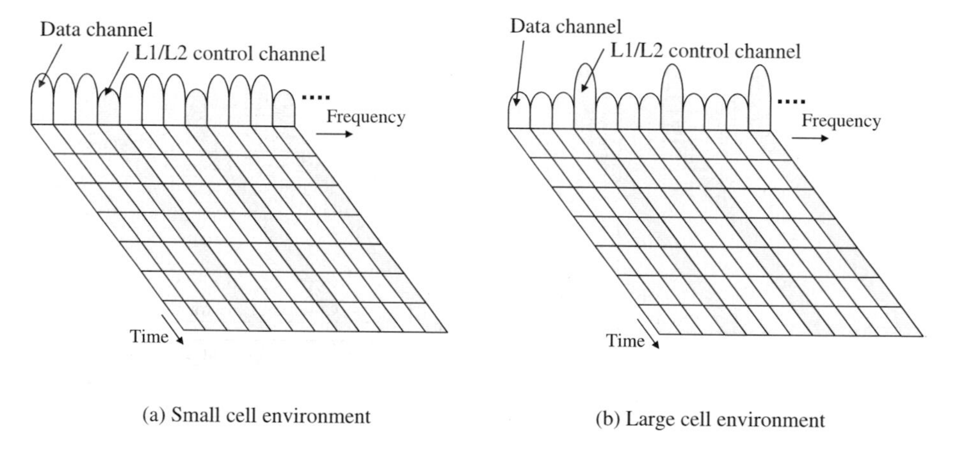


*Figure 12.* Power balancing in FDM multiplexing

complexity at the UE receiver. However, the number of symbols available to the shared data channel is also changed according to the number of scheduled sets of UE since the number of symbols for the L1/L2 control channel is dependent on the number of scheduled sets of UE both for TDM and FDM. Therefore, the control of the transport format for the shared data channel according to the configuration of the L1/L2 control channel is required for both TDM and FDM.

### *4.4.3 Channel coding scheme for L1/L2 control channel*

The coding schemes, i.e., joint or separate coding, in the downlink L1/L2 control channel listed below have a major impact on the design of the downlink L1/L2 control channel structure.

- Joint or separate coding of downlink transmission-related Cat. 1 information (control information related to scheduling (resource assignment))
- Joint or separate coding between downlink transmission-related Cat. 1 information and Cat. 2 and 3 information (control information related demodulation and decoding) within the same UE.
- Joint or separate coding between downlink transmission-related control information and uplink transmission-related information

In general, joint coding is advantageous from the viewpoints of the number of control signaling bits and the channel coding gain. Separate coding is advantageous from the viewpoint of the effect of link adaptation such as transmission power control (TPC) and the adaptive modulation and coding channel rate (AMC), the effect of beam-forming or pre-coding, and frequency diversity via channel dependent scheduling. Here, we focus on joint or separate coding for the downlink L1/L2 control channel for downlink transmission related information. **Figure 13** shows the possible channel coding schemes for L1/L2 control

| Option | Cat. 1 information for multiple users | Cat. 1 and Cat. 2/3 information | |
|---|---|---|---|
| 1 | Joint | Joint | Cat. 1 and 2/3 for multiple sets of UE |
| 2 | Joint | Separate | Cat. 1 for multiple sets of UE<br>Cat. 2/3 for UE 1<br>Cat. 2/3 for UE 2<br>Cat. 2/3 for UE 3 |
| 3 | Separate | Joint | Cat. 1 and 2/3 for UE 1<br>Cat. 1 and 2/3 for UE 2<br>Cat. 1 and 2/3 for UE 3 |
| 4 | Separate | Separate | Cat. 1 for UE 1 / Cat. 2/3 for UE 1<br>Cat. 1 for UE 2 / Cat. 2/3 for UE 2<br>Cat. 1 for UE 3 / Cat. 2/3 for UE 3 |

*Figure 13.* Channel coding scheme for L1/L2 control information

information. In principle, the following tradeoffs exist between the joint and separate coding schemes for downlink transmission-related control information.

- **Total number of control signaling bits and overhead associated with channel coding**

The joint coding scheme can reduce the overall number of control signaling bits for multiplexed sets of UE. Moreover, the size of the overhead such as the cyclic redundancy check (CRC) code associated with each coding block can be decreased in the joint coding scheme rather than the separate coding scheme.

- **Channel coding gain**

The joint coding scheme can provide a higher channel coding gain than the separate coding scheme, since the number of information bits accommodated within one coding block becomes larger.

- **Reception quality using link adaptation**

The separate coding scheme has a high affinity to UE-dependent link adaptation such as TPC and AMC for the L1/L2 control channel. We proposed a CQI-based TPC and consider applying the TPC to the L1/L2 control channel to mitigate the fluctuation in the received level due to instantaneous fading. The application of AMC to the L1/L2 control channel was also proposed. Thus, the required average received signal energy per bit-to-noise power spectrum density ratio ($E_b/N_0$) of the L1/L2 control channel in a multipath fading channel can be decreased by employing the separate coding scheme rather than the joint coding scheme due to the user-dependent precise link adaptation. In the joint coding scheme, the required transmission power may be significantly increased, since TPC compensates for the worst CQI among sets of UE to which the shared L1/L2 control information should be correctly decoded.

- **Effect of channel dependent scheduling gain**

The separate coding scheme has a high affinity to UE-dependent channel dependent scheduling since the control signaling bits can be transmitted from the assigned RBs.

It is shown that the combination of separate coding of the downlink transmission-related Cat. 1 information and separate coding between downlink transmission-related Cat. 1 information and Cat. 2 and 3 information require fewer radio resources than joint coding since the difference in the impact of the accuracy of link adaptation is much greater than that in the total number of control signaling bits and channel coding gain.

## 4.5 MBMS

MBMS transmissions are performed in the following two ways: Multi-cell transmissions and single-cell transmissions. Moreover, in the case of multi-cell transmission, tight inter-cell (Node B) synchronization, in the order of substantially less than the CP duration, can be optionally applied to enable UEs to simultaneously receive MBMS transmissions from multiple cell sites, so called SFN or Single Frequency

Network reception. To support multi-cell reception also in case of large cells with substantial propagation times, a long CP with the length of 16.67 μsec can be used in case of MBMS transmission.

### 4.5.1 *Multiplexing of MBMS traffic*

MBMS transmissions can be carried out on a separate carrier, with only MBMS transmission. Alternatively, MBMS transmission and unicast data traffic can share the same carrier.

In a system where Unicast and MBMS traffic are multiplexed into the same transmission band, TDM and/or FDM multiplexing are used. With TDM multiplexing, the MBMS traffic is transmitted in specific sub-frames using the same transmission band as the Unicast traffic. Moreover, TDM multiplexing of different MBMS streams is preferred in order to minimize the reception time of a specific MBMS stream. This enables lower power consumption for MBMS capable UE as shown in **Figure 14**.

Additional FDM multiplexing between MBMS traffic and unicast traffic is also supported and is especially required in case of an overall cell transmission bandwidths larger than the mandatory UE reception bandwidth capability.

### 4.5.2 *Reference signal for MBMS*

A reference signal for channel estimation of the MBMS channel is necessary in the MBMS sub-frame. In this case, channel gains are estimated using multiple reference signals with different time delays from multiple Node Bs providing MBMS service. Focusing on only MBMS traffic provisioning, it is sufficient to accommodate the

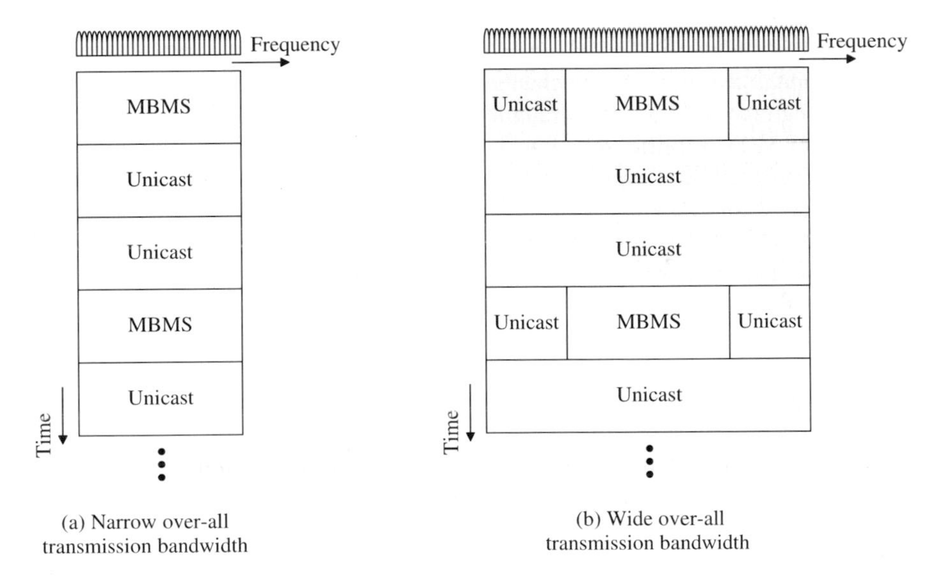


*Figure 14.* Multiplexing of MBMS traffic

shared data channel providing MBMS traffic and cell-common scrambled reference signal in an MBMS sub-frame. However, the downlink L1/L2 control information such as the scheduling grant and ACK/NACK bit in hybrid ARQ is necessary for uplink data transmission. Therefore, the cell-specific L1/L2 control channel and cell-specific scrambled reference signal for decoding control signaling bits and channel-quality measurement must be multiplexed in the MBMS sub-frame. Thus, comprehensive designs of cell-common scrambled and cell-specific scrambled reference signals are required in the MBMS sub-frame. The following two candidates are considered for reference signal structures associated with the scrambled code for the MBMS sub-frame.

- **Cell-common scrambled reference signals associated with additional cell-specific scrambled reference signals**

In this structure, a cell-common scrambled code is multiplied to the reference signals, which are used for channel estimation of MBMS traffic symbols. Moreover, the additional cell-specific scrambled reference signals are necessary for cell-specific channel-quality measurement and for channel estimation of the L1/L2 control channel. It should be noted that different cell-common scrambled codes are used for different MBMS streams.

- **Cell-specific scrambled reference signals associated with the same cell-specific scrambled modulation between reference signals and data**

The cell-specific scrambled reference signal for channel estimation of MBMS traffic was proposed. In this method, when applying the same scrambling modulation in the frequency domain to the data symbols in the same MBMS sub-frame, the cell-specific scrambled reference signal is used for channel estimation of the MBMS traffic symbols without detecting the scrambling code information in each cell. The cell-specific scrambled reference signals are simultaneously used for cell-specific channel-quality measurement and channel estimation for the L1/L2 control channel. Thus, although the cell-specific scrambled reference signal can be used for channel estimation of MBMS traffic as well as L1/L2 control information, the interpolation of channel gains in the frequency and time domains cannot be applied since the set of UE does not know the scramble code assigned to each cell. Therefore, the reference signal with the same scrambled modulation as the original reference signal must be repeated to improve the channel estimation accuracy.

The cell-common and cell-specific scrambled reference signals were compared from the viewpoint of the achievable PER performance of the MBMS traffic symbols. It was reported that almost the same PER was achieved between the cell-common scrambled reference signal including the addition of the overhead for a cell-specific reference signal for the uplink L1/L2 control channel and the cell-specific scrambled reference signal including repetition of the reference signal, although the former exhibited slight superiority to the latter under the high mobility conditions due to the interpolation of channel gains both in the time and frequency domains. Accordingly, the cell-common scrambled reference signal is considered to be the current working assumption for reference signals in the MBMS sub-frame.

## 4.6 Shared Data Channel

### *4.6.1 Localized and Distributed Transmissions*

In OFDM-based radio access, the granularity in the frequency domain is much narrower than that of DS-CDMA based 3G systems. Thus, frequency domain in addition to time domain (hereafter simply frequency domain) channel-dependent scheduling was proposed and its effectiveness in increasing the user throughput was investigated in several papers. In case of frequency-domain channel-dependent scheduling, the entire wideband signal is divided into resource blocks (RB) of continuous sub-carriers, (i.e. localized resource blocks). By dynamically assigning each RB to a UE with instantaneously good channel conditions at the corresponding frequency band, the user and cell throughput can be increased. This is also often referred to as multi-user diversity.

In a high mobility case, however, the control loop of channel-dependent scheduling cannot track the instantaneous fading variation and, consequently, the multi-user diversity effect can not be utilized to the same extent as in case of low mobility. In this case, discontinuous transmission in the frequency domain with a comb-shaped spectrum is more appropriate in order to, instead, achieve high-quality reception using frequency diversity over the entire transmission band. Thus discontinuous transmission is necessary to achieve high throughput and affinity to simultaneous localized transmission in the same sub-frame. The distributed OFDMA transmission is roughly categorized into sub-carrier-level transmission and RB-level transmission as shown in **Figure 15(a)** and **15(b)**, respectively. Sub-carrier-level distributed transmission provides a large frequency diversity gain due to the mapping of the transport channel to the physical channel at the sub-carrier level. Sub-carrier-level distributed transmission is achieved by puncturing bits of the localized transmission within the same RB. Therefore, the punctured-bit information for distributed transmission in addition to the RB information is necessary to demodulate the localized-transmission-UE. This brings about an increase in the number of control signaling bits for simultaneous transmitting sets of UE employing localized transmission. Moreover, the bandwidths or the transport block size of coexisting localized transmissions are changed according to the number of simultaneously assigned sub-carrier-level distributed transmissions.

Meanwhile, in the RB-level distributed transmission, the sets of UE employing localized transmission do not need information pertaining to the resource allocation of the sets of UE employing distributed transmission and vice versa. RB-level distributed transmission achieves simple radio resource assignment using the same RB unit as in localized transmission in a unified manner, although there is a slight sacrifice of the achievable frequency diversity effect. Thus, the RB-level distributed transmission seems more practical than the sub-carrier-level transmission from the viewpoint of simple radio resource assignment and the control signaling format. For low data-rate traffic such as VoIP, however, RB-level distributed transmission cannot obtain a sufficient frequency diversity effect because only one

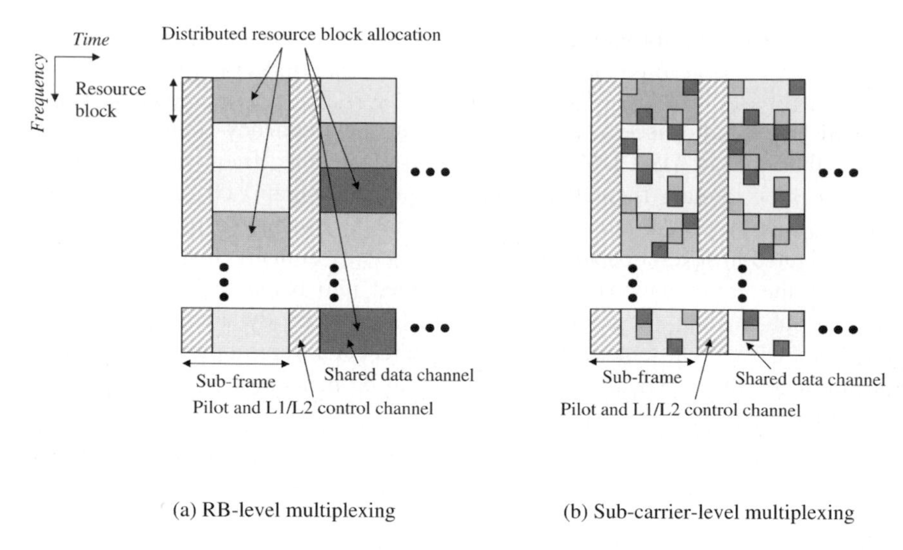


*Figure 15.* Sub-carrier-level and RB-level distributed OFDMA transmissions

RB size is sufficient to accommodate the payload size of such low-rate traffic. Therefore, block-wise RB-level distributed OFDMA transmissions with $N_D$-block divisions were proposed in order to obtain the frequency diversity effect even for low-rate traffic (here $N_D$ indicates the number of VRBs within one PRB). In the block-wise RB-level distributed transmission, one or a few RBs are further divided into a few blocks each. The divided blocks with the size of $1/N_D$ of one RB are multiplexed to generate one RB that comprises $N_D$ original RBs. In the block-wise RB-level distributed transmission, A RB that is assigned to distributed transmission, is segmented into $N_D$ units (the RB is called a VRB. The $N_D$ units of $N_D$ sets of UE using distributed transmission are multiplexed into the one PRB. Therefore, in particular for small size traffic, which requires only one or a few PRBs in the original BR-level distributed transmission, we can increase the number of PRBs into the one distributed transmission that is multiplexed by employing the proposed $N_D$-block division. The $N_D$ virtual distributed transmissions are multiplexed into the one PRB and virtual granularity in the frequency domain of resource assignment for the distributed transmission is identical to that for localized transmission. Furthermore, we do not allow multiplexing of distributed and localized transmissions in the same PRB. Meanwhile, in the sub-carrier-level distributed transmission, distributed transmission is multiplexed into the same PRB of the localized transmission by puncturing. Therefore, unlike sub-carrier-level distributed transmission, we can assign radio resources both to the distributed and localized transmissions in a unified manner in the RB-level distributed transmission. Moreover, in the scheme, the localized transmission channel can be decoded independently of the distributed transmission without applying control signaling on the resource assignment of distributed transmission, and vice versa. Meanwhile, the puncture pattern information of the distributed transmission is necessary for localized transmission to decode the own channel. Therefore, the block-wise RB-level distributed transmission can achieve simpler resource assignment than sub-carrier-level distributed transmission.

### *4.6.2 Frequency domain channel-dependent scheduling*

The Node B scheduler (for Unicast transmission) dynamically controls which time and frequency domain resources are allocated to a certain UE at a given sub-frame. Downlink control signaling informs UEs what resources and respective transmission formats have been allocated. The scheduler can instantaneously choose the best multiplexing strategy from the available methods, e.g. localized or distributed transmissions in the frequency domain. The flexibility in selecting RBs and multiplexing sets of UE will influence the available scheduling performance. Scheduling is tightly integrated with link adaptation and hybrid ARQ. The decision of which user transmissions to multiplex within a given sub-frame may for example be based on

- QoS parameters and measurements
- Payloads buffered in the Node-B ready for scheduling
- Pending retransmissions
- CQI reports from the sets of UE
- UE capabilities
- UE sleep cycles and measurement gaps/periods
- System parameters such as bandwidth and interference level/patterns

The UE is able to measure and report to the Node B the channel quality of one RB or a group of RBs, in the form of the CQI or Channel Quality Indicator. The achievable performance depends on the trade-off relation between the uplink signaling overhead and gains by frequency and time domain channel-dependent scheduling, and link adaptation taking varying channel-conditions and type of scheduling into account. Therefore, the granularity of the CQI reporting in the time and frequency domain is adjustable in terms of sub-frame units and RB units, and set on a per UE or per UE-group basis. CQI feedback from the UE, which indicates the downlink channel quality, can be used at Node B at least for the following purposes:

- Time/frequency selective scheduling
- Selection of modulation and coding scheme
- Interference management
- Transmission power control for physical channels, e.g., L1/L2-control signaling channels.

In a low mobility environment, when the round trip delay of frequency and time domain channel-dependent scheduling can track the instantaneous fading variation, high gain is achieved. However, in high mobility for a UE the gain in the channel-dependent scheduling is not obtained. In this situation, frequency hopping is beneficial to achieve high quality reception by randomizing the fluctuation of the received signal level in the frequency and time domains, in addition to distributed transmission.

### *4.6.3 Adaptive modulation and coding (AMC)*

Link adaptation (AMC: adaptive modulation and coding) with various modulation schemes and channel coding rates is applied to the shared data channel. Link adaptation uses either localized (for frequency selectivity) or distributed (for frequency diversity) transmission modes. The selection of localized or distributed transmission may be based on the service type, UE speed, packet size, or other factors. In the following discussion, we assume that one channel-coded block (stream) is transmitted from one user using multiple RBs. Note that in the case of localized transmission, the RB is defined as the minimum transmission bandwidth. The support of multiple channel-coded streams may also be necessary for segmenting large IP packets, etc.

In localized OFDMA transmission, three link adaptation methods are considered.

- RB-dependent adaptive modulation and RB-common channel coding rate method
- RB-common adaptive modulation and channel coding rate method
- RB-dependent adaptive modulation and channel coding rate method

In the RB-dependent adaptive modulation and channel coding rate method, the reliability of one coded bit differs among different RBs when the channel coding rate is changed according to the CQI within one coded block. Accordingly, the achievable channel coding gain is reduced compared to the other two methods. It was reported that the gain in the achievable throughput using the RB-dependent adaptive modulation from the RB-common modulation is small such as approximately 3%. This is because interleaving over multiple RBs assigned to one set of UE works well, and even the common modulation scheme is not optimum at each RB. Therefore, the same coding and modulation is applied to all groups of RBs belonging to the same L2 PDU scheduled to one user within one TTI and within a single stream. This applies to both localized and distributed transmission. **Figure 16** shows a block diagram for the RB-common modulation method. The operation of the method is as follows.

- CRC bits are attached at every L2 PDU and channel encoding is performed. The coding rate is common to all RBs.
- Channel interleaving is performed after hybrid ARQ transmission operation such as coding rate control.
- The interleaved coded block is segmented into multiple RBs.
- preading, e.g., repetition, may be performed as part of the channel coding or as part of the data modulation.
- A common modulation scheme is assigned to all assigned RBs regardless of the CQI information reported for the respective RBs. The modulation scheme and coding rate are decided from the average CQI information for all RBs assigned to the same sub-frame.
- Scrambling may be employed after adaptive modulation.

Note that when multiple RBs are assigned, they are not required to be adjacent. Clearly, the number of control signaling bits in RB-common modulation is reduced compared to that for RB-dependent modulation. Furthermore, in the case of multiple

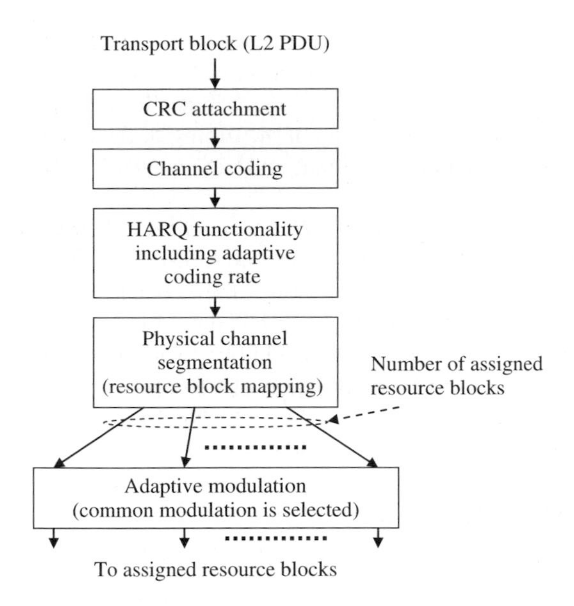


*Figure 16.* Resource block-common adaptive modulation and resource block-common channel coding rate scheme (for localized and distributed transmission modes)

channel-coded streams, the schemes in **Figure 16** may be separately applied to several different groups of RBs.

### *4.6.4 Hybrid ARQ*

E-UTRA Hybrid ARQ is based on so called Incremental Redundancy, with Chase combining being a special case. Furthermore, an N-channel Stop-and-Wait Hybrid-ARQ protocol is used.

Hybrid ARQ is categorized into synchronous and asynchronous processing.

- Synchronous hybrid ARQ: The (re)transmissions for a certain hybrid ARQ process are restricted to occur at know sub-frame timings. Thus, explicit signaling of the hybrid ARQ process number is not necessary, since the process number is derived from e.g., the sub-frame number.
- Asynchronous hybrid ARQ: The (re) transmissions for a certain hybrid ARQ process can occur at any sub-frame. Therefore, explicit signaling for the hybrid ARQ process number is necessary.

The various forms of hybrid ARQ can be further classified as adaptive and non-adaptive schemes from the viewpoint of transmission attributes, e.g., RB allocation, modulation scheme, transport block size, and duration of retransmission. Each scheme has the following control channel requirements.

- Adaptive hybrid ARQ: Transmitter may change some or all of the transmission attributes used in each retransmission from those of the initial transmission due to, e.g., a change in the propagation channel conditions. Therefore, the associated

control information is to be transmitted together with the retransmitted RB(s) such as the modulation scheme, RB application, and duration of transmission.
- Non-adaptive hybrid ARQ: The changes in the transmission attributes for retransmissions are known to both the transmitter and receiver at the timing of the initial transmission. Therefore, the associated control information does not need to be retransmitted.

Assuming these categorizations, the HS-DSCH in WCDMA uses the adaptive, asynchronous hybrid ARQ scheme, while E-DCH in WCDMA uses the non-adaptive, synchronous hybrid ARQ scheme. Furthermore, the asynchronous, adaptive ARQ is the current working assumption in the E-UTRA downlink, similar to HD-DSCH. Therefore, Node B transmits the process number to a target UE associated with the UE ID, which is conveyed by modulation of CRC bits.

## 4.7 SCH and Cell Search Procedure

### *4.7.1 Cell search in OFDM based radio access*

In the Evolved UTRA, a set of UE must acquire the best cell with the minimum path loss between the cell and the target UE. The cell acquisition process is called cell search. Similar to WCDMA, one-cell frequency reuse is the baseline in the Evolved UTRA, although the interference coordination using partial fractional frequency reuse is under discussion. In the cell search procedure, the UE acquires time and frequency synchronization with a cell and detects the Cell ID of that cell. The cell search procedure is categorized into the initial cell search and a neighboring cell search for handover. In this section, we mainly focus on the initial cell search. The assumptions for the cell search process in the Evolved UTRA are given below.
- The Evolved UTRA and UTRAN employ the 3G spectrum including additional spectra. This is also true in WCDMA. However, in addition, the GSM spectrum such as the 900-MHz frequency band is considered, which has a narrower frequency spectrum than 5 MHz. Therefore, the UE must search multiple frequency spectra in the initial cell search process.
- Similar to WCDMA, a set of UE must acquire a carrier frequency to connect to the target cell, which is on the frequency raster separated by every 200 kHz (or possibly a multiple of 200 kHz).
- Support of multiple transmission bandwidths from 1.25 to 20 MHz was decided. Meanwhile, a set of UE does not have any knowledge regarding the transmission bandwidth of the target cell. Thus, the UE must find the transmission bandwidth of the target cell from the candidates from 1.25 to 20 MHz.
- OFDM-based radio access was adopted in the downlink. Therefore, the SCH structure and cell search procedure, which is appropriate to OFDM, are necessary.

Cell search in the Evolved UTRA is performed employing two physical channels transmitted in the downlink, the SCH and BCH. Additional usage of a reference signal was proposed to detect, e.g., the cell ID. The primary purpose of the SCH is to acquire the carrier frequency and received timing, i.e., at least the SCH symbol timing of the downlink signal in the target cell. The primary purpose of the BCH is to broadcast a certain set of cell and/or system-specific information similar to the current UTRA BCH transport channel. In addition to the SCH symbol timing and carrier frequency information, the UE must acquire at least the following cell-specific information.

- The overall transmission bandwidth of the cell
- Cell ID
- Radio frame timing information when this is not directly given by the SCH timing, i.e., if the SCH is transmitted more than once every radio frame
- Information regarding the antenna configuration of the cell (number of transmitter antennas)
- CP length information regarding the sub-frame in which the SCH and/or BCH are transmitted

Each set of information is detected by using the SCH, reference symbols, or the BCH.

### *4.7.2 SCH structure*

The SCH in the downlink is necessary to detect the sub-frame timing, frame timing, cell-ID, etc. There are two different options for SCH structures: Hierarchical SCH and non-hierarchical SCH.

**(1) Hierarchical SCH**

The hierarchical SCH comprises a primary SCH (P-SCH) and secondary SCH (S-SCH) similar to the SCHs in WCDMA. The P-SCH has the same sequence, which is common to all cells. The S-SCH has a cell-specific, i.e., Node B-specific, sequence. The hierarchical SCH is suitable for cross-correlation based SCH-symbol timing detection (SCH-replica based detection). The SCH symbol timing of the target cell is detected by taking the correlation between the received signal and cell-common P-SCH replica in the time domain. It is required that the P-SCH code have a good auto-correlation property in the time domain.

In order to reduce the level of computational complexity inherent to the cross-correlation based detection, a layered (hierarchical) code structure in the time domain, which is adopted for the SCH in WCDMA, is beneficial. The P-SCH can be used as a reference to detect the S-SCH in the frequency domain after the SCH symbol timing detection. The S-SCH is used to detect the cell ID group, radio frame timing, and other control information conveyed by the S-SCH. For the S-SCH, many codes should be defined in order to carry many control information bits, and it is desirable that the detection of the S-SCH code be simple.

**(2) Non-Hierarchical SCH**

In non-hierarchical SCH, the SCH is mapped every $N$ ($N$ is greater than one) sub-carriers in the frequency domain within the same OFDM symbol duration. In this case, the same signal waveform is repeated N times within one OFDM symbol duration. Therefore, by detecting the auto-correlation of the repeated signal waveforms, the SCH symbol timing is estimated. The application of the Generalized Chirp Like (GCL) sequence to the non-hierarchical SCH was proposed. In the non-hierarchical SCH, auto-correlation detection for all possible GCL sequence candidates can be simply implemented by performing one discrete Fourier transform (DFT). The auto-correlation based detection using a non-hierarchical SCH enables a lower level of computational complexity than the cross-correlation based detection using a hierarchical SCH. In contrast, it was reported that the SCH symbol timing detection probability by auto-correlation based detection becomes worse than that for cross-correlation based detection particularly under low received SINR conditions. This is because the influence of the background noise and interference is greater in auto-correlation based detection than in cross-correlation based detection.

The 3GPP agreed to employ the hierarchical SCH in which the cell search time performance is prioritized especially for handover cell search, which is conducted under very low SINR conditions.

### *4.7.3 Multiplexing of SCH*

#### (1) SCH Transmission Bandwidth

The SCH structure is based on the constant bandwidth of 1.25 MHz regardless of the overall transmission bandwidth of the cell, at least for the initial cell search as shown in **Figure 17**. This is a result of the simple channel structure and simple acquisition operation being prioritized because the SCH is the first channel that a set of UE acquires without information about the transmission bandwidth.

#### (2) Multiplexing in Time domain

A small number of SCH symbols per radio frame is desirable in order to reduce the overhead. Furthermore, from the viewpoint of the timing detection performance against the background noise and interference, the signal energy of the SCH should be concentrated on a small number of OFDM symbols. Meanwhile, time diversity

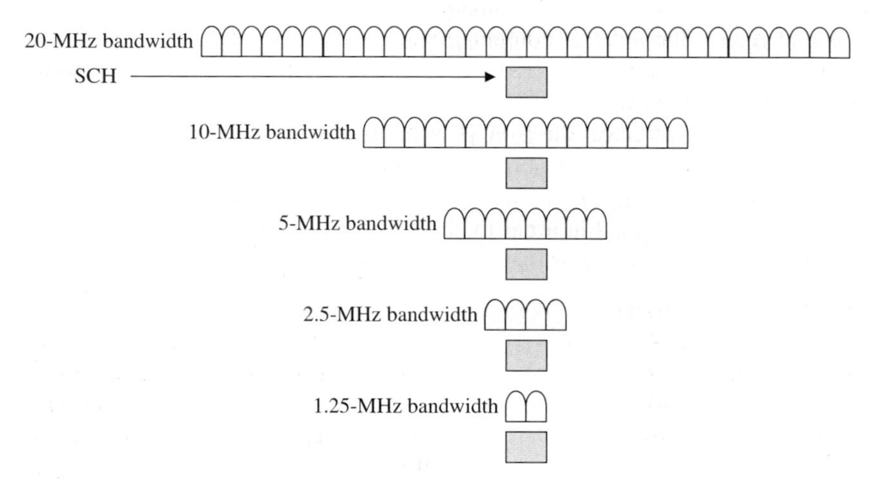


*Figure 17.* SCH transmission bandwidth

employing multiple SCH symbols is very effective in achieving fast cell search by improving the detection probability of the SCH particularly in a high mobility environment. As a result, it was decided that the SCH and BCH should be transmitted one or multiple times every 10-msec radio frame. The optimum number of SCHs per radio frame is to be specified from the cell search time performance including inter-radio access technology (RAT) measurement for various mobility conditions, and the impact on the TDD mode.

Next, we focus on the multiplexing of the SCH symbol in the sub-frame duration. The use of two types of CPs, short and long, was adopted mainly for Unicast and the MBMS. Assuming this condition, the SCH symbol mapping at the last OFDM symbol within a sub-frame was proposed as shown in **Figure 18**. In the SCH symbol mapping, the transmission timing of the SCH signal becomes identical regardless of the CP length. As a result, a set of UE can detect all SCH symbols for cell search without identifying the CP length of the SCH symbol.

**(3) Multiplexing in frequency domain**

As shown in **Figure 17**, the SCH is transmitted from the central part of the given spectrum regardless of the overall transmission bandwidth of a cell. Furthermore, the center frequency of the center sub-carrier of the SCH, i.e., over the entire transmission bandwidth of each cell site, is designed to satisfy the 200-kHz raster condition regardless of the entire transmission bandwidth of the cell site. By employing SCH multiplexing, a set of UE can detect the SCH on the 200-kHz raster and carrier frequency simultaneously irrespective of the total transmission bandwidth of the target cell. The system information and cell-unique information in the BCH is also transmitted from the central part of the total transmission bandwidth. Note that the current working assumption for the UE capability for the minimum reception bandwidth is 10 MHz. Thus, the SCH multiplexing in **Figure 17** was decided when the total transmission bandwidth of a cell site is 10 MHz or narrower. Three options for SCH multiplexing in the frequency domain were proposed when the total transmission bandwidth is 20 MHz.

Using the SCH multiplexing in the frequency domain, a set of UE first detects the central part of the spectrum of the best cell. Then, the UE moves to the

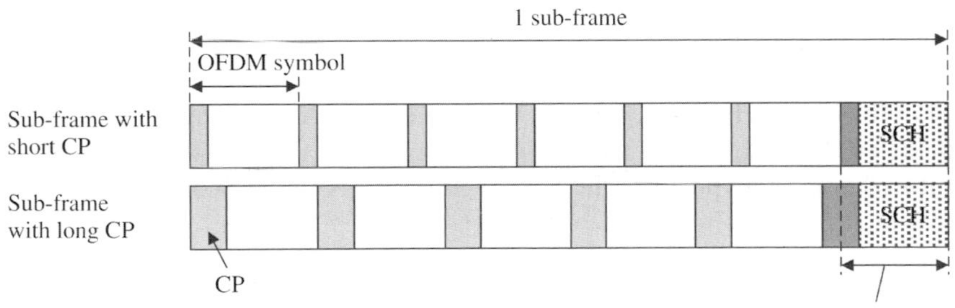


*Figure 18.* SCH symbol mapping

transmission bandwidth for actual communications. This transmission bandwidth for communications is assigned through the BCH from the Node B after the initial cell search. In the SCH multiplexing, the sub-carrier spacing does not have to satisfy the integer relationship with the 200-kHz frequency raster. Furthermore, we can avoid inefficient transmissions of the BCH signaling from any fractional transmission bandwidth. Since the transmission bandwidth for actual communications is assigned by the BCH from the cell site, each cell site can manage the transmission bandwidth assignment of the UE. As a result, unbalanced utilization of the spectrum of each cell site can be avoided.

Next, we focus on the multiplexing of the P-SCH and S-SCH in the hierarchical SCH structure. In the cross-correlation based detection, three types of P-SCH and S-SCH multiplexing methods are considered: FDM, CDM, and TDM. In the S-SCH correlation detection, coherent detection of the S-SCH using the P-SCH as a reference can improve the detection probability performance. From this viewpoint, FDM or CDM is desirable since the P-SCH is transmitted close to each sub-carrier of the S-SCH both in the time domain and in the frequency domain. Meanwhile, TDM is more advantageous than FDM in that the P-SCH does not suffer from interference from the S-SCH at the SCH timing detection in the time domain. Moreover, TDM can generate a larger number of S-SCH sequences compared to FDM since all the sub-carriers within a 1.25-MHz bandwidth can be used for the S-SCH transmission. Considering the advantages and disadvantages of the respective multiplexing methods between the P-SCH and S-SCH, TDM based multiplexing was decided as the working assumption in the 3GPP specification.

The number of codes for the S-SCH depends on the code length. Assuming the SCH transmission bandwidth of 1.25 MHz, the number of S-SCH codes per SCH symbol becomes less than approximately 70 because the maximum number of sub-carriers allocated to the 1.25-MHz SCH is 75. If the S-SCH code directly identifies the cell ID as proposed, the number of S-SCH codes is too small when assuming a 1.25-MHz transmission bandwidth. To solve this problem, $M$ (typically two)-code interleaved mapping of the (S)-SCH using FDM within one OFDM symbol was proposed as shown in **Figure 19**. The resultant number of codes becomes $(N_{sym}/M)^M$ ($N_{sym}$ indicates the number of S-SCH sub-carriers). For instance, when the code length is $N_{sym}$ = 64, the number of total codes could be 1024 when $M$ = 2, which is much larger than that when $M$ = 1. By using the interleaved mapping of $M$ codes, the frequency diversity effect is gained for the respective codes. In the auto-correlation based detection, coherent detection cannot be applied to detect the SCH. Thus, a SCH code sequence is assigned to the differential phase of the contiguous SCH sub-carriers in the auto-correlation based method whereas, we can assign it to the absolute phase of the S-SCH sub-carriers in the SCH-replica based method.

**(4) Cell search procedure**

In the method described, the cell ID (cell-specific scrambling code) is directly identified only by the SCH without using the reference signal. To enable flexible cell ID assignment, many cell IDs such as 512 for WCDMA, are

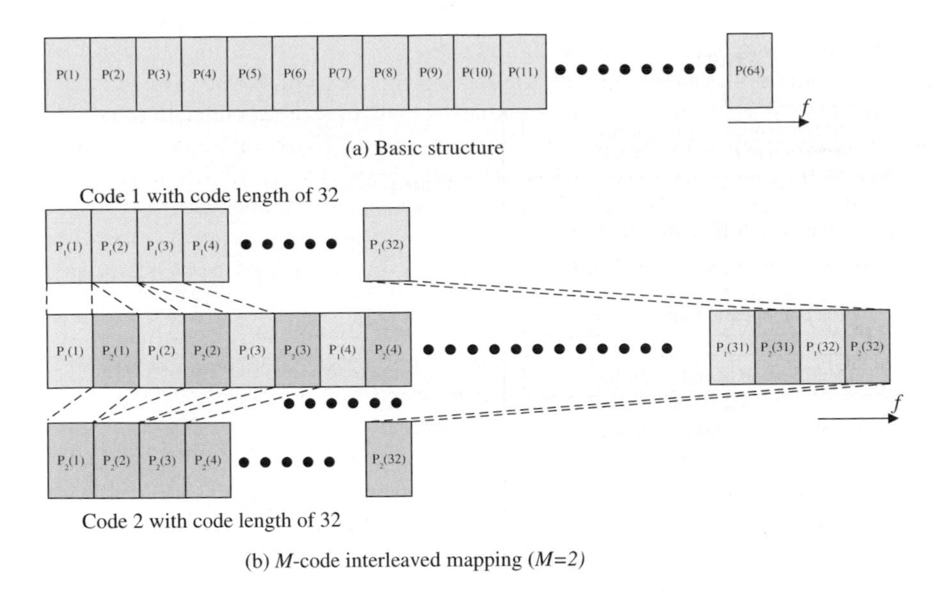


*Figure 19.* Proposed (S-)SCH mapping in frequency domain (example of basic code length of 16)

necessary. However, the number of cell IDs generated is limited by the SCH sequence length. Moreover, according to the increase in the number of cell IDs defined in the system, the cell search time becomes longer. The grouping of cell IDs was proposed, which is similar to WCDMA. The overall cell search procedure including the possible cell ID grouping is as follows as indicated in **Figure 20**.

- **SCH symbol timing detection**

First, the SCH symbol timing is detected following the detection of the OFDM symbol timing (FFT window timing). In the hierarchical SCH structure, the SCH symbol timing is detected from the correlation between the received signal and the SCH replica in the time domain. The repeated signal waveform arising from the decimation of the SCH sub-carrier in the frequency domain is detected in the auto-correlation based detection with the non-hierarchical SCH structure. Simultaneously, the carrier frequency of the target cell, which is on the 200-kHz frequency raster defined in the 3G system, is detected. The detection and compensation of the frequency drift can also be performed in this step.

- **Cell ID or Cell ID group index detection**

From the SCH symbol timing detected, the S-SCH or direct SCH sequence in the frequency domain is detected in the hierarchical or non-hierarchical SCH structure. The SCH sequence indicates the cell ID or cell ID group index when the grouping of cell IDs is used similarly to that in WCDMA. In addition to the cell ID or cell ID group index, the following information can be conveyed by the SCH sequence:

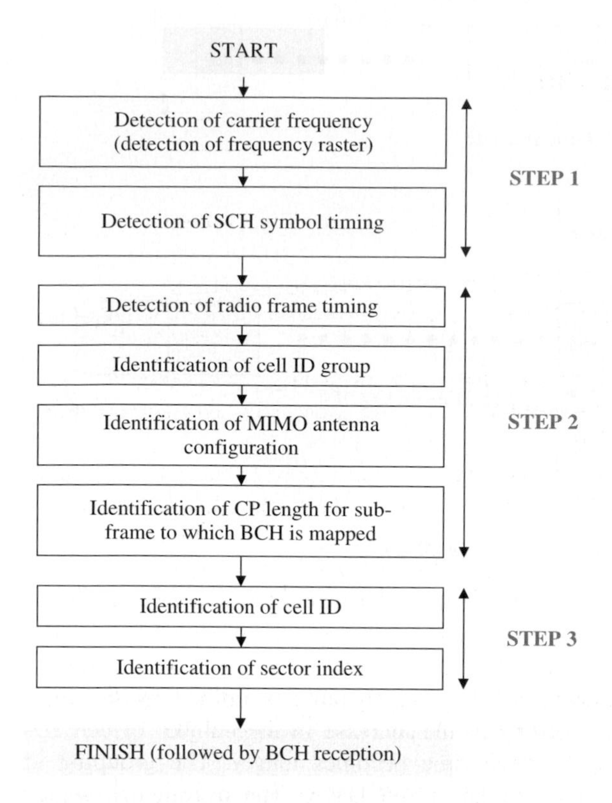


*Figure 20.* Flow of initial cell search procedure in Evolved UTRA

radio frame timing, MIMO antenna configuration, and CP length of the sub-frame with primary BCH, etc.

When the cell ID grouping is applied, the cell ID belonging to the group detected by SCH sequence in frequency domain and the sector index are identified. The detection of the cell ID is performed by taking the correlation between the received signal after FFT and the reference signal symbol replica. Even without cell ID grouping, the use of reference signal is beneficial for detecting sector index and for verification of cell ID.

## 5. UPLINK PHYSICAL CHANNEL

Uplink physical channels are categorized into the contention-based channel and scheduled based channel. Furthermore, the random access channel (RACH) is included in the contention-based channel. The reference signal, L1/L2 control channel, and shared data channel basically belong to the scheduled channel. In this section, we explain the structures and features of the E-UTRA uplink physical-channel structure related to

- Reference signals
- Random Access Channel (RACH)
- L1/L2 Control Channel
- Shared Data Channel

## 5.1 Uplink Reference Signal

Two uplink reference signals are defined in the E-UTRA uplink for two purposes.

- Demodulation reference signal: Used for channel estimation for uplink coherent demodulation/detection at a Node B
- Sounding reference signal: Used for channel-quality measurements for uplink frequency- and/or time-domain channel-dependent scheduling, link adaptation, etc.

The uplink demodulation reference signal, sounding reference signal, and data/control symbols are TDM-multiplexed using different LBs as show in **Figure 21**.

The transmission bandwidth of the demodulation reference signal is equal to the transmission bandwidth of the shared data channel or L1/L2 control channel. The demodulation reference signals are mapped onto the 4th and 11th LBs within 1-msec TTI (see also Fig. 4). Since only localized FDMA is used for the uplink shared data channel, the demodulation reference signals from different UEs are basically orthogonally multiplexed using localized FDMA within a TTI. However, when MIMO is used in the uplink (both in single-user and multiuser MIMO), orthogonal reference signals are also employed between multiple transmit antennas in the code domain as described in Section 4.1.1.

The bandwidth of the sounding reference signal is basically set wide to achieve frequency-domain channel dependent scheduling. However, if the UE at the cell edge transmits a wideband sounding reference signal, the CQI measurement error

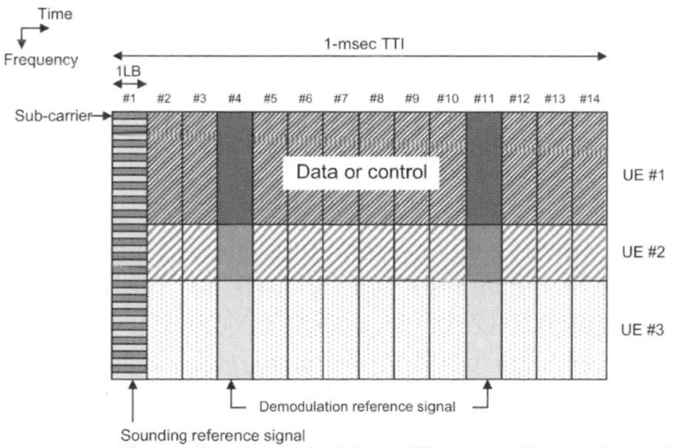


*Figure 21.* Orthogonal uplink reference signal structure

is increased due to the decreased received signal power density and the inter-cell interference caused by the sounding reference signal is increased. Thus, it is necessary to support multiple sounding reference signals with different transmission bandwidths.

Since the uplink transmissions are time-aligned within a CP, orthogonal sounding reference signals occupying the same LB and the same frequency block can be generated to accommodate different UEs within the same Node B in the uplink using either CDMA with the same spectrum or FDMA using a different spectrum as explained in the following section.

### *5.1.1 Orthogonal reference signals in code domain*

In the uplink, synchronization of the inter-UE timing within a sample-duration is not maintained at the Node B input. Thus, if a random sequence is used for the reference-signal sequence, the reference signals will suffer from multi-access interference (MAI) and MPI. Therefore, a sequence with a good auto-correlation property is necessary such as the CAZAC sequence. However, the number of sequences is small when the CAZAC sequence is used as the reference signal sequence because the scrambled code cannot be used in conjunction with the CAZAC sequence. The number of available CAZAC sequences is limited according to the length of the CAZAC sequence, i.e., proportional to the sequence length. For instance, the number of Zadoff-Chu sequences is the number of integers relatively prime to the sequence length "$N$." Thus, assuming $N$ is a prime number, the number of sequences becomes $(N-1)$ at maximum. Note that in a cellular system with one-cell frequency reuse, a sufficient number of reference signal sequences, i.e., at least the same number of sequences as cell IDs in WCDMA such as 512, is necessary to achieve flexible reference signal sequence assignment. By employing a cyclic shift based CAZAC code generation method, multiple CAZAC sequences are generated by cyclic shift of the original CAZAC sequence as shown in **Figure 22**. The cyclic shift based CAZAC sequences provide a good cross-correlation property to mitigate mutual interference within the duration of the cyclic shift value. The Zadoff-Chu sequence with the code length of $N$ (here we assume $N$ is an odd number), $a_u(k)$, is represented as

$$(4) \qquad a_u(k) = \exp\left(\frac{-j2\pi u}{N} \cdot \frac{k(k+1)}{2}\right), k = 0, 1, \ldots, \tilde{N}1.,$$

where $u$ denotes the sequence index, which is a prime relation to the $N$ value. When $N$ is a prime number, the number of available sequences is ($N$-1). When cyclic shift is applied, the number of orthogonal CAZAC sequences becomes $L = \lfloor N/\Delta \rfloor$ for the sequence length of $N$. The resultant CAZAC sequence with cyclic shift, $a_{u,l}(k)$, is given in the next equation using the cyclic shift index $l$ ($l$ = 0, 1, ..., $\tilde{L}1$).

$$(5) \qquad a_{u,l}(k) = a_u\{(k + l \cdot \Delta) \bmod N\}, l = 0, 1, \ldots, \tilde{L}1.$$

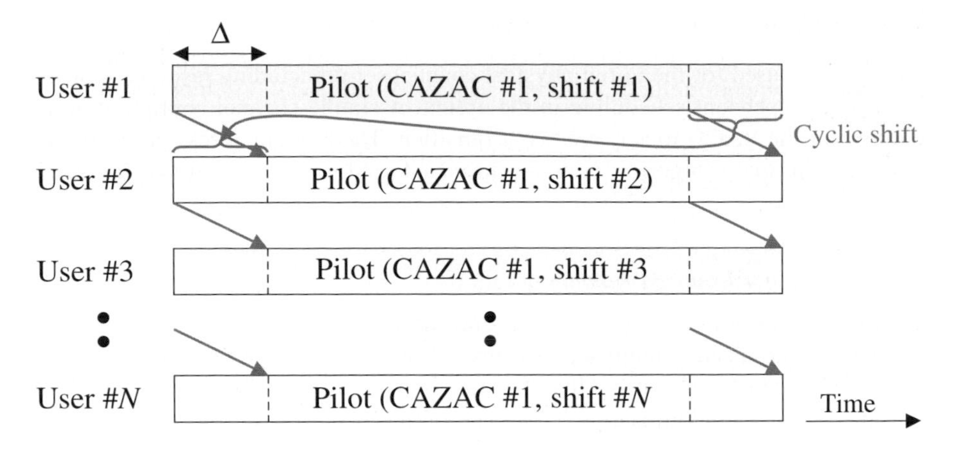


*Figure 22.* Cyclic shifted CAZAC sequence generation

The cyclic shift value, $\Delta$, to achieve an orthogonal sequence is decided from the total received timing errors among simultaneously receiving sets of UE. Moreover, the total number of received timing errors arises from the timing errors of the maximum delayed paths and transmission timing errors, which synchronize the received timings of the downlink reference signals. Accordingly, by using the cyclic shift based CAZAC sequences, the orthogonality among sets of UE in the same transmission bandwidth is achieved. The cyclic shift based CAZAC sequences from the one original CAZAC sequence is used with priority. Thus, when all cyclic shift based CAZAC sequences from the one original sequence are used up, the cyclic shift based CAZAC sequences from the second sequence are employed.

### *5.1.2 Orthogonal reference signals in frequency domain*

It is difficult to achieve orthogonality using CDMA among simultaneous sets of UE with different transmission bandwidths. Thus, FDMA is employed for multiplexing reference signals particularly with different transmission bandwidths. Employing distributed transmission provides channel gain over the assigned transmission bandwidth. Therefore, distributed transmission is used for multiplexing multiple reference signals using FDMA.

Furthermore, the orthogonal reference signal with the combination of CDMA and FDMA was proposed as shown in **Figure 21**. When distributed FDMA transmission is used, the sequence length of the reference signal becomes short in the time domain. Thus, in the method, the multiplexing using distributed FDMA transmission is restricted only to sets of UE with different transmission bandwidths. Instead, CDMA using the cyclic shifted CAZAC sequences are used for multiplexing sets of UE with the same transmission bandwidth. Thus, by combining FDM and CDM multiplexing, orthogonality among sets of UE with different and identical transmission bandwidths is flexibly achieved.

## 5.2 RACH

The RACH is used for the initial physical-channel set-upincluding resource request for channel-dependent scheduling in the uplink. A simple channel configuration is required for the RACH to achieve fast acquisition. The random access procedure is classified into two categories: Non-synchronized random access and synchronized random access.

### *5.2.1 Non-synchronized random access*

**(1) Purposes of non-synchronized random access**

The non-synchronized random access is used when the UE is not time synchronized with simultaneously accessing UEs. The random-access procedure is based on the transmission of a random-access burst. The purposes of non-synchronized RACH are as follows.

- Received timing measurement for time alignment so that the UE transmission timing is adjusted within the CP duration. Note that in the Evolved UTRA uplink, the orthogonality among simultaneous accessing users in the frequency domain is maintained as long as the received timings are aligned within the CP duration.
- Request for UE ID that is given by the Node-B
- Uplink resource request for sending an L2/L3 message such as the Radio Resource Control (RRC) message, Non-Access Stratum (NAS) message, or detailed scheduling request information

**(2) Resource assignment for non-synchronized RACH**

**Figure 23** shows the resource assignment for the non-synchronized RACH. The non-synchronized RACH is transmitted as a contention-based channel in the time/frequency resource, assigned by the Radio Resource Management (RRM) configuration. Prior to attempting non-synchronized random access, the UE shall synchronize with the downlink transmission. Thus, the transmission timing of the non-synchronized RACH can be determined based on the received timing of the downlink signal such as the downlink reference signal. The appropriate transmission power of the non-synchronized RACH can be determined through open-loop transmission power control so that distance-dependent path loss and shadowing variation are compensated. It is possible to vary the random access burst transmit power between successive bursts using power ramping with a configurable step size including a zero step size for both FDD and TDD or per-burst open loop power determination for the TDD case only. Frequency hopping at each RACH attempt is beneficial to improve the RACH detection.

**(3) Structure of non-synchronized RACH**

The non-synchronized RACH transmission consists of only a preamble, which achieves accurate detection of the RACH attempts including RACH receive timing. The required control signaling is implicitly carried by the preamble signature sequence. A preamble sequence having a good auto-correlation property such as

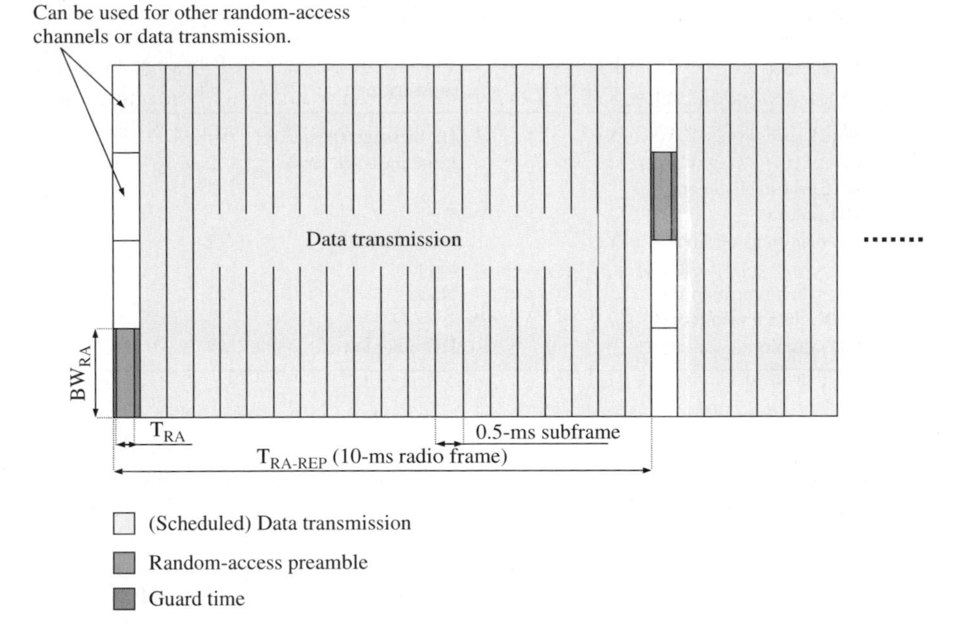


*Figure 23.* Resource assignment for non-synchronized RACH

a CAZAC sequence is used to achieve good detection probability. A small non-synchronized RACH control message is desirable to achieve a high detection probability and for a short delay of the physical channel setup. It was decided that the number of available information bits for the non-synchronized RACH is limited to four to six bits from the viewpoint of sufficient coverage provisioning.

An efficient transmission method for the required control signaling bits is necessary using the limited accommodated bit size. Thus, the grouping of the purposes for the non-synchronized RACH was proposed to achieve efficient required control signaling bits. **Table 3** shows the grouping of the purposes associated with the respective purposes. In the grouping scheme, purposes such as initial access, RRC re-establishment, and handover failure are grouped into the same purpose group. Moreover, uplink data transmission, uplink synchronization for downlink data reception, and inter-Node B handover are each assigned to a separate group. Thus, according to the grouping scheme, we need only a two-bit group index for the random access purpose in addition to a random ID for the non-synchronized RACH. Furthermore, for the purposes of initial access, RRC re-establishment, and handover failure, the UE requests the Node B to acquire the cell-specific UE ID, i.e., the Cell specific-Radio Network Temporary Identifier (C-RNTI). Meanwhile, for the purposes of the uplink data transmission, the uplink synchronization for downlink data transmission, and completion of handover, the UE can use the C-RNTI, which has been assigned from the connected Node B.

*Table 3.* Purposes of Non-synchronised RACH and Its Grouping

| Purpose | UE has UE ID (C-RNTI)? | Contents of message part | Purpose group index |
|---|---|---|---|
| Initial access | No | L3 message (basically large message size) | #1 |
| RRC re-establishment | | | |
| HO failure | | | |
| UL data transmission | Yes | UL scheduling request | #2 |
| UL synchronization for DL data transmission | | None | #3 |
| HO complete | | HO complete | #4 |

Therefore, from the group index of the random access purpose, the Node B can identify the need for the C-RNTI feedback and size of the message to be conveyed using the shared channel.

**(4) Procedure of non-synchronized RACH**

**Figure 24** shows the procedure using the non-synchronized RACH with the proposed two-bit group index of the random access purpose. The non-synchronized random access is processed in one-step as follows.

- UE transmits the non-synchronized RACH.
  - The preamble sequence is selected according to the Random ID selected by the UE randomly. Otherwise, a random ID can be selected according to the rough CQI value or UE power headroom using a one-to-one corresponding manner.
  - The purpose group index is also indicated by the selected preamble sequence.
- Node B sends a response to the non-synchronized RACH via the downlink.

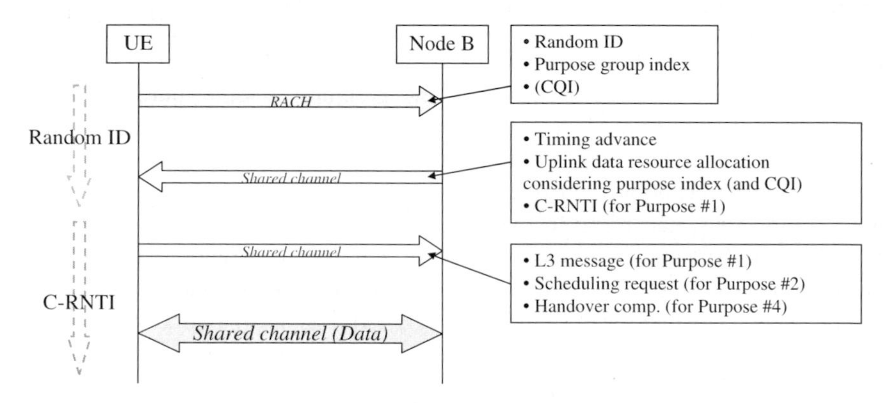


*Figure 24.* Non-synchronized random access procedure using RACH that transmits random ID and purpose group index

  - Timing adjustment is determined by the received timing of the non-synchronized RACH.
  - Uplink shared radio resources are allocated to the UE. The amount of the shared radio resources is determined by the expected message size that is indicated by the purpose group index and CQI value if available.
  - If the received purpose group index is #1, the Node B allocates new C-RNTI to the UE.
- The UE sends the control message information using the allocated uplink shared radio resources.
  - C-RNTI can be used to manage this shared channel transmission in the same way as that in the normal shared data channel transmission.
  - When the entire message is not transmitted using the initially assigned radio resources, additional shared radio resources are allocated by the Node B based on more accurate CQI.
  - hybrid ARQ can be applied.
- After finishing the RACH message transmission, actual data transmission starts.

### *5.2.2 Synchronized random access*

Synchronized random access is used by the UE to request resources for uplink data transmission during the interval in which received timing alignment is maintained. One of the objectives of the synchronized random access procedure is to reduce the overall latency. Synchronized random access and data transmission are also time and/or frequency multiplexed.

**Figure 25** shows the scheduling request in the active state. When the uplink scheduling grant is transmitted in the downlink L1/L2 control channel, the uplink shared data channel is transmitted from a set of UE using the assigned RB. Meanwhile, in the discontinuous reception (DRX) mode without uplink timing synchronization, the non-synchronized RACH is transmitted to establish uplink timing synchronization followed by a scheduling request procedure. Synchronous random access is employed in the active mode, during which the uplink timing synchronization is maintained; however, the uplink scheduling grant to assign RBs for the shared data channel is not transmitted over, e.g., several-hundred-milliseconds to several seconds.

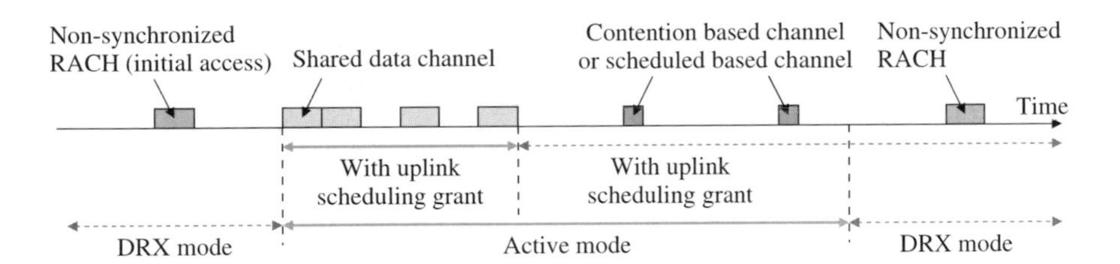


*Figure 25.* Scheduling request in active state

Therefore, during the active mode without shard data transmission, the following two functions are necessary.

- Transmission of scheduling requests
- Synchronization timing update, i.e., tracking of timing synchronization

The contention-based or scheduled based physical channels are considered to achieve these functions.

- Contention-based channels

In the contention-based mode, the synchronized random access channel is used for both purposes. In the case, the synchronized RACH for timing synchronization is transmitted asynchronously among sets of UE with a short delay. Establishing timing synchronization is not always guaranteed due to misdetection by contention with other channels.

- Schedule-based channels

In the scheduled-based mode, reference signal or CQI signal is used for synchronization timing update, and the L1/L2 control channel is used for scheduling requests instead in the schedule-based mode. Timing synchronization is accurately maintained without contention with other physical channels whereas, the physical channel for timing synchronization must be transmitted at the pre-assigned frequency and time resource, which leads to a delay longer than that for a contention based channel.

## 5.3 Uplink L1/L2 Control Channel

L1/L2 control signaling is essential to channel-dependent scheduling, link adaptation, and hybrid ARQ, etc. In the E-UTRA uplink, efficient and high-quality transmission of L1/L2 control signaling must be established while maintaining the low-PAPR feature of the single-carrier transmission in order to achieve a wide coverage area. Similar to the downlink, the transmission bandwidth capability of the UE should be taken into account in the design of the L1/L2 control signaling channel.

### *5.3.1 Uplink L1/L2 Control Information*

The control signaling information conveyed by the L1/L2 control channel is as follows.

- Data-associated signaling, which is transmitted with uplink data transmission
  - Transport format (MCS information) of an uplink data channel, if the UE determines the transport format (if it is decided by Node B, this information is not necessary)
  - Hybrid ARQ related information of an uplink data channel, e.g., redundancy version and new data indicator
- Data-non-associated signaling, which can be transmitted regardless of the uplink data transmission states
  - ACK/NACK in response to downlink data transmission

- Downlink CQI for the purpose of downlink scheduling and link adaptation

In addition to the two types of data-non-associated signaling, scheduling request for uplink data transmissions can be transmitted by the L1/L2 control channel in the active mode.

### *5.3.2 Multiplexing of L1/L2 control channel*

**(1) TDM-based low-PAPR transmission of L1/L2 control information**

There are three multiplexing combinations for the uplink reference signal, data, and L1/L2 control signaling within a sub-frame for a single set of UE.

- Multiplexing of the reference signal, data, and data-associated L1/L2 control signaling
- Multiplexing of the reference signal, data, data-associated, and data-non-associated L1/L2 control signaling
- Multiplexing of the reference signal and data-non-associated L1/L2 control signal in the data-associated control channel

It is necessary to multiplex the L1/L2 control signaling channel and shared data channel for the same UE within the same sub-frame. If the L1/L2 control signaling channel is multiplexed into the shared data channel using FDM (multicarrier) or CDM (multicode), the PAPR is increased. Therefore, as shown in **Figure 26**, TDM-based multiplexing of the L1/L2 control signaling channel into other physical channels was approved. The low PAPR property can be maintained by implementing one IFFT that takes advantage of the TDM-based feature.

**(2) Multiplexing of L1/L2 control signaling**

As was specified, E-UTRA supports multiple transmission bandwidths from 1.25 to 20 MHz. Accordingly, the L1/L2 control information must be transmitted in a unified manner to all types of UE that have different capabilities pertaining to the minimum maximum-transmission bandwidth. Therefore, similar to the downlink, we propose a block-wise transmission scheme for the L1/L2 control channel as shown in **Figure 27** that has the following features. The allocated system bandwidth of each cell is segmented into one or several frequency blocks.

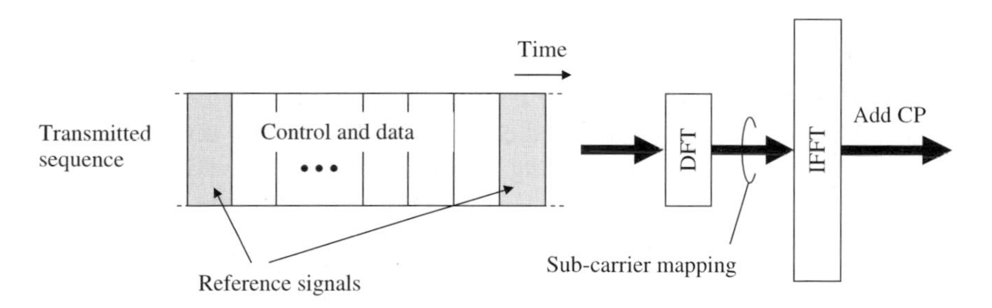


*Figure 26.* Transmission principle of TDM-based multiplexing

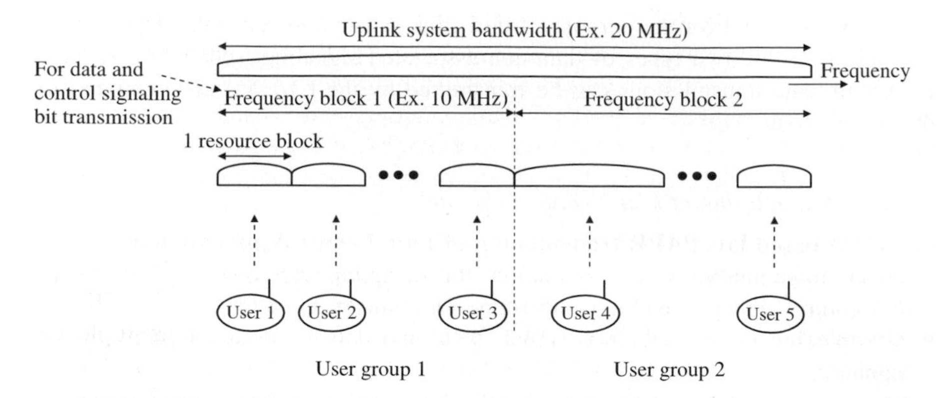


*Figure 27.* Example of block-wise transmission for L1/L2 control channel

- The pre-determined transmission bandwidth of the L1/L2 control channel should be equal to or less than the minimum maximum-transmission bandwidth in the UE capability.
- To each set of UE, at least one of the frequency blocks is assigned by Node B. The UE with a high capability uses multiple consecutive frequency blocks. Within the assigned frequency block, the UE transmits the L1/L2 control information, the reference signals for scheduling and demodulation, and the data channel transmission using one or multiple RBs within the pre-assigned frequency block.
- Frequency block assignment for each set of UE can be changed adaptively or by using a pre-determined manner. The purpose for changing the frequency block assignment is to average the traffic load of each frequency block and to avoid fatal transmission error due to, for example, severe frequency selective interference from other cells.

There are two methods to multiplex the data-non-associate L1/L2 control signaling bits from the UE without uplink shared data channel transmission: TDM within a sub-frame or FDM. In the TDM case, data-non-associated L1/L2 control information such as ACK/NACK and CQI signals are multiplexed into one block duration by TDM in a sub-frame, in which different assigned sets of UE transmit the shared data channel. In the same sub-frame, the reference signal of the data-non-associated UE is multiplexed with those of other sets of UE in the same block duration by distributed FDMA or CDMA. However, when TDM multiplexing within a sub-frame is used in the limited resource allocation to L1/L2 control signaling bits, the achievable coverage is decreased for a set of UE located at the cell boundary. Meanwhile, when the data-non-associated control signaling bits for sets of UE that transmit only the L1/L2 control signaling is multiplexed exclusively in a semi-statically assigned time-frequency region, i.e., FDM, wide coverage is provided, since the whole sub-frame duration is used for ACK/NACK and CQI transmission. However, the delay will be increased due to a long transmission interval since assigning an exclusive time-frequency region for the ACK/NACK and CQI to every sub-frame will severely degrade the signaling overhead in the uplink.

Therefore, to address the disadvantages of the above mentioned multiplexing methods, the multiplexing of ACK/NACK and CQI signals using separated multiple narrowband exclusive time-frequency regions with frequency hopping was proposed as indicated in **Figures 28**. In this figure, multiple narrowband time-frequency regions exclusively used for the data-non-associated L1/L2 control information such as ACK/NACK and CQI transmission are prepared at each sub-frame. Between the separated multiple time-frequency regions, frequency hopping is used to achieve frequency diversity (intra-sub-frame or inter-sub-frame frequency hopping). By using the narrowband exclusive time-frequency regions for ACK/NACK and CQI transmission, the overhead of the time-frequency resources for the ACK/NACK and CQI is reduced. The ACK/NACKs and CQIs of different sets of UE are multiplexed using the frequency/time/code domain or a hybrid of them within the assigned time-frequency region. The number of time-frequency regions for ACK/NACK and CQI transmission and the bandwidth of that at each sub-frame can be re-configurable according to the number of ACK/NACK and CQI signals.

### *5.3.3 Link adaptation for the L1/L2 control channel*

At least the distance-dependent path loss and shadowing variation must be mitigated in the uplink in order to guarantee a high-quality PER. In scheduling-based packet access, the time-frequency resources are assigned to different sets of UE every sub-frame. The inter-Node B interference varies with a wide dynamic range every sub-frame, since the location of an interfering set of UE changes. Therefore, the effect of fast link adaptation, which compensates for the instantaneous fading variation, is unclear.

TPC based on CQI measurement (CQI-based TPC) was proposed. The transmission power is controlled based on the CQI measurement results, which are originally required for AMC and channel-dependent scheduling for the uplink data channel. The required transmission power (or required transmission power relative to the reference signal/data channel) is directly calculated based on the CQI value

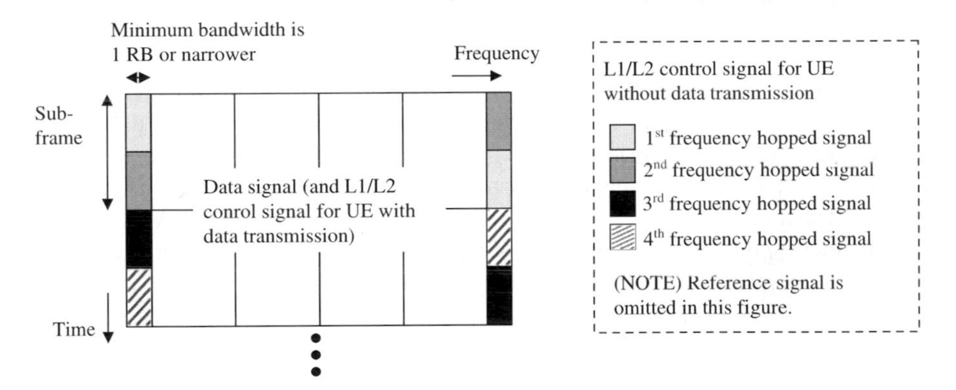


*Figure 28.* ACK/NACK and CQI multiplexing method using separated multiple narrowband exclusive time-frequency regions with frequency hopping

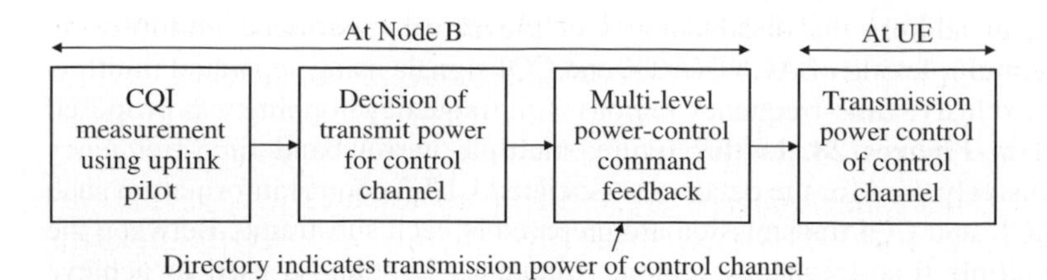


*Figure 29.* General description of fast transmission power control using multi-step power-control commands based on the CQI value for uplink

and fed back to the UE using multi-step power-control commands as shown in **Figure 29**. Therefore, CQI-based TPC does not need the history of its own transmission power. For the uplink data-associated control signaling channel, the MCS information for the uplink data channel is utilized as the power-control command. Therefore, no additional feedback command is required in the downlink. For the uplink data-non-associated control signaling channel such as ACK/NACK of the downlink shared data channel, the CQI of the downlink, and uplink scheduling request, multi-step power-control commands are fed back to the UE.

Furthermore, when the uplink L1/L2 control channel is transmitted with the data channel, in addition to the CQI-based TPC, a rough and slow adaptive modulation and coding rate including the repetition factor (AMC) can be applied. In this case, the relationship between the MCS used for the L1/L2 control channel and the data channel should be fixed. The use of AMC especially to control the coding rate and repetition factor in addition to the CQI-based TPC alleviates the limitation to the maximum transmission power of the UE, extends the coverage area, and improves the radio resource efficiency.

## 5.4 Shared Data Channel

### *5.4.1 Localized and Distributed Transmissions*

In general, according to the increase in the transmission bandwidth, the channel quality is improved owing to the increasing frequency diversity. In the uplink, nevertheless, we must consider the restricted transmission power of a set of UE due to the battery usage. When the transmission bandwidth is excessively increased for the data rate, the received power spectrum density is reduced. Accordingly, the channel quality becomes worse since the increasing channel estimation error exceeds the improvement in the frequency diversity effect.

Next, we consider the orthogonal multiplexing of simultaneously accessing sets of UE in the frequency domain. Unlike in the downlink, the orthogonality using the distributed FDMA transmission is not perfectly achieved due to the residual synchronization and frequency offset error particularly for a narrow comb, i.e., sub-carrier, bandwidth. From the aforementioned two factors, the localized FDMA transmission is considered for the shared data channel employing the adaptive transmission bandwidth according to the offered data rate.

### 5.4.2 Channel-dependent scheduling

In the uplink shared data channel, a certain frequency resource in a certain sub-frame, i.e., RU, is dynamically assigned to the target UE for uplink data transmission by the Node B scheduler. Downlink control signaling informs the set(s) of UE of what resources and the respective transmission formats that have been allocated. The selection of the sets of UE that are multiplexed within a given sub-frame is based on the following criteria.

- QoS parameters and measurements
- Payloads buffered in the UE ready for transmission
- Pending retransmissions
- Uplink channel quality measurements
- UE capabilities
- System parameters such as bandwidth and interference level/patterns

Among these, channel-dependent scheduling using the uplink CQI is effective in increasing the user and cell throughput associated with the combined usage with AMC including 16QAM modulation. Moreover, frequency domain channel-dependent scheduling is beneficial in improving the user and cell throughput performance levels in the uplink using SC-FDMA radio access as well as in the downlink OFDM radio access. To achieve frequency domain channel-dependent scheduling, a reference signal (pilot channel) with a wider transmission bandwidth than that for the shared data channel is necessary for CQI measurement before transmitting the shared data channel. However, with the limited maximum transmission power of a set of UE, an excessively wide transmission bandwidth for a reference signal for CQI measurement brings about increased CQI measurement error, and degrades the gain in the frequency domain channel-dependent scheduling for a set of UE located near the cell boundary.

Therefore, an adaptive transmission bandwidth of a reference signal for CQI measurement according to the path loss between a set of UE and a Node B was proposed. In the adaptive transmission bandwidth scheme, the UE, which is in a transmission-power-limited situation near the cell boundary, transmits a narrow-bandwidth reference signal to reduce the CQI measurement errors. Thus, the achievable throughput is improved since the improvement in channel-dependent scheduling by improving the CQI measurement accuracy exceeds the loss in frequency diversity effect. However, reference signals with different transmission bandwidths for CQI measurement are multiplexed using distributed FDMA. Moreover, a reference signal for demodulation, i.e., channel estimation for a shared data channel is multiplexed simultaneously within the same short block duration. Therefore, according to an increase in the number of reference signals multiplexed simultaneously using the distributed FDMA transmission, the sequence length of the reference signals becomes short. Accordingly, the number of reference signal sequences generated is decreased. Therefore, to avoid the decrease in the number of reference signal sequences, group-wise frequency resource allocation of the reference signals for CQI measurement was proposed associated with an adaptive transmission bandwidth in the frequency domain channel-dependent scheduling. **Figure 30** shows the group-wised frequency allocation assuming two transmission bandwidths for reference signals for CQI measurement when frequency domain channel-dependent scheduling is applied. The transmission bandwidth of a reference signal for CQI measurement is assigned by a Node B according to the path loss between a set of UE and the Node B. A set of UE with a low path loss uses a reference signal for CQI measurement with a wide transmission bandwidth such as 5 MHz, whereas a set of UE with a high path loss, i.e., located near the cell boundary, employs a narrow transmission bandwidth such as 1.25 MHz for CQI measurement. Different reference signals with different transmission bandwidths for CQI measurement are multiplexed into different frequency bands using localized FDMA transmission, i.e., by group-wised resource allocation in the frequency domain. Therefore, a reference signal for demodulation is multiplexed with only one reference signal for CQI measurement to minimize the reduction in the number of reference signal sequences accommodated by distributed FDMA. Reference signals for CQI measurement with the same transmission bandwidth are multiplexed using CDMA employing the cyclic-shifted sequences originating from the same CAZAC sequence within the same short block.

Similar to the downlink, the gain of the frequency and time domain channel-dependent scheduling is reduced in a high mobility environment, since the feedback loop in the channel-dependent scheduling cannot track the instantaneous channel variation. In such a situation, frequency hopping is beneficial in randomizing the fluctuations in the received signal level. Frequency hopping can mitigate the fluctuation in the interference to the surrounding cells compared to the channel-dependent scheduling.

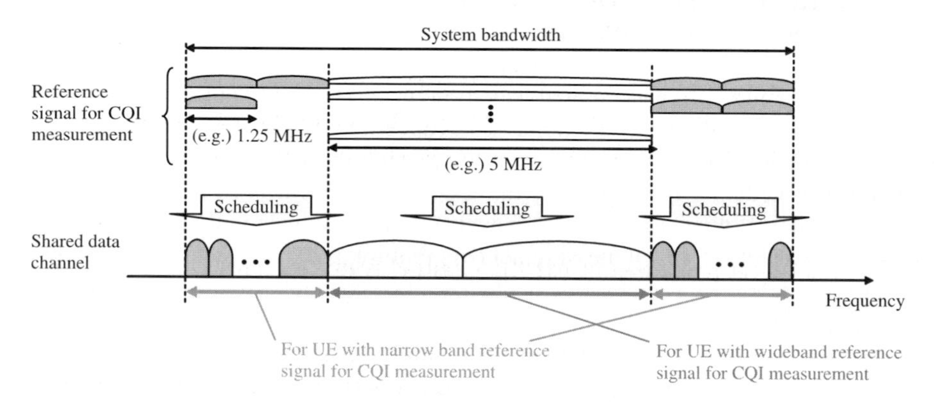


*Figure 30.* Group-wise frequency resource allocation of reference signals for frequency domain channel-dependent scheduling

### *5.4.3 Link adaptation*

Link adaptation for the shared data channel is necessary in order to guarantee the required minimum transmission performance of each set of UE such as the user data rate, packet error rate, and latency, while maximizing the system throughput. Three types of link adaptation are performed according to the channel conditions, the UE capability such as the maximum transmission power and the maximum transmission bandwidth, and the required QoS such as the data rate, latency, and packet error rate. The three link adaptation methods are an adaptive transmission bandwidth, adaptive transmission power control (TPC), and adaptive modulation and channel coding (AMC). The transmission bandwidth is decided mainly from the data rate requirement. Slow TPC may be used so that the distance-dependent path loss and shadowing variation are compensated. Moreover, the transmission power of a set of UE is also controlled to suppress the interference to neighboring cells particularly when the UE is located near the cell boundary. Then, the optimum MCS is selected from the CQI information reported every TTI so that the required packet error rate is satisfied against the instantaneous multipath fading channel.

### *5.4.4 Hybrid ARQ*

Incremental Redundancy is used as hybrid ARQ as well as in the downlink. Chase Combining is a special case of Incremental Redundancy and is thus implicitly supported as well. The N-channel Stop-and-Wait protocol is used for uplink hybrid ARQ similar to the downlink. Furthermore, the synchronous, non-adaptive ARQ is the current working assumption in the uplink. The retransmitted packet is transmitted from the sub-frame with the pre-decided delay from the initial transmitted packet. Therefore, the UE does not have to report the UE ID and process number to the Node B because the packet retransmission from the target UE is known to the Node B.

# 6. MIMO CHANNEL TRANSMISSION (MIMO MULTIPLEXING/DIVERSITY)

The MIMO channel transmissions such as MIMO multiplexing and MIMO diversity are considered in the E-UTRA. In particular, OFDM based radio access in the downlink has affinity to MIMO channel transmissions due to the robustness against time dispersion.

## 6.1 High level principles of MIMO channel transmissions

**(1) Downlink**

The baseline antenna configuration for MIMO channel transmissions in the downlink is two transmission antennas at the cell site and two reception antennas at the UE. Higher-order MIMO channel transmissions are also supported with the maximum antenna configurations of $4 \times 4$.

**(2) Uplink**

The baseline of the UE transmission is single-antenna transmission. Moreover, the baseline antenna configuration for single-user MIMO in the uplink is two transmission antennas at a set of UE and two reception antennas at the cell site.

## 6.2 MIMO multiplexing

**(1) Downlink**

In the Unicast mode, the following techniques are under discussion and are to be decided.

**• Single-user and Multi-user MIMO**

Both single-user and multi-user MIMO are supported in the downlink (note that multi-user MIMO is also called spatial division multiple access (SDMA)). In the single-user MIMO, Spatial division multiplexing (SDM) of multiple modulation symbol streams is solely assigned to a single set of UE using the same time and frequency (code) domain resources. Meanwhile, the modulation symbol streams are simultaneously assigned to different sets of UE using SDMA. The application of single-user or multi-user MIMO to a set of UE is determined by Node B in either a dynamic or semi-static manner.

**• Number of code words**

Multiple code words that use the same time and frequency domain resources and are independently channel-encoded are considered, in which CRC is attached to each code word. A single code word is included as a special case. The maximum number of code words per RB transmitted from the Node B is 1, 2, or 4. The maximum number of code words that can be received at a set of UE is fixed depending only on the UE capability. The number of code words per set of UE is restricted to, e.g., 2, from the viewpoint of the control signaling overhead.

**• Link adaptation**

Link adaptation is applied independently for each code word when multiple code words are transmitted to a set of UE with the control interval of 0.5 msec to a few milliseconds.

**• Pre-coding**

Multi-user MIMO is supported only with pre-coding. In pre-coding, the antenna domain MIMO signal processing is converted into the beam domain processing. Feedback information from a set of UE is necessary, which indicates the equivalent channel gain from each transmission antenna. Both codebook based and non-codebook based pre-coding are considered. In the codebook based scheme, the pre-coding vector(s) is selected from codebook(s) to reduce the control signaling overhead associated with the minimum size of the codebook. Regardless of the codebook usage, the feedback signaling should be minimized. The updating interval of the pre-coding vector is sufficiently short so that it can track the instantaneous channel variation such as 0.5 msec or longer. Since the pre-coding vector is updated to track the instantaneous channel variation, the achievable performance gain, i.e., the generated beam associated with each set of UE, is degraded when the feedback of the channel gain from a set of UE cannot track the channel variation in a high mobility environment. Therefore, pre-coding is suitable for local areas with a relatively short ISD and with heavy traffic in the coverage area.

Adaptive beam forming with a narrow antenna separation is considered as an alternative method of pre-coding to achieve SDMA. Adaptive beam forming generates a directive narrow beam toward the direction of each UE. The feedback of the channel gain from a set of UE is not necessary, since it tracks only the direction of the UE (note that the direction of the UE is estimated from the direction of arrival of the received signal in the uplink). Therefore, adaptive beam forming is appropriate to increase the coverage area for large cells, which support low-to-high mobility sets of UE.

**Rank adaptation**

Rank adaptation is supported, in which an antenna subset of MIMO layers is selected according to the channel condition of each set of UE. When the rank is one, transmit diversity is used. The number of code words transmitted to a set of UE is controlled through rank adaptation when the UE supports multiple code words.

Antenna selection is also considered, in which only the number of transmission antennas is selected according to the channel conditions of the UE using the same MIMO scheme.

The MIMO channel transmission for MBMS is also discussed in the downlink. In an MBMS channel, transmitting a feedback signal from multiplexed sets of UE is not feasible. Thus, the potential candidate for MIMO channel transmission is an open loop scheme. In terms of transmit diversity, it seems that the additional diversity gain is small, since the large frequency diversity gains is already obtained due to delayed signals from multiple cell sites by soft-combining reception at a set of UE.

**(2) Uplink**

Both single-user and multi-user MIMO is supported in the uplink. A specific example of SDMA in the uplink is 2-by-2 multi-user MIMO, where two sets of UE transmit an uplink signal on a single antenna and share the same time and frequency domain resource allocation. There the sets of UE employ orthogonal reference signals for a Node B to separate respective streams from two sets of UE.

## 6.3 Transmit diversity

**(1) Downlink**

Transmit diversity is considered mainly for application to the control channels. In particular, open loop transmit diversity is considered for the downlink control channels. Various numbers of transmission antennas at a cell site must be supported. Similarly, various types of UE with different capability antenna configurations are supported. Thus, a unified transmit diversity scheme for the control channels is necessary regardless of the antenna configuration at the cell site and regardless of the UE capability from the viewpoint of simplicity. The following open loop transmit diversity schemes are considered for the cell-specific control channels.

- Block-code based transmit diversity (STBC, SFBC)
- Time (or frequency) switched transmit diversity
- Cyclic delay diversity (CDD)
- Pre-coded transmission using selected pre-coding vector(s) including selection transmit diversity

**(2) Uplink**

In the uplink, the complexity of the UE transmitter is considered together with the achievable transmit diversity gain. The UE transmitter configuration with one RF chain, i.e., transmitter circuitry, and two antennas is considered as promising candidates for uplink transmit diversity. In the configuration, selection diversity with either open loop control or closed loop control is applied. The final adoption and scheme are decided from the substantial gain considering the performance loss such as insertion loss of the RF switch.

## 7. INTER-CELL INTERFERENCE MITIGATION

The improvements in frequency efficiency and user throughput near the cell boundary are very important requirements in the E-UTRA. Therefore, to achieve this, the following three techniques are discussed to mitigate inter-cell interference.

- Inter-cell-interference randomization
- Inter-cell-interference cancellation
- Inter-cell-interference co-ordination/avoidance

In addition to the above, the use of an adaptive beam-forming antenna transmitter/receiver at the cell site is a general method that can be considered as a means to mitigate inter-cell-interference.

In terms of inter-cell-interference randomization, the cell-specific scrambled code is applied to reference signals and other physical channels with repetition coding. The inter-cell-interference is suppressed in line with the processing gain.

In the inter-cell-interference cancellation, the following two methods are considered. First is spatial suppression based on, e.g., the MMSE criterion employing multiple antennas. Second is based on the detection and subtraction of inter-cell interference. In inter-cell-interference cancellation, the suppression gain, which is beyond what can be achieved by just taking advantage of the processing gain in inter-cell-interference randomization, is obtained.

In inter-cell-interference coordination, resource allocation is restricted through coordination between cells. The coordination of resource allocation at each cell is achieved by the configuration for the common channels and by scheduling for the scheduled based channel such as the shared data channel. The inter-cell-interference coordination restricts the time and frequency resources that are available to the resource manager or the transmission power that can be applied at a certain time and the frequency resources. The restrictions on the allocation of resources or transmission power will provide the possibility for improvement in the received

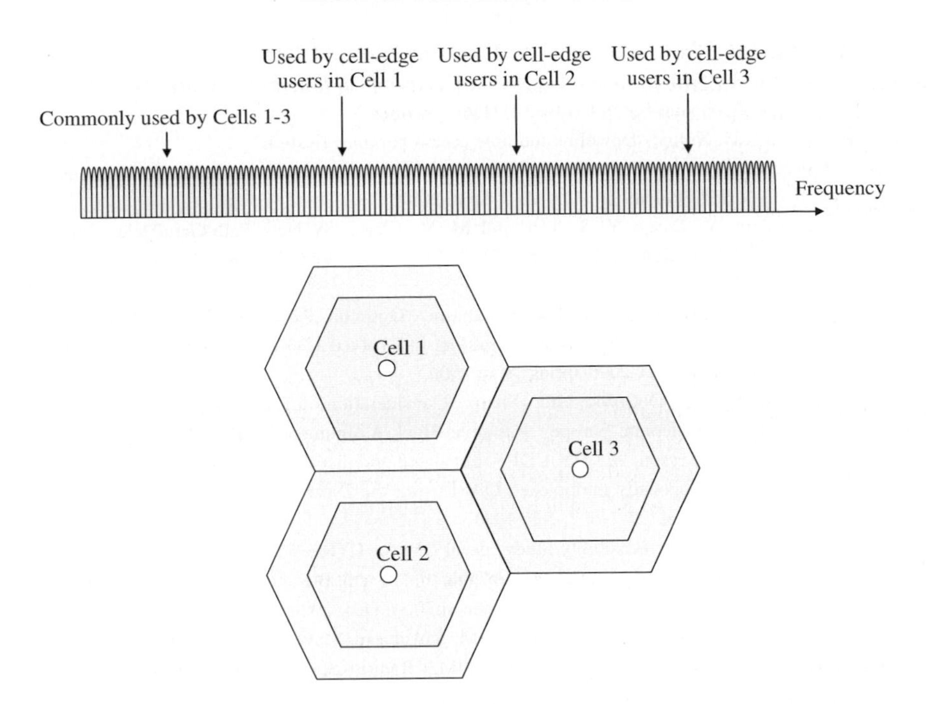


*Figure 31.* Principle of inter-cell-interference coordination

SINR and the achievable user throughput at the cell edge based on the corresponding time and frequency resources in a neighboring cell. **Figure 31** illustrates an example of the coordination of inter-cell-interference. As shown in the figure, one-cell frequency reuse is used in the central part near the cell site of each cell where the influence of other-cell interference is slight. Meanwhile, fractional frequency reuse is introduced in regions near the cell edge to avoid the influence of other-cell interference.

The coordination between cells is mainly achieved by a scheduler at each cell site independently, not by additional tight inter-Node B communication and/or additional UE measurements and reporting. In this case, certain inter-communications between different Node Bs will be required in order to set and reconfigure the scheduler restrictions. Two cases are considered: Static and semi-static interference coordination. The reconfiguration of the restrictions is performed on a time scale corresponding to days or seconds in the static or semi-static interference coordination cases, respectively.

## REFERENCES

[1] 3GPP, RP-040461, "Proposed Study Item on Evolved UTRA and UTRAN," Dec. 2004.
[2] 3GPP, TR 25.913, v7.3.0, "Requirements for Evolved UTRA and UTRAN"
[3] 3GPP, TR 25.912, v7.0.0, "Feasibility Study for Evolved UTRA and UTRAN"

[4] 3GPP, TR 25.814, v7.0.0, "Physical Layer Aspects for Evolved UTRA"
[5] 3GPP, R1-050587, NTT DoCoMo, Fujitsu, Intel Corporation, Mitsubishi Electric, NEC, Sharp, "OFDM Radio Parameter Set in Evolved UTRA Downlink"
[6] 3GPP, R1-050384, Nokia, "Downlink multiple access parameterisation"
[7] M. Batariere, K. Baum, and T. P. Krauss, "Cyclic Prefix Length Analysis for 4G OFDM Systems," in Proc. IEEE VTC2004-Fall, pp. 543– 547, Sept. 2004.
[8] L. J. Greenstein, V. Erceg, Y. S. Yeh, and M. V. Clark, "A New Path-Gain/Delay-Spread propagation model for digital cellular channels," IEEE Trans. Veh. Technol., vol. 46, no. 2, pp. 477–485, May 1997.
[9] S. Nagata, Y. Ofuji, K. Higuchi, and M. Sawahashi, "Optimum Resource Block Bandwidth for Frequency Domain Channel-Dependent Scheduling in Evolved UTRA Downlink OFDM Radio Access," in Proc. IEEE VTC2006-spring, May 2006.
[10] 3GPP, R1-061667, NTT DoCoMo, NEC, Sharp, "Consideration on Resource Block Size"
[11] 3GPP, R1-061795, Qualcomm Europe, "Resource Block Assignment for E-UTRA"
[12] 3GPP, R1-060095, Ericsson, "E-UTRA DL – Localized and distributed transmission"
[13] B. Hirosaki, "An orthogonally multiplexed QAM using the Discrete Fourier Transform," IEEE Trans. On Commun., Vol. 29, No. 7, July 1981.
[14] 3GPP, TR25.892, v1.1.0, "Feasibility Study for OFDM for UTRAN Enhancement"
[15] R. Dinis, D. Falconer, C.T. Lam, and M. Sabbaghian, "A multiple access scheme for the uplink of broadband wireless systems," in Proc. Globecom2004, Dec. 2004.
[16] T. Kawamura, Y. Kishiyama, K. Higuchi, and M. Sawahashi, "Investigations on Optimum Roll-off Factor for DFT-Spread OFDM Based SC-FDMA Radio Access in Evolved UTRA Uplink," in Proc. ISWCS2006, Sept. 2006.
[17] 3GPP, R1-050584, Motorola, "EUTRA Uplink Numerology and Design"
[18] 3GPP, R1-050588, NTT DoCoMo, Fujitsu, Mitsubishi Electric, Sharp, "Radio Parameter Set for Single-Carrier Based Radio Access in Evolved UTRA Uplink"
[19] 3GPP, R1-051185, Ericsson, "PAR-reducing modulation for E-UTRA SC-FDMA uplink," Oct. 2005.
[20] 3GPP, R1-060293, Nokia, "UL modulation scheme"
[21] T. Kawamura, Y. Kishiyama, K. Higuchi, and M. Sawahashi, "Comparisons of 16QAM Modulation Schemes Considering PAPR for Single-Carrier FDMA Radio Access in Evolved UTRA Uplink", in Proc. IEEE ISSSTA2006, Sept. 2006.
[22] 3GPP, R1-050853, NTT DoCoMo, Fujitsu, "Common Pilot Channel Structure for OFDM Based Radio Access in Evolved UTRA Downlink"
[23] Y. Kishiyama, K. Higuchi, and M. Sawahashi, "Intra-Node B orthogonal pilot channel structure for OFDM radio access in Evolved UTRA downlink," in Proc. IEEE VTC2006-Spring, May 2006.
[24] 3GPP, R1-061665, NTT DoCoMo, Fujitsu, NEC, Sharp, Toshiba Corporation, "Broadcast Channel Structure for E-UTRA Downlink"
[25] 3GPP, R1-060302, NTT DoCoMo, Fujitsu, Mitsubishi Electric, NEC, Panasonic, Sharp, Toshiba Corporation, "Broadcast Channel Structure for E-UTRA Downlink"
[26] 3GPP, TS 25.211, v7.0.0, "Physical channels and mapping of transport channels onto physical channels (FDD)"
[27] A. Morimoto, Y. Kishiyama, K. Higuchi, and M. Sawahashi, "Efficient Synchronous Transmissions within the Same Node B for Paging Channel in OFDM Based Evolved UTRA Downlink" in Proc. ICCS2006, Oct. 2006.
[28] 3GPP, R1-061672, NTT DoCoMo, Fujitsu, Mitsubishi Electric, NEC, Sharp, Toshiba Corporation, "Coding Scheme of L1/L2 Control Channel for E-UTRA Downlink"

[29] 3GPP, R1-051145, NTT DoCoMo, Fujitsu, NEC, Panasonic, Sharp, Toshiba Corporation, "CQI-based Transmission Power Control for Control Channel in Evolved UTRA"
[30] 3GPP, R1-051331, Motorola, "E-UTRA Downlink Control Channel Design and TP"
[31] 3GPP, R1-060182, Toshiba Corporation, "MBMS Structure for Evolved UTRA"
[32] 3GPP, R1-060779, NTT DoCoMo, Mitsubishi Electric, NEC, Sharp, Toshiba Corporation, "Investigations on Pilot Channel Structure for MBMS in E-UTRA Downlink"
[33] E. Costa, E. Filippi, H. Haas, S. Ometto, and E. Schulz, "Adaptive sub-band allocation in MC-CDMA/FDM systems," in Proc. International OFDM Workshop 2002, Sept. 2002.
[34] B. Classon, P. Sartori, V. Nangia, Z. Xiangyang, and K. Baum, "Multi-dimensional adaptation and multi-user scheduling techniques for wireless OFDM systems," in Proc. IEEE ICC2003, May 2003.
[35] S. Nagata, Y. Ofuji, K. Higuchi, and M. Sawahashi, "Block-Wise Resource Block-Level Distributed Transmission for Shared Data Channel in OFDMA Evolved UTRA Downlink" in Proc. WPMC2006, Sept. 2006.
[36] 3GPP, R1-060039, NTT DoCoMo, Fujitsu, Mitsubishi Electric, NEC, QUALCOMM Europe, Sharp, Toshiba Corporation, "Adaptive Modulation and Channel Coding Rate Control for Single-antenna Transmission in Frequency Domain Scheduling in E-UTRA Downlink"
[37] 3GPP, TS.25.212, v.6.8.0, "Multiplexing and channel coding"
[38] 3GPP, R1-051329, Motorola, "Cell Search and Initial Acquisition for OFDM Downlink"
[39] 3GPP, R1-061187, NTT DoCoMo, Fujitsu, Mitsubishi Electric, NEC, Panasonic, Toshiba Corporation, "Comparison on Cell Search Time Performance between SCH-Replica Based and Auto-Correlation Based Detections in E-UTRA Downlink"
[40] 3GPP, R1-050590, NTT DoCoMo, "Physical Channels and Multiplexing in Evolved UTRA Downlink"
[41] 3GPP, R1-051412, Nokia, "Cell Search procedure for initial synchronization and neighbour cell identification"
[42] 3GPP, R1-061664, NTT DoCoMo, Fujitsu, Mitsubishi Electric, NEC, Toshiba Corporation, "Neighboring Cell Search Method for Connected and Idle Mode in E-UTRA Downlink"
[43] 3GPP, R1-061662, NTT DoCoMo, Fujitsu, NEC, Toshiba Corporation, "SCH Structure and Cell Search Method for E-UTRA Downlink"
[44] D. C. Chu, "Polyphase codes with good periodic correlation properties," IEEE Trans. Inform. Theory, vol. IT-18, pp. 531–532, July 1972.
[45] 3GPP, R1-050822, Texas Instruments, "On Allocation of Uplink Pilot Sub-Channels in EUTRA SC-OFDMA"
[46] 3GPP, R1-060046, NTT DoCoMo, NEC, Sharp, "Orthogonal Pilot Channel Structure in E-UTRA Uplink"
[47] 3GPP, R1-060296, Nokia, "Random access message – text proposal"
[48] Y. Kishiyama, K. Higuchi, and M. Sawahashi, "Investigations on Random Access Channel Structure in Evolved UTRA Uplink," in Proc. ISWCS2006, Sept. 2006.
[49] 3GPP, R1-061184, NTT DoCoMo, Fujitsu, Mitsubishi Electric, NEC, Sharp, Toshiba Corporation, "Random Access Channel Structure for E-UTRA Uplink"
[50] 3GPP, R1-061114, Panasonic, "Random access design for E-UTRA uplink"
[51] 3GPP, R2-060996, Qualcomm Europe, "Access Procedures"
[52] 3GPP, R1-060320, NTT DoCoMo, NEC, Sharp, Toshiba Corporation, "L1/L2 Control Channel Structure for E-UTRA Uplink"
[53] 3GPP, R1-061675, NTT DoCoMo, Sharp, Toshiba Corporation, "Data-non-associated L1/L2 Control Channel Structure for E-UTRA Uplink"

[54] Y. Ofuji, K. Higuchi, and M. Sawahashi, "Frequency Domain Channel-Dependent Scheduling Employing an Adaptive Transmission Bandwidth for Pilot Channel in Uplink Single-Carrier – FDMA Radio Access," in Proc. IEEE VTC2006-Spring, May 2006.

[55] 3GPP, R1-061679, NTT DoCoMo, Fujitsu, NEC, Sharp, Toshiba Corporation, "Group-wised Frequency Resource Allocation for Frequency Domain Channel-dependent Scheduling in SC-Based E-UTRA Uplink"

[56] 3GPP, R1-050694, Alcatel, "Multi-cell Simulation Results for Interference Coordination in new OFDM DL"

[57] 3GPP, R1-051123, Qualcomm Europe, "Further description of dynamic FFR for OFDM based E-UTRA downlink,"

# INDEX